Dmitry Ivanov · Boris Sokolov

Adaptive Supply Chain Management

Springer

Dr. Dmitry Ivanov
Chemnitz University of Technology
Department of Economics and
Business Administration
09107 Chemnitz
Germany
dmitry.ivanov@wirtschaft.tu-chemnitz.de
www.ivanov-scm.com

Prof. Dr.-Ing. Boris Sokolov
Saint Petersburg Institute of Informatics
and Automation of the Russian Academy
of Science (SPIIRAS)
14th Line VO 39
St. Petersburg 199178
Russia
sokol@iias.spb.su

ISBN 978-1-4471-5788-5 ISBN 978-1-84882-952-7 (eBook)
DOI 10.1007/978-1-84882-952-7
Springer London Dordrecht Heidelberg New York

British Library Cataloguing in Publication Data
A catalogue record for this book is available from the British Library

Cover design: eStudioCalamar, Figueres/Berlin

Printed on acid-free paper

Springer is part of Springer Science+Business Media (www.springer.com)

Adaptive Supply Chain Management

With special gratitude to our dear families who endured the work on this book with encouraging smiles and to so many sincere friends and colleagues

Foreword

Research on decision-making support for supply chain management (SCM) has been conducted from different perspectives so far. Though considerable advancements have been achieved in operations research, control theory, evolutionary algorithms, and agent-based systems, their isolated application frequently led to inevitable problem simplifications. In the worst cases, the artificial transformation of a real problem to a specific problem to apply particular solution techniques could be stated as examples. The comprehensive and near-real-world frameworks for decision-making support in SCM require solid skills from the investigators in both business administration, information technologies, and mathematical modelling. Fortunately, the joint knowledge of the authors of this book provides the required skills at a very high level, which could ensure the advancements in SCM decision-making support as described in this book. Unlike many other researchers who consider SCM as a new application domain for existing research methods, the authors of this book have been continuously developing special decision-making support techniques with regard to specific features of supply chains for the last few years. While proposing these advancements today, the book provides guidelines and challenges to conduct research on SCM in the near future. This book will be very helpful for students and post-graduates and will enable them to understand better the possibilities and challenges in conducting first-class research on complex production-logistics systems that are commonly known as supply chains at the present time.

Joachim Kaeschel, Prof. Dr.habil.
Chair of Production Economics and Industrial Organization
Chemnitz, Germany

Decision-making support systems for supply chain management (SCM) have been extensively developed during the past two decades. In this regard, remarkable progress within operations research, control theory, and artificial intelligence has been achieved. However, partial SCM problems have usually been considered as a "thing in itself" without reflecting the tight interlinking of the related sub-problems with regard to the global supply chain performance. To tackle this weakness, the authors present a multi-disciplinary approach for integrated decision-making in the SCM domain. They show that this multi-disciplinary research into supply chain management allows consideration of the whole problem semantics in an interlinked form, and not only as the optimization of local problems. This is enabled by enriching operations research methods with dynamics, non-linearity, stability and adaptivity considerations as well as by realising discreteness and subjectivism of decision-making in dynamic control models. The combined application of operations research, control sciences and artificial intelligence induces a synergy effect to bring the decision-making support systems for SCM closer to real management problems. In this and many other aspects, the academic contribution of this book is very high. The vision and concrete tools presented for supporting decision-making in the domain of SCM are path-breaking and worth thorough studying by young researchers, PhD students, and also professionals and advanced scientists.

Prof. Dr.-Ing. Herbert Kopfer
Chair of Logistics
University of Bremen
Germany

Supply chain management (SCM), once developed from the information technologies perspective and further treated from the business administration perspective, now requires comprehensive engineering frameworks to incorporate business models, technological infrastructures, information coordination and decision-making models. Hence, multi-disciplinary research approaches are in strong demand. In this setting, this book opens up new perspectives for conducting research into SCM from integrated perspectives to amplify concepts and models that are not consistent and limited to very specific cases. The authors put the main emphasis in this book not only on the development of new mathematical constructions but first of all on system thinking and new perspectives to consider SCM problem semantics and the art of applied mathematical modelling for complex business systems. Therefore, this monograph aims to cover the gap between the mathematical modelling and the management system thinking. With novel approaches and models, the book makes a number of valuable contribution to operations research, systems analysis and control sciences. E.g., cconventionally isolated static and dynamic problems of supply chain planning and scheduling are considered as a whole and a new integrated problem of supply chain planning and execution has been revealed, stated and solved due to the mutual enriching of operations research and optimal control techniques. This book may be especially useful for PhD students and post-docs who are interested in advanced and practice-relevant techniques in supply chain management.

Alexandre Dolgui, Ph.D., Dr.Hab.
Professor
Director of the Centre for Industrial Engineering and Computer Science
Saint Etienne National Institute of Science and Technology
Saint Etienne, France

Supply chain management (SCM) is a relatively new research field. However, the insights gained from recent research already allow us to talk about new research developments within the SCM domain. These developments concern both new conceptual business frameworks for supply management with regard to advanced information technologies and social responsibilities and new mathematical approaches to quantitative supply chain models and comprehensive decision-making support. These new ways of thinking are reflected in this book. The book undertakes a comprehensive theoretical rethinking of a complex of multi-disciplinary problems in the SCM domain. It recognizes the central role of people in decision making on supply chain planning and execution and reflects this in the modelling approaches developed, e.g., by bringing people and their subjective decision making into systems and control science. The modern challenges in SCM that have been taken up by the authors are thoroughly worked through and answered with constructive methods. The book *Adaptive Supply Chain Management* may potentially motivate new research in the SCM domain.

Viktor Sergeev, Prof. Dr.Sc. (econ.)
Chair of Supply Chain Management
President International Logistics Center
High School of Economics
Member of European Logistics Association (ELA) Board
Moskau, Russia

The present economic crisis is having a profound negative impact on most supply chains. Existing supply chains have been configured with the old assumptions of unlimited economic growth and utter stability of environmental conditions. However, this approach was wrong. Sustainable supply chain development requires not taking environmental stability as a given invariable state but considering supply chains and the environment as a whole. Finding the necessarily stable conditions for supply chain development should be based on the simultaneous balancing of changes in the environment of supply chains and in the supply chains themselves. This is the only way to remain profitable and competitive in real complex supply chain execution environments. This is the stand taken in this book. It is very gratifying that modern research on SCM takes up this practical challenge and proposes methodical guidelines for decision-making frameworks with the focus on ensuring supply chain profitability through stability. As a director for operations logistics of an internationally active logistics and supply chain provider, I can encourage the research community to take this viewpoint at the forefront of future investigation. Future business models in SCM will be based on the ensuring supply chain profitability as balanced with the supply chain stability to answer the challenges of uncertainty rather than on the unlimited profit growth. This will require proper scientific tackling. And this book is intended to illustrate this.

Ako Djaf, Dipl.-Wirt.-Ing.
Director Operations Logistics
Schenker Russia

Preface

We study what we can see,
but what we see is not always what exists.
Paulo Coelho

The term "*supply chain management*" (SCM) was coined in the 1980–90s. Presently, SCM is considered as the most popular strategy for improving organizational competitiveness along the entire value chain in the twenty-first century.

A *supply chain* (SC) is a network of organizations, flows and processes wherein a number of various enterprises (suppliers, manufacturers, distributors and retailers) collaborate (cooperate and coordinate) along the entire value chain to acquire raw materials, to convert these raw materials into specified final products, and to deliver these final products to customers.

SCM studies human decisions in relation to cross-enterprise collaboration processes to transform and use the SC resources in the most rational way along the entire value chain, from raw material suppliers to customers, based on functional and structural integration, cooperation, and coordination throughout.

SCs influence the world economy and are influenced by it. The economic environment has changed significantly since the autumn of 2008. Hence, the necessity for new viewpoints on SCM has become even more obvious. The former paradigm of total and unlimited customer satisfaction has naturally failed because of the limited resources for this satisfaction.

In these settings, the duality of the main goals of SCM – maximizing the service level and minimizing costs – should be enhanced by the third component – maintaining SC stability. This triangle goal framework will build the *new SCM paradigm* that can be formulated as the maintenance of stability and the harmonization of value chains with possibly full customer satisfaction and cost-efficient resource consumption for ensuring the performance of production-ecological systems at the infinite time horizon. Therefore, new conceptual frameworks and mathematical tools for decision-making support are needed.

In taking the level of the engineering frameworks and mathematical models to the forefront of this study, the research logic includes the following main components (see Fig. 1). The SC design starts with the *system formation*. In this research stream, a wide variety of organizational issues are investigated in relation to the collaboration motivation, organizational structures, trust, etc. This level is out of the scope of this study. However, we will reflect the organizational issues in Chap. 2.

The first step in building an SC is *structural design*. Within the SC, a number of structures (organizational, functional, informational, technological and financial) are to be formed to ensure a backbone for the achievement of the system's goals.

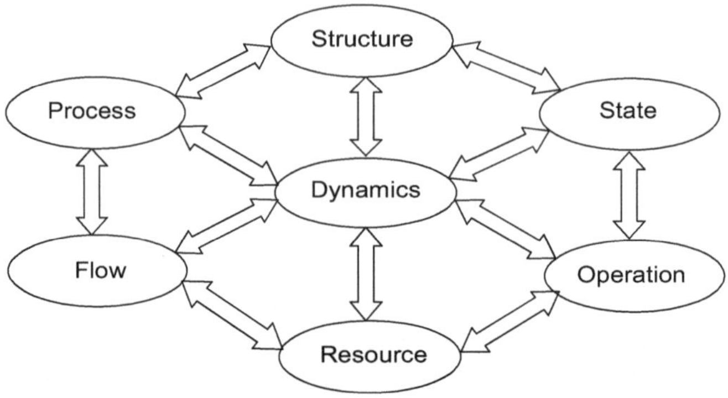

Fig. 1 Main components of system research on SCM

The *processes* and *flows* within the structures are implemented through the resource provision and consumption. These *resources* may be related to materials, time, information, people and finances. To utilize the resources, a number of *operations* is to be planned and executed. The execution happens not statically but dynamically, and is unwound over different time horizons. Through the structure and operations dynamics, SCs may appear in different *states*. The above-mentioned components are tightly interlinked.

The *primary objective* of this book is to reflect these conjunctions both at the conceptual and formal levels to contribute to the existing knowledge in the SCM domain in developing a framework, mathematical models and software prototypes for handling uncertainty and dynamics in SCs, taking into account high-dimensional and computational complex problems, the explicit interconnection of the planning and execution decisions, and decision-making by people to ensure SC adaptability, stability and crisis-resistance.

The *motivation for this contribution* consists of several topics that can be captured under the general problem of the handling of uncertainty and dynamism while planning and executing SCs.

First, the composite objective of maximizing both the *SC stability and economic performance* can be considered as a timely and crucial topic in modern SCM. The profit losses through non-purposeful (e.g., demand fluctuations) and purposeful (e.g., terrorism or thefts) perturbation impacts can amount to 30% of the annual turnover. The current economic decline and its impacts on SCs confirm the necessity for rethinking the SC optimization vision of an unlimited profitability growth. Striving for maximal profitability in the hope of an unperturbed environment and unlimited economic growth led to tremendous collapses and losses in SCs. The crisis provides the ultimate evidence that one of the main tasks of SCM is to balance profitability and stability to remain competitive in the perturbed economic environment. Besides, the stability indicator meets the SCM nature to a greater extent. Increases in sales and cost reductions may be related to operational logistics improvements at local knots of SC. But the stability of the whole SC is even the direct performance indicator of SCM.

Second, within SCs, there are lots of problems that have been conventionally treated in isolation from each other but which are indeed tightly *interlinked.* In our opinion, these explicit interconnections may potentially provide new quality of decision-making support for SCM. For example, planning and scheduling are an integrated management function. SCs consist of different structures. The planning and execution of SCs are based on the same decision-making procedures that do, however, differ in their decision-making speed. Hence, the planning and execution stages of SCM can be considered within an adaptation framework. In following these assumptions, we will consider dynamics in SCs at the structural–operational level with concrete processes and parameters rather than at the strategic level, which usually suffers from a lack of links to concrete processes and a high abstraction degree. We will consider SCs as multi-structural systems and propose an SC structure–operations dynamics approach to establish adaptive feedback loops at the tactical and operational decision-making levels.

Third, we assume that SCs as complex systems are described by *different models.* In modelling SC structure and operations dynamics, a number of *particular features of the SCM domain* should be taken into account. The processes of SC execution are non-stationary and non-linear. It is difficult to formalize various aspects of SC functioning. Besides, the SC models have high dimensionality. There are no strict criteria of decision-making for SCM and no *a priori* information about many SC parameters. The SC execution is always accompanied by negative perturbation impacts. These negative perturbation impacts initiate the SC structure dynamics and predetermine a sequence of positive control compensating for the perturbations. Unlike the automatic systems, adjustment control decisions in SCs are taken by managers and not by automatics. This results in individual interests, risk perceptions and time delays (from minutes to months) between disruption identification and taking adjustment measures.

To answer these challenges, the research approach in this study is based on the *combined application* of modern optimal control theory, operations research (OR), systems analysis and artificial intelligence. In particular, recent advancements in applying different OR approaches in the SCM domain as well as advancements in modern control theory and applied mathematics, namely the theory of structure dynamics control (SDC) and multi-model complexes will be considered.

Fourth, another particular feature of this book is the developing of *generic model constructions* and guidelines for near real-world problem identification and the application of a problem-specific solution method (or a combination of methods) rather than dealing with clear identifiable problems and specific problem cases with a known desirable outcome in a known environment. Such problem localizations may lead to unrealistic simplifications where the connection of the model to reality fails. Real problems are different and involve multiple decision makers and different interests and value sets (e.g., individual risk perception). Of course, the continuous improvements in the solution of partial SCM problems are very important. There is a wealth of literature on these problems and the optimization potential. However, this literature indicates even more case-study applications but as yet almost no new methodical approaches.

In this research, we will not concentrate on the solutions to partial benchmarking problems of SCM that are known and localized in the literature. We will put the emphasis on the fact that these partial problems and specific cases are tightly interlinked with each other with regard to different SCM levels, different SC structures, and in SC dynamics. Even these links will be at the forefront of our considerations to develop a *problem semantic* for the SCM domain. In the situation when the further optimization of known referenced problems can hardly provide a significant increase in SC performance, even the investigation into the links between different problems may lead to a new breakthrough in SCM research.

In this book, we will consider the modelling level at a higher degree of abstraction and develop generic methodical constructs that can be localized in concrete environments with the help of certain methodical guidelines. Finally, the proposed approach will reflect the fact that SCs evolve as a result of subjective decisions taken by people on the basis of compromised iterative decision making procedures within an information environment.

Book Organization

The contents of the book are organized in 15 chapters' format as follows.

Chapter 1. Evolution of Supply Chain Management (SCM)
The chapter starts by describing the role of SCM in enterprise management. Subsequently, the predecessors and establishment of SCM are discussed. Particularly, market and enterprise management paradigm developments in 1960–2010, objective economic grounds of SCM development, and the development of SCM are presented. Finally, the issues of SCM and related disciplines are discussed. We analyse the interrelations of logistics and SCM as well as arguing the multidisciplinary nature of SCM. The chapter highlights the key role of SCM in modern enterprise management and concludes that this role will increase in the coming years due to further globalization, customer orientation and advancements in information technology (IT). Finally, the main research directions on SCM are discussed.

Chapter 2. Drivers of Supply Chain Management
In this chapter, the issues in agile, flexible and responsive SCs and related categories of postponement, virtual enterprises (VE) and coordination are considered. Different interrelations of these categories are examined. We also present the state of the art of adaptive SCM (A-SCM). Subsequently, the basics of the conceptual vision of the A-SCM approach at the organizational level are considered. We start with the main definitions, and then we consider the A-SCM framework as composed of elements drawn from SCM, VE, agile/responsive SCs, and sustainable SCM. In the A-SCM framework, we do not set off different value-adding chain management strategies with each other, but consider them as an integrated framework. All three value chain drivers – products and their life cycles, customers and their orders, and suppliers/outsourcers – are enhanced by combining the elements from SCM, agility, and sustainability. Moreover, these drivers are interlinked within a unified information space.

Chapter 3. Decision-making Support for Supply Chain Management
This chapter deals with decision-making support for the SCM domain. The first part of the chapter is devoted to basic approaches to modelling SCs. We consider the model-based decision-making support as being composed of mathematical and informational models as well as of different hybrid models. In the class of mathematical models, we distinguish the research paradigms of OR, control theory and agent-based approaches. We also consider the main solution techniques: optimization, simulation, statistics, heuristics and hybrid models. Subsequently, we consider the information modelling of SCs subject to business process reengineering models, IT-driven models, and the use of modern IT techniques and methods for the integration of SC decision-making models. In the second part of the chapter, we analyse the information systems-driven decision-making support. Basic IT with regard to different application areas within the SCM domain are presented.

Finally, we consider possibilities to develop integrated modelling frameworks (IMF) from informational and mathematical perspectives.

Chapter 4. Challenges in Research on Modern and Future Supply Chains
This chapter deals with modern developments and challenges in SCM from both the practical and theoretical points of view. We highlight the following main challenges for SCM: compromising potential SC economic performance and SC stability; capturing uncertainty and dynamics; handling the multi-structural nature of SCs; ensuring interrelations and optimality of decisions at different management levels; conducting multi-disciplinary research on SCM; establishing links to different stages of the product life cycle, related enterprise management functions and the environment. Finally, we consider challenges in the further development of IT and organizational aspects for SCM. We conclude by summarizing 12 main misunderstandings of SCM that we have experienced in our teaching and consulting work so far.

Chapter 5. Uncertainty, Risk and Complexity
In the focus of this chapter are the issues of uncertainty, risk and complexity in SCs. We discuss the origins of uncertainty, risk and complexity and provide proper classifications. The uncertainty factors are divided into environmental uncertainty, human thinking and decision-making uncertainty. The distinguishing purposeful and non-purposeful perturbation influences also form the basis of the proposed classifications. Subsequently, issues in SC complexity are presented. We conclude this chapter by analysing the practical issues of uncertainty in SCs. This chapter shows the interrelations between uncertainty and complexity management. The interlinking of uncertainty, risk, disturbances and deviations are discussed. Finally, constructive arguments to consider SCs as complex systems are discussed.

Chapter 6. Handling Uncertainty in Supply Chains
As indicated in the recent literature, there are two types of risk affecting SCs: (1) risks arising from the problems of coordinating supply and demand and (2) risks arising from disruptions to normal activities. According to this classification, we will continue to consider the issues of uncertainty and risk in this chapter. This chapter analyses purposeful perturbation influences from the point of view of SC security, and non-purposeful perturbation influences from the point of view of SC vulnerability. We describe different kinds of both purposeful and non-purposeful perturbation influences. Subsequently, managerial impacts to handle uncertainty in SCs are addressed. In particular, leverages of SC reliability and flexibility are analysed.

Chapter 7. STREAM: Stability-based Realization of Economic Performance and Management
In this chapter we develop the conceptual basics of the approach to balancing SC economic performance and stability as the primary objective in SC planning and optimization. The developed concept is named STREAM (Stability-based Realization of Economic Performance and Management). The concept STREAM, as the name implies, is based on the idea that the SC's potential economic performance will be realized through the SC's stability. The conceptual model of STREAM is

based on conceptualizing the subject domain from uniform SCM and system-cybernetic positions by means of the interconnected considerations of (1) control and perturbation influences in SCs and (2) verbally describable properties of an SC as a business process (for example, security and flexibility) and theoretically attributed properties of an SC as a complex system (for example, adaptability and resilience). Finally, general algorithms of SC (re)planning under uncertainty are presented.

Chapter 8. Quantitative Modelling of Supply Chains
This chapter is devoted to the modelling approaches in the SCM domain. The chapter starts with an analysis of OR on SCM that can be divided into three primary approaches to conducting SC modelling. These are optimization, simulation and heuristics. Subsequently, control theory application in the SCM domain is discussed. Finally, the approaches of complex adaptive systems (CAS) and multi-agent systems (MAS) are analysed. A critical analysis of the advantages and limitations of different modelling techniques concludes this chapter. The chapter highlights the main features and application areas of OR, control theory and agent-based models in the SCM domain.

Chapter 9. DIMA – Decentralized Integrated Modelling Approach
In this chapter, the basics of the SC multi-disciplinary treatment in the DIMA (Decentralized Integrated Modelling Approach) methodology are presented. The main principles of the DIMA are SC elements' activity, multiple modelling, integration and decentralization. We consider these principles in detail in the course of the chapter. We introduce the concept of an "active modelling object" (AMO) as part of the generic model constructions. Integration is considered from four perspectives: the integration of various modelling approaches and frameworks, the integration of planning and execution models, the integration of decision-making levels, and the implementation of integration throughout "conceptual model → mathematical model → computation". The integration and combined application of various models is implemented by means of multi-model complexes and qualimetry of models.

Chapter 10. Structure Dynamics Control and Multi-model Analysis
One of the main features of SCs is the multi-structural design and changeability of structural parameters because of objective and subjective factors at different stages of the SC life cycle. In other words, SC structure dynamics is constantly encountered in practice. In this chapter, we present the concept and the models of SC structure dynamics. The common conceptual basis facilitates the construction of a complex of unified dynamic models for SC control. The models describe the functioning SC along with the collaboration processes within them. The unified description of various control processes allows the simultaneous synthesis of different SC structures. The proposed approach allows us to establish a dependence relation between the control technology applied to SCs and the SCM goals. This statement is exemplified by an analysis of SC functional abilities and goal abilities. It is important that the presented approach extends new scientific and practical results obtained in the modern control theory for the SCM domain.

Chapter 11. Adaptive Planning of Supply Chains
In this chapter, we discuss the kernel of planning and scheduling as an integrated management function and provide a classification of planning tasks. Then we consider the method of adaptive planning. Subsequently, we present a general conceptual framework of the adaptive planning and scheduling with the models' adaptation. The main purpose of the adaptation framework is to ensure parameter tuning of the dynamic scheduling model with regard to changes in the execution environment. In the proposed framework, the plans' adaptation is connected with the models' adaptation. Within the framework, a special controller concept is presented. Finally, we consider SC planning levels and their reflections. We develop a framework of decision-making consistency in SCM on the basis of adaptive planning principles.

Chapter 12. Modelling Operations Dynamics, Planning and Scheduling
In this chapter, we present mathematical models and algorithms for operations dynamics planning and scheduling. The basics of the research approach are discussed. A complex of dynamic models for integrated planning and scheduling is presented. This complex is composed of dynamic models for collaborative operations control, resource control and flow control. Subsequently, we consider algorithms for optimal SC operations control and develop our own one. The proposed approach is based on the fundamental scientific results of modern control theory and systems analysis in combination with the optimization methods of OR. We formulate the planning and scheduling as optimal control problems, taking into account the discreteness of decision-making and the continuity of flows with the use of special techniques, e.g., by transferring the non-linearity from the dynamic models into the left part of the differential equations in constraints. The modelling procedure is based on an essential reduction of a problem dimensionality that is under solution at each instant of time due to connectivity decreases. For the computations, the dynamic Lagrange relaxation, transformation of the optimal control problem to the boundary problem and maximization of Hamiltonians with the use of Pontryagin's maximum principle are used.

Chapter 13. Supply Chain Reconfiguration and Model Adaptation
In this chapter, we consider issues of SC reconfiguration and models' adaptation. We classify different SC reconfiguration issues within the control loop. The considerations presented lead us from a narrow traditional interpretation of complex systems' reconfiguration to a wide interpretation within a new applied theory of SDC. In the first phase of re-configuration, the forming (generation) of allowable multi-structural macro-states is performed. In other words, a structural-functional synthesis of a new SC should be fulfilled in accordance with an actual or forecasted situation. In the second phase, a single multi-structural macro-state is selected, and adaptive plans of SC transition to the selected macro-state are constructed. Subsequently, a mathematical model of the SC reconfiguration and algorithms of parametric and structure adaptation are presented.

Chapter 14. Supply Chain Global Stability and Manageability
In this chapter the mathematical model complex of SC stability analysis is presented. These formal models present at the mathematical level the conceptual model of the global stability. In its development, stability comes to be interpreted in different ways beginning with the classical BIBO stability up to the non-quantified "conceptual" stability concepts. We consider as stability the SC property to approach the real SC performance to the planned one under the interacting SC processes in the real perturbed execution environment with regard to the variety of execution and goal criteria. The SC stability analysis addresses the problem of the direct connection of SC stability and economic performance. The model is based on the dynamic interpretation of the SC functioning process and uses for the first time the method of attainable sets (AS) for the SCM domain.

Chapter 15. Experimental Environment
In this chapter the concept of the integrated experimental environment developed and its partial components are considered. A vision of a special software environment, which contains a simulation and optimization "engine" of SC planning, a Web platform, an ERP system and an SC monitor, is presented. For experiments, we elaborated two software prototypes: (1) SNDC – Supply Network Dynamics Control and (2) SCPSA – Supply Chain Planning and Stability Analysis. We provide some case examples with experimental results that reflect the models of the previous chapters.

Target Audience

The book is targeted to a broad range of professionals involved in SCM. It is structured to appeal to audiences seeking a discussion on conceptual business models, generic methodical principles and modelling approaches to modern concepts in SCM, as well as those interested in applied SCM problems from the decision-making support point of view.

The main target group consists of graduate students, professors and research associates in SCM, logistics management, industrial engineering, systems engineering, management science, decision analysis, operations management and applied OR, and practitioners and researchers working in the fields of SCM and operations management who aim to combine mathematical aspects of problem solving with the use of modern business concepts and IT solutions. This book may be used for teaching in graduate and professional development courses. It also provides valuable reference material for research in the area of SCM, logistics management, production and operations management. The professional societies interested in these areas are:

- European Operations Management Association (EurOMA)
- International Federation of Operations Research and Management Science (INFORMS)

- Production and Operations Management Society (POM)
- International Federation of Automatic Control (IFAC)
- Council of SCM Professionals (CSCMP)
- Institute of Electrical and Electronics Engineers (IEEE)
- Institute of Industrial Engineers (IIE)
- Decision Science Institute (DSI)
- German Operations Research Society (GOR)
- American Production and Inventory Control Society (APICS)
- Society of German Professors on Business Administration (VHB)
- German-Russian Logistics Society (DR-LOG)
- Society of Collaborative Networks (SOCOLNET)

Acknowledgements

We gratefully acknowledge all those who have helped us in bringing this book to publication. First and foremost, we have greatly benefited from the wealth of a vast array of published materials on the subject of SCs and associated topics, including SCM and SC modelling.

We would like to thank the reviewers of this book. The content of this book has benefited immensely from their valued insights, comments, and suggestions. Our sincere thanks to Prof. Dr-Eng. R. Yusupov, Prof. Dr-Eng. M. Okhtilev, Prof. Dr-Eng. A. Arkhipov, Prof. Dr J. Kaeschel and Prof. Dr-Eng. V. Kalinin for their invaluable contributions to this book. We would also like to acknowledge gratefully the contribution of Prof. Dr-Eng. M. Schenk, Prof. Dr-Eng. Y. Tolujev and M.Sc. Tobias Reggelin to Sect. 8.2 and the contributions of Dr. S. Potrjasaev and Dr. A. Malyugin to programming our models and prototype software development. With regard to the manuscript preparation, we thank Mrs A. Ikonnikova and Mrs E. Solodova for their technical assistance.

The research described in this paper is partially supported by grants from the Russian Foundation for Basic Research (grants 07-07-00169, 08-08-00403, 09-07-00066), the Saint Petersburg Scientific Center RAS, and the Department of Nanotechnologies and Information Technologies of the Institute of Systems Analysis RAS (2.3).

We also acknowledge the International Journal of Production Research, the European Journal of Operational Research, the International Journal of Manufacturing Technology and Management, the International Journal of Integrated Supply Management, the International Journal of Agile Systems and Management, the European Conference in Modelling and Simulation (ECMS), the IFAC Symposium on Information Control in Manufacturing INCOM, the International Conference on Virtual Enterprises (PRO-VE), and the German-Russian Logistics Workshops (DR-LOG) for prior publication of selected material that has been expanded into this book.

Finally, we wish to thank Mr Anthony Doyle, Senior Editor for Engineering, his able assistant Ms Claire Protherough, and the entire Springer production team for their assistance and guidance in successfully completing this book.

Last but not least – we cordially thank our families who supported us enormously during the work on this book.

About the Authors

Dr. Sc. (econ), Dr rer.pol. Dmitry Ivanov, born in 1978, is a research associate at the chair of production economics and industrial science at Chemnitz University of Technology, Germany. He studied production management and industrial organization at Saint Petersburg State University of Technology and Design. He graduated in 2000 with distinction. In 2006, he gained the degree of Dr.rer.pol. at Chemnitz University of Technology. In 2008, he gained the degree of Doctor of Economics Science at Saint Petersburg University of Economics and Finance (FINEC).

From 1999 to 2001 he worked as a project engineer in an ERP systems consulting house and as a research associate at Saint Petersburg State University of Technology and Design. In 2001–2002 he studied and graduated with distinction on an international post-graduate programme at Chemnitz University of Technology in Germany. From 2003 to 2005 he worked as a research associate at Saint Petersburg State Polytechnical University. At this time he was also a project manager for subcontracting networks founded by the Saint Petersburg Government. In 2005 he won a German Chancellor Scholarship and spent 1 year at Chemnitz University of Technology. Since 2006 he has worked at Chemnitz University of Technology.

He is the (co)-author of more than 150 scientific works. He is a presidium member of the Russian National Supply Chain Council. He leads the German–Russian Logistics Society DR-LOG. His research interests lie in the area of adaptive and agile supply chains, applied optimal control theory, OR and business information systems. His works have been published in various academic journals, including International Journal of Production Research, European Journal of Operational Research, International Journal of Manufacturing Technology and Management, International Journal of Integrated Supply Management, International Journal of Agile Systems and Management, etc.

Homepages: www.ivanov-scm.com; www.dr-log.org

Prof. Dr-Eng. Boris Sokolov, born in 1951, is a deputy director at the Russian Academy of Science, Saint Petersburg Institute of Informatics and Automation. He received his M.Sc., PhD and Dr. Sc.Eng. in 1974, 1983 and 1993 respectively. From 1966 to 1999 he developed his military career at the Mozaisky Academy in Leningrad/St Petersburg, from cadet to colonel. Since 2000 he has been a professor in St Petersburg State University of Aerospace Instrumentation. Professor Sokolov is the author of a new scientific lead: optimal control theory for structure dynamics of complex systems. Since 1985, he has been continuously developing this new scientific school. By now his numerous followers got their PhD and Dr. Science degrees under his supervision.

The research interests of Prof. Sokolov are as follows: basic and applied research in mathematical modelling and mathematical methods in scientific research, optimal control theory, and mathematical models and methods of decision-making support in complex technical-organizational systems under uncertainties and multi-criteria. In 2008, he became an honoured scientist of Russia.

He is the author and co-author of 3 books on systems and control theory and of more than 180 scientific works published in various academic journals, including European Journal of Operational Research, International Journal of Manufacturing Technology and Management, Journal of Computer and Systems Sciences International, Differential Equations, and Automation and Remote Control.

Homepage*:* www.spiiras-grom.ru

Contents

Abbreviations

3PL	Three-part logistics
4PL	Four-part logistics
AC	Adaptive control
AMO	Active modelling object
APS	Advanced planning systems
ARIS	Architecture of information systems
A-SCM	Adaptive supply chain management
AS	Attainable set
BIBO	Bounded-in-bounded-out
BOM	Bill of materials
BTO	Build-to-order
CALS	Continuous acquisition and life cycle support
CAS	Complex adaptive system
CPFR	Collaborative planning, forecasting, and replenishment
CS	Control system
DAMG	Dynamic alternative multi-graph
DIMA	Decentralized integrated modelling approach
DSS	Decision support systems
ECR	Efficient consumer response
EDI	Electronic data interchange
EMP	Event management plan
EPC	Event-process chain
EPI	Event probability index
ERP	Enterprise resource planning
FA	Functional abilities
FD	Function of diagnostics
FIFO	First-in-first-out
FMCG	Fast moving consumer goods
GA	Goal abilities
GSF	General systems framework
IDEF	Integration definition for function modelling
IDSS	Integrated decision-support systems
IMF	Integrated modelling framework
ISO	International organization for standardization
IT	Information technologies
JIS	Just-in-sequence
JIT	Just-in-time

LIFO	Last-in-first-out
MAS	Multi-agent system
MES	Manufacturing execution system
MILP	Mixed integer linear programming
MIP	Mixed integer programming
MPC	Model predictive control
MSMS	Macro-structural macro-states
NP	Nondeterministic polynomial
OEM	Original equipment manufacturer
OLAP	On line analytical processing
OPP	Order penetration point
OR	Operations research
PCI	Peripheral component interconnect
PID	Proportional–integral–derivative
PLM	Product life cycle management
QI	Quality index
QR	Quick response
RFID	Radio frequency identification
SC	Supply chain
SCD	Supply chain design
SCEM	Supply chain event management
SCM	Supply chain management
SCMo	Supply chain monitoring
SCOR	Supply chain operations reference
SCP	Supply chain planning
SCPSA	Supply chain planning and stability analysis
SDC	Structure dynamics control
SNDC	Supply network dynamics control
SS	Simulation system
STREAM	Stability-based realization of economic performance and management
TCO	Total cost of ownership
UML	Unified modelling language
VE	Virtual enterprise
VMI	Vendor-managed inventory
WIP	Work-in-progress
WMS	Warehouse management systems
WWW	World-wide-web
XML	Extensible markup language

Chapter 1
Evolution of Supply Chain Management (SCM)

The secret of success is for a man to be ready
for his opportunity when it comes.
Benjamin Disraeli

We know what we are, but know not what we may be.
William Shakespeare

1.1 Role of SCM in Enterprise Management

Only a few years have passed since enterprise management and organizational structure have been considered from the functional perspective: marketing, research and development, procurement, warehousing, manufacturing, sales, and finance. The modern value creation logic challenges other schemes (see Fig. 1.1).

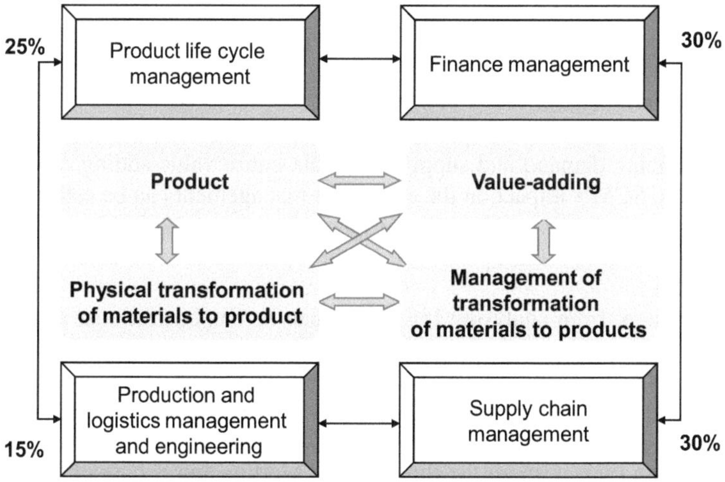

Fig. 1.1 The main elements of enterprise management

The basic element of entrepreneurship is the creation of *added value*. This is the basis for all further considerations. In normal business conditions, this value is connected with a product or a service. To be more precise, the added value creation is dispersed over the whole value chain, from raw materials to product dissemination and consumption. The product life cycle management (PLM) is the first component of enterprise management. Its impact on enterprise growth is different in different branches and industrial environments. However, a related value of about 25% can be estimated.

The second enterprise management component is *finance*. The financial flows that accompany the material flows need to be handled efficiently. This concerns both the direct financial flows for product creation and indirect financial flows such as stock exchange activities, investments, etc. The impact of financial management on enterprise management may add up to 35%, especially if an enterprise is presented at stock exchanges. Financial management is out of the scope of this book.

The physical production of a product is based on local (but tightly interlinked!) product transformation (*manufacturing*) and transition (*logistics*) processes. The logistics' impact on the enterprise management usually amounts to 10–15%.

In the modern customer-driven economy, a product must not only be produced but also marketed. This means that a product is to be produced according to customers' requirements. Besides, minimum costs for product creation are usually desired. To achieve this, on the one hand, manufacturing and logistics process optimization is required. On the other hand, a continuous balance of demands and supplies is needed. This balance can be ensured by means of integrating and balancing the local processes along the entire value-adding chain. The last aspect is even the kernel idea of SCM.

Our empirical experiences and the insights of the existing literature show that even the aspect of the optimization of the links between manufacturing and logistic processes in a value-adding chain has a greater impact on the business optimization in modern SCs. This is due to (1) even more complex SCs and an uncertain environment and (2) the significantly longer-lasting research on manufacturing and logistic optimization.

SCM is one of the key components of enterprise management and is responsible for balancing demand and supply along the entire value-adding chain (Christopher 2005). SCM's impact on the enterprise management can be estimated as up to 30%. From decisions on the SC configuration arise up to 80% of the total SC costs (Harrison 2005) and up to 75% of the operational costs in SCs (Wannenwetsch 2005).

In conclusion, from analysing Fig. 1.1, it can be emphasized that all the enterprise management drivers should be tightly interlinked for maximum *performance* (effectiveness and efficiency). PLM can be interlinked with SCM, i.e., through the suppliers' and customers' participation in new product development and engineering. A better synchronization of material, informational, and financial flows will have a positive impact on all the three flows (Mertins and Schallock 2009). Marketing can have a profound impact on adjusting SC imbalances with regard to over-inventories (Christopher 2005, Rudolph and Drenth 2007). The simultaneous optimization of manufacturing and logistics processes and the links between these processes also brings positive effects with regard to shareholders' satisfaction (Olle 2008).

Let us consider SCM as a system within enterprise management (see Fig. 1.2).

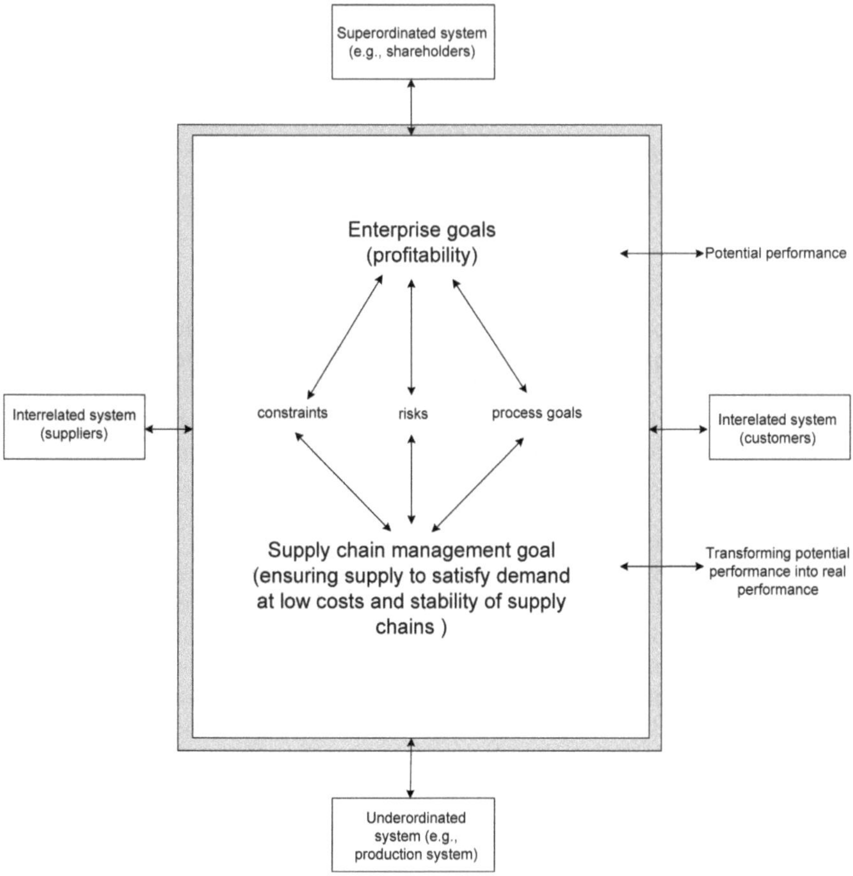

Fig. 1.2 System of SCM

SCM is subject to the goals of the superordinated strategic management level and should be harmonized with the enterprise competitive strategy (Fischer 1997, Chopra and Meindl 2007), given constraints (i.e., financial constraints) and risks (the achievement of potential SC goals is always subject to a certain amount of risk due to uncertainty).

SCM within an enterprise is based on both enterprise-internal activities and interactions with the external interrelated systems. These systems are customers and suppliers. SCM is in turn the goal-set system for the underordinated systems, e.g., a production system.

1.2 Predecessors and Establishment of SCM

1.2.1 Market and Enterprise Management Paradigm Developments 1960–2010

Over the last 50 years, a transition from the producers' market to the customers' markets has occurred. This transition began in the 1960s with an increasing role of marketing in the conditions of *mass production* of similar products to an anonymous market. This period is known as the economy of scale. After filling the markets with products, the quality problems came to the forefront of enterprise management. In the 1970s, total quality management (TQM) was established.

The increased quality caused the *individualization* of customers' requirements in the 1980s. This was the launching point for the establishment of the economy of the customer. This period is characterized by efforts for optimal inventory management and a reduction in production cycles.

In the 1980–1990s, handling a high product variety challenged enterprise management. Another trend was the so-called *speed effect*. The speed of reaction to market changes and cutting time-to-market became even more important. Consequently, the optimization of internal processes simultaneously with external links to suppliers was rooted in the concepts of lean production and just-in-time.

Throughout the 1990s, companies concentrated on development approaches to core competencies, outsourcing, innovations and collaboration. These trends were caused by globalization, advancements in IT and integration processes into the world economy. Particularly in the 1990s, the paradigm of SCM was established.

1.2.2 Objective Economic Grounds of SCM Development

The development of SCM was driven in the 1990s by three *main trends*: customer orientation, markets globalization and establishing an information society. These trends caused changes in enterprise competitive strategies and required new adequate value chain management concepts (see Fig. 1.3).

First, to remain competitive, enterprises concentrated on *product individualization* and maximum meeting of customers' requirements. *Flexibility* and *responsiveness* came to be the key factors in supply management. Second, in the 1990s, new markets in Asia, Eastern Europe, and South America were extensively acquired and the production facilities actively shifted to these regions. Thirdly, remarkable advances in *IT* and establishing the World Wide Web (WWW) provided the basis for innovative business concepts. On the whole, the focus turned to the consideration of entire value-adding chains, all the elements and links within them and outside the own enterprise to ensure business profitability and competitiveness. This launched the mass establishment of SCM.

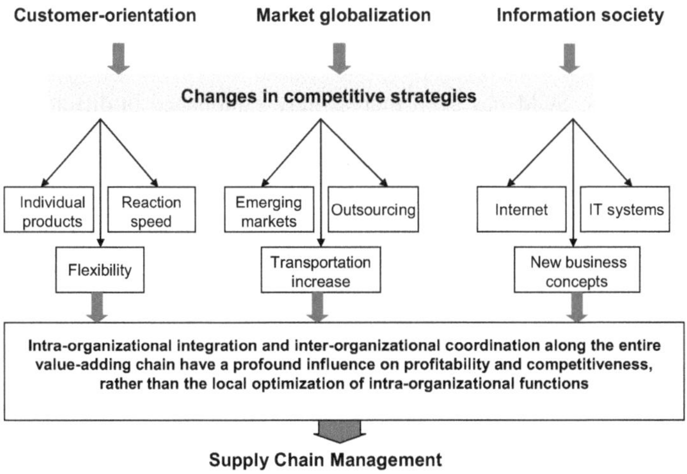

Fig. 1.3 Objective grounds of SCM development

The practice of SCM has provided enough evidence that intra-organizational integration and inter-organizational coordination along *the entire value-adding chain* have a profound influence on profitability and competitiveness, rather than the local optimization of intra-organizational functions.

1.2.3 Development and Merit of SCM

The first use of the term "SCM" is commonly related to the article "SCM: Logistics Catches up with Strategy" by Oliver and Weber (1982). They set out to examine material flows from raw material suppliers through SC to end consumers within an integrated framework that has been named SCM.

The origins of SCM can also been seen in early works on postponement (Alderson 1950), system dynamics and the bullwhip effect (Forrester 1961), inter-firm cooperation (Bowersox 1969), optimal multi-echelon inventory management (Geoffrion and Graves 1974), just-in-time (JIT), and lean production. The first book on SCM appeared in 1992 (Christopher 1992).

In practice, SCM became important in the 1990s in retail networks, the automotive industry, electronics and textiles. Trends of outsourcing, increased competition pressure, the establishment of new organizational forms in conditions of globalization, integration and IT development, expanding logistic services – all of these have driven the development of SCM.

The practical realization of ideas of balancing and synchronizing demand and supply along the entire value-adding chain has been enabled by business information systems and the Internet. IT provided a new level of coordination capabilities in SCs and enabled a breakthrough in SC responsiveness and flexibility. Information technology, on one hand, serves as an environment to support SCM. On the other hand, it has been the enabler of much advancement in SCM. The coordination and integration made SCs much more than simple inter-organizational cooperation.

In recent years, SCM has been increasingly established in different branches, such as aerospace, automotive, pharmaceutical, telecommunications, textiles and clothing, retail, 3PL, FMCG (fast moving consumer goods), construction and material, health care, food and beverages and high-tech manufacturing. The research on SCM has a different emphasis, i.e., in the automotive industry, the issues of alignment of market demands and supply and costing are at the forefront of discussions. For the FMCG sector, stockless processes, agility and mobile (IT) are the most important issues. For retail and clothing, the issues of outsourcing, uncertainty, collaboration and reverse logistics have a greater impact. In aerospace, the stage of product utilization and service is of a crucial nature.

Regarding the *merit and performance of SCM*, the following figures can be shown as examples. The increase in sales and reduction in costs in the value-adding chain due to SCM amounts to 15–30%. Partial effects such as inventory reductions, an increase in service level, SC reliability and flexibility, a reduction in transaction costs, etc., amount to 10–60%. These effects occur due to balancing supplies along the entire value-adding chain to ensure mutual iterative balances of production and logistics processes subject to full customer satisfaction. As customer orientation, globalization and IT advancements are still the ongoing trends in markets, the importance of SCM will become ever greater. Hence, SCM will be further developed and there will be ever more investments in SCs (Christopher 2005).

1.2.4 Organizational Aspects of SCM

SCM, as the term implies, is primarily directed to the inter-organizational level. Another successful application of SCM depends to a very large extent on the intra-organizational changes. Even the collaborative processes with an extended information systems application are managed by people who work in different departments: marketing, procurement, sales, production, etc. The interests of these departments are usually in conflict with each other. Hence, not only outbound synchronizations but also internal organizational synchronization are encompassed by SC organization.

Some levels of *SC organization* can be distinguished (Christopher 2005, Jahns 2008, Werner 2008). These are open market negotiations, cooperation, communication/integration, coordination and collaboration (see Table 1.1).

Table 1.1 Main categories of SC organization

Category	Content	Organization level
Open market negotiations	Supply on the basis of commerce offers	Lowest
Cooperation	Long-term contracts with suppliers and customers	Low
Communication/ Integration	Building channels and links within and outside the enterprise	Middle
Coordination	Building information interchange within the integration channels	High
Collaboration	Joint business strategies, collaborative promoting, sales and order forecasts, technological know-how sharing, process synchronization	Highest

If an enterprise handles its supplies and demands based on long-term contracts with suppliers and customers, this is *cooperation*. From cooperation, channels and links of an informational (i.e., e-mail, fax, ERP system) and physical (i.e., transportation) nature are initiated. If these channels are systematically used for information interchange (i.e., sharing data about demand planning or inventory levels or shipment control with RFID chips), this is *coordination*. Finally, if, along with the integration and coordination, your enterprise attracts suppliers and customers to new product development and design, co-creates joint business collaborated in joint promotion actions and sales forecasts, and takes part in know-how sharing, this is *collaboration*. Actually, only a few SCs in the world have achieved the highest collaboration and synchronization level. Between 15% and 20% of SCs are at the stage of advanced coordination, and about 50% can be placed between communication, integration and simple coordination.

The variety of possible SC organizational structures can be very large. However, while identifying and structuring possible SC organizations, the following five components can be considered as orientations for the identification of SC structures. These are competition, supplied and replenished products, production, and export–import relations (see Fig. 1.4).

In each of the five components, two key analysis categories are distinguished. These are as follows:

- competition: market share and the number of customers;
- supplied and replenished products: the number of different products and the number of different variants of each product;
- production: manufacturing depth: the number of technological processes and the number of operations in each of the processes; and
- export–import: how many products are outsourced abroad and what part of these products is sold in the country of outsourcing.

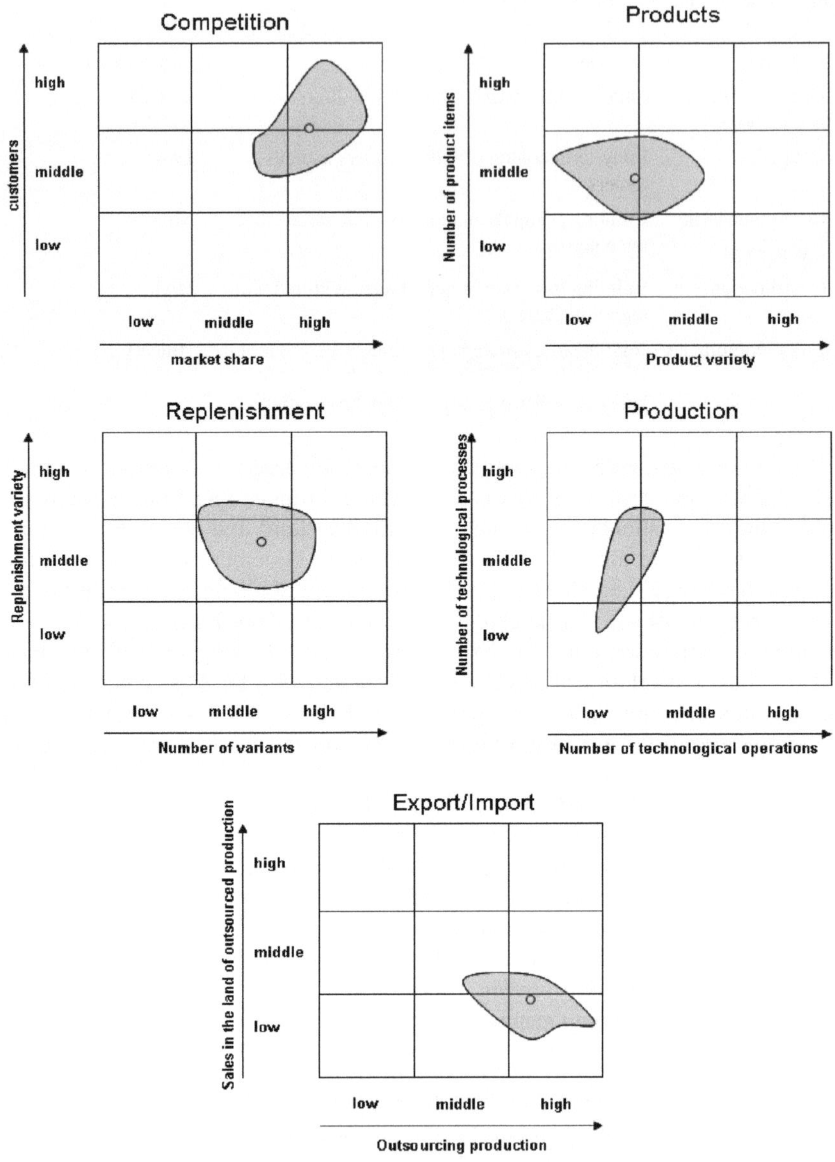

Fig. 1.4 Components for SC structures identification

At the first stage, enterprise data are placed within each of the five matrices. At the second stage, the data aggregation takes place to reveal SC structures, material, information, and financial flows. Some examples are presented in Fig. 1.5.

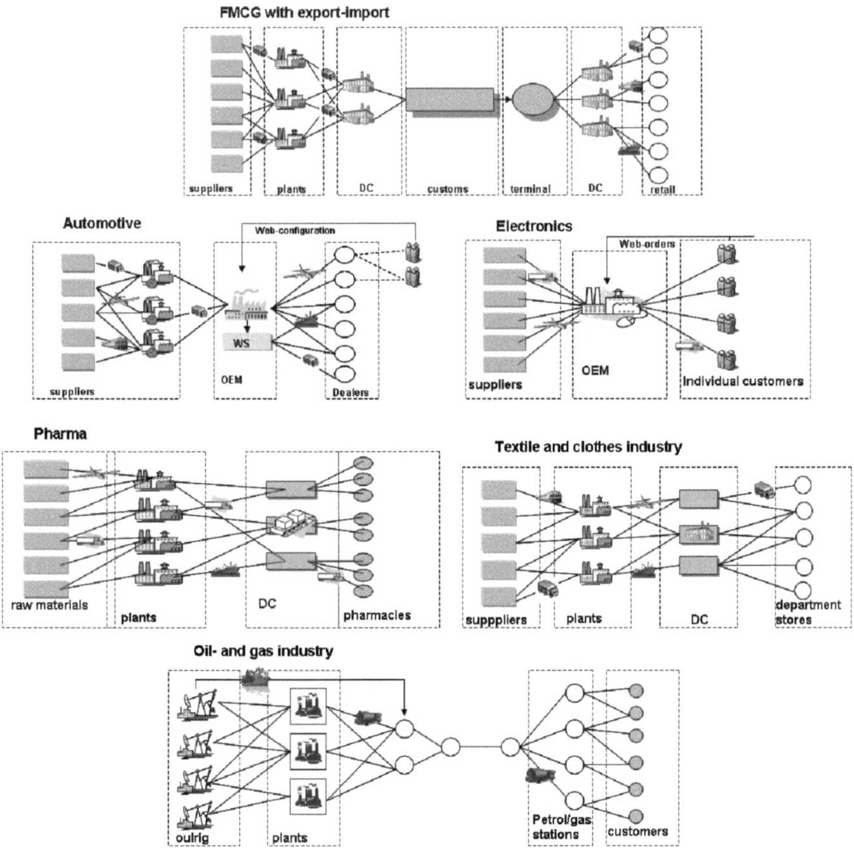

Fig. 1.5 Examples of SC structures

After the SC structures have been identified, the business processes should be analysed and proper information architecture should be selected. These aspects will be considered in the following section in this chapter.

Finally, let us consider the main organizational aspects of SC paradigm as compared with conventional self-oriented business making vision. In Table 1.2, we summarize the modern views on the main organizational aspects of SC paradigm (Christopher 2005, Mangan *et al.* 2007, Becker 2008).

Table 1.2 The main organizational aspects of the SC paradigm

Focuses of a conventional business vision	Focuses of a SCM business vision
Suppliers and own production/services	Focus on customers
Make-to-stock	Make-to-demand
Safety stocks	Coordination
Local optimization of transport, manufacturing, inventories	SC optimization
Functional thinking	Process thinking
Penalties for supply breaks	100% keeping supply terms
Capacity loading optimization	Flexibility and customer satisfaction
Operative planning on the basis of middle-term plans	Operative planning on the basis of current demand
Permanent lack of necessary materials	Monitoring availability and stock levels of materials
Optimizing direct costs	Optimizing total SC costs
Optimizing usage of containers	Building batches as JIT and just-in-sequence
Volume maximization	Cycle time optimization
Enterprises compete	SCs compete

As can be observed from Table 1.2, SC organization is a very important and complex problem. It requires thorough changes in enterprise organization and inter-enterprise links' reorganization. IT and innovative business concepts can have a profound effect on your enterprise but, first of all, to establish flexible SCs, flexible people are needed who will implement the IT usage and the management.

1.3 SCM and Related Disciplines

1.3.1 Logistics and SCM

The interrelation of logistics and SCM is a "hot spot" in many discussions. Actually, the elaboration of a unique viewpoint on this aspect should not be counted on. Sometimes, these discussions appear very similar to discussions on interrelations of theatre and cinema in the 1940–1950s. Nevertheless, both the theatre and the cinema exist now. So both the logistics and SCM will exist in the future.

In modern literature, four main viewpoints of the interrelation of logistics and SCM can be classified (Mangan *et al.* 2007). These are:

- logistics as a part of SCM;
- SCM as a part of logistics;

- SCM instead of logistics; and
- logistics and SCM are independent and have some intersection points.

Without standing for one of these viewpoints, let us discuss our own understanding of logistics and SCM (see Fig. 1.6).

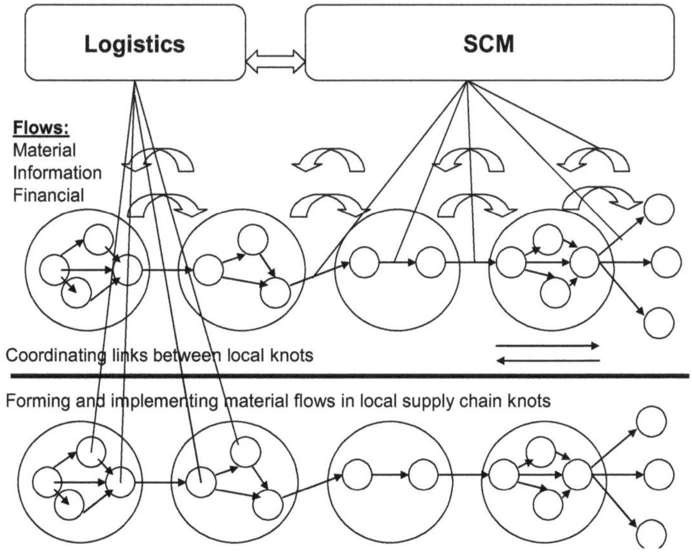

Fig. 1.6 Interrelation of logistics and SCM

In analysing the existing research literature and empirical case studies, the following can be concluded: logistics deals mostly with local functions for implementing the physical transition of material flows and SCM deals with the value-adding chain as a whole and concentrates on the managerial links between the local functions for implementing the physical transition of inbound and outbound material flows

In Fig. 1.6, the spheres of physical material flows and the management of these flows with information and financial flows are distinguished. To explain it very simply – the circles in Fig. 1.6 are the subject of logistics and the managerial (informational and financial) links between these circles are the subject of SCM. Logistics is attracted to optimizing the realization of physical transitions; SCM is attracted to the management level. In other words, logistics takes care of providing the right goods, in the right place, at the right time, in the right volume, in the right package, in the right quality, with the right costs, and SCM takes care of balancing the supplies along the entire value-adding chain subject to the full customer satisfaction.

In Fig. 1.7, an extract from a SC is presented to depict the above-mentioned issues.

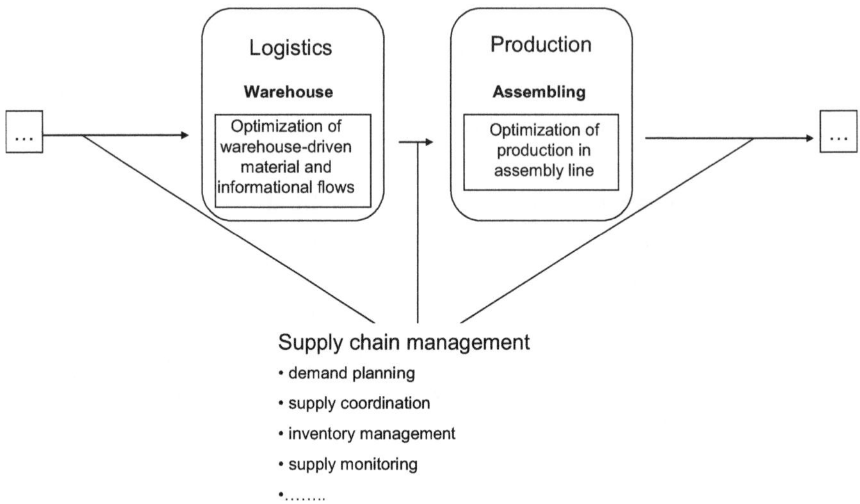

Fig. 1.7 An extraction from a value chain and functionalities of logistics, manufacturing, and SCM

As examples of *logistics problems*, warehouse management, transportation optimization, procurement quantity optimization, local inventory management, cross-docking design, inter-modal terminals design, etc. can be named. Accordingly, manufacturing deals with optimizations in assembly lines, production cells, etc. As examples of *SCM problems*, distribution network design, demand forecasting, collaborative inventory management, supply coordination, supply monitoring, and controlling can be identified. In practice, the logistics and SCM problems interact and are tightly interlinked. This is impossible to consider logistics and SCM in isolation from each other. SCM and logistics mutually enriches themselves. SCM is a very important part of logistics. In its turn, logistics is a very important part of SCM.

1.3.2 The Multi-disciplinary Nature of SCM

SCM is interlinked with logistics, operations management, strategic management, marketing, industrial organization, production management and informatics. Some examples follow. Cooperation is the basis of SCM. However, the issue of cooperation is within the scope of industrial studies. Cooperation process flows in turn belong to the logistic and SCM competences. SCM integrates the strategic goals of production (process flexibility, productivity and efficiency) and logistics (providing the production and customers with products, low logistic costs and a high logistic service level).

Besides cooperation, coordination belongs to the most important SCM components. The basis of the coordination is IT. The functionalities of IT can be different, i.e., SC planning, SC monitoring, data interchange, radio frequency identification (RFID), trace and tracking, etc. Hence, SCM is interlinked with informatics and engineering. The above-mentioned interrelations are presented in Fig. 1.8.

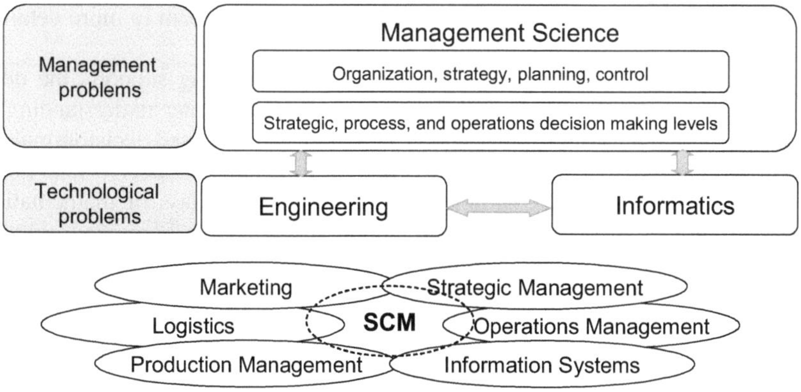

Fig. 1.8 SCM as a multi-disciplinary framework

Only a few years have passed since SCM was considered as an extension of logistics and purchasing management. As stated in Christopher (2005), modern SCM is a wider concept than just logistics. Moreover, SCM has been extensively developed into an independent research and management domain. For the last few years, the focus of SCM has shifted to the management level. Organization, strategies, planning and control at different decision-making levels are the subjects of management science. To implement the managerial functions, information and engineering technologies are needed. In its turn, the level of the existing technologies enables or disables management concepts. Hence, SCM should be considered as a multi-disciplinary framework of management science, engineering science and information science.

1.3.3 The Main Directions of Research on SCM

When considering the literature on SCM, it is apparent that three major research streams exist. One contains conceptual business research, another is modelling, and the third deals with software engineering and information tools (Simchi-Levi *et al.* 2004, Chandra and Grabis 2007, Gunasekaran and Ngai 2009).

In recent years, the concepts of *business networks* have been developed increasingly. A large number of concepts of network organization and management, such as extended enterprise, agile SCs, responsive SCs, virtual enterprise, etc. have been proposed (Christopher and Towill 2001, Camarinha-Matos and Afsarmanesh 2004, Ross 2004, Yusuf *et al.* 2004, Gunasekaran *et al.* 2008). Coordination concepts and strategies for SCM have also been developed, such as collaborative planning, forecasting, and replenishment (CPFR), vendor-managed inventory (VMI), and SC monitoring (SCMo). In this stream, researchers try to grasp managerial structure issues related to business networking. The studies give extensive guidelines for a deeper understanding of the basic principles and advantages of new organizational forms. We will consider this research stream in more detail in Chap. 2.

The second research direction is *modelling*. SC modelling supports the decision-making process through computational tools and a greater understanding of SC operating characteristics (Chatfield *et al.* 2009). Model-based decision-making support is composed of mathematical and informational models (Fiala 2005, Chandra and Grabis 2007, Chatfield *et al.* 2009). In the class of mathematical models, a number of decision-making support research paradigms and solution methods can be distinguished. The main research paradigms are OR, control theory and agent-based approaches. The main solution techniques are optimization, simulation, statistics, heuristics, and hybrid models (Beamon 1998, Swaminathan *et al.* 1998, Tayur *et al.* 1999, Fox *et al.* 2000, Shen *et al.* 2001, de Kok and Graves 2004, Simchi-Levi *et al.* 2004, Koechel and Nielandaer 2005, Surana *et al.* 2005, Chopra and Meindl 2007, Chandra and Grabis 2007, Sarimveis *et al.* 2008). Information models describe the problems from an information processing perspective. Into the models of this class fall business process reengineering models, coordination-driven models, and data warehouse and knowledge-base models (Chandra and Grabis 2007). As emphasized by Simchi-Levi *et al.* (2003), Chandra and Grabis (2007), Kuehnle (2008), Chatfield *et al.* (2009) and Ivanov (2009), SC problems are tightly interlinked with each other and have multi-dimensional characteristics that require the application of different integrated frameworks of decision-making support.

Research on *management information systems* can be divided into theoretical and empirical paths. The latter contains a wealth of different application cases in diverse SC environments (Subramanian and Iyigungor 2006, Ketikidis *et al.* 2008) The key ingredients of IT in SCs are the use of the Internet and Web-based service portals, integrated information/knowledge within ERP software, and decision-support systems (DSS) that utilize proven algorithms for various strategic, tactical, and planning problems in specific industry domains (Fiala 2005, Chandra and Grabis 2007). The theoretical research on IT in SCM discusses improvements in information sharing (Chen *et al.* 2000, Dejonckheere *et al.* 2003, Chandra *et al.* 2007), developing a taxonomy for IT in the SCM domain (McLaren and Vuong 2008), and the use of modern IT techniques and methods for integration of SC decision-making models (Chandra and Grabis 2008, Chatfield *et al.* 2009).

The mathematical model-based, information model-based, and IT-supported decision-making components are tightly interlinked with each other. Mathematical models support decision-making through quantitative aspects. Information models describe and depict both SCs and decision-making processes. Information systems support the decision makers in practice.

In line with the above classification into three research streams, we will organize the layout of this book. The key point of our considerations will be directed to the modelling level. However, the levels of conceptual business research and IT will also be considered. In Chap. 2 we will consider the existing conceptual business framework in the SCM domain and develop a vision of the conceptual framework for A-SCM. In Chap. 3 we will consider the SC modelling and information systems domains. Chapter 4 will discuss the main challenges in research on SCM with regard to all the three research streams. Subsequently, we will propose our approaches to the stated challenges. Chapters 5 and 6 will be devoted to the uncertainty, risks and complexity in SCs. Based on these considerations, the concept STREAM will be presented in Chap. 7. Chapter 8 will analyse the approaches to the quantitative modelling of SCs and reveal their advantages and shortcomings. These considerations will be further developed in Chap. 9 where the multi-disciplinary modelling approach to SC modelling, named DIMA, will be presented. The following chapters of the book will be devoted to answering the challenges of SC multi-structural design, structure, and operation dynamics, reconfiguration, adaptation, and stability estimation at the modelling level. In Chap. 10 we will present the approach to SC structure dynamics control as a whole. In Chap. 11–14 the main frameworks and models that have been generalized in Chap. 10 will be considered in detail. Chapter 11 will be devoted to SC adaptive planning. In Chap.12 a complex of dynamic models for SC scheduling will be presented. In Chap. 13 the reconfiguration and models' adaptation problems will be addressed. Models of SC global stability analysis will be presented in Chap.14. The developed experimental environment and some experimental examples will be presented in Chap. 15.

References

Alderson W (1950) Marketing efficiency and the principle of postponement. Cost and Profit Outlook 3:15–18

Beamon BM (1998) Supply chain design and analysis: models and methods. Int J Prod Econ 55(3):281–294

Becker T (2008) Prozesse in Produktion und Supply Chain optimieren. Berlin, Springer

Bowersox DJ (1969) Physical distribution development, current status, and potential. J Market 33:63–70

Camarinha–Matos LM, Afsarmanesh H (2004) The emerging discipline of collaborative networks. In: Camarihna-Matos L (Ed) Virtual enterprises and collaborative networks. Kluwer Academic Publishers, Boston

Chandra C, Grabis J (2007) SC configuration. Springer, New York

Chandra C, Grabis J (2008) Information technology support for integrated supply chain modeling. Hum Syst Manag 27(1):3–13

Chandra C, Grabis J, Tumanyan A (2007) Problem taxonomy: a step towards effective information sharing in supply chain management. Int J Prod Res 45(11):2507–2544

Chatfield DC, Harrison TP, Hayya JC (2009) SCML: An information framework to support supply chain modeling. Eur J of Oper Res 196(2):651–660

Chen F, Drezner Z, Ryan J K and Simchi–Levi D (2000) Quantifying the bullwhip effect in a simple supply chain: the impact of forecasting, lead times, and information. Manag Sci 46(3):436–443

Chopra S, Meindl P (2007) Supply Chain Management. Strategy, planning and operation. Pearson Prentice Hall, New Jersey

Christopher M (1992) Logistics and Supply Chain Management. Pitman Publishing, London
Christopher M (2005) Logistics and Supply Chain Management: Creating Value-Adding Networks. Financial Times Prentice Hall, Dorchester
Christopher M, Towill DR (2001) An integrated model for the design of agile supply chains. Int J Phys Distrib Oper Manag 31:235–244
Dejonckheere J, Disney SM, Lambrecht MR and DR Towill (2003) Measuring and avoiding the bullwhip effect: a control theoretic approach. Eur J Oper Res 147(3):567–590
Fiala P (2005) Information sharing in supply chains. Omega 33:419–423
Fisher ML (1997) What is the right supply chain for your product? Harv Bus Rev march–april: 83–93
Forrester J (1961) Industrial dynamics. MIT Press Cambridge
Fox MS, Barbuceanu M, Teigen R (2000) Agent–oriented supply chain management system. Int J Flex Manuf Syst 12:165–188
Geoffrion A, Graves G (1974) Multicommodity distribution system design by Benders decomposition. Manag Sci 29(5):822–844
Gunasekaran A, Ngai NWT (2009) Modeling and analysis of build-to-order supply chains. Eur J Oper Res 195(2):319–334
Gunasekaran A, Kee–hung Laib and TCE Cheng (2008) Responsive supply chain: a competitive strategy in a networked economy. Omega 36(4):549–564
Harrison TP (2005) Principles for the strategic design of supply chains. In: Harrison TP, Lee HL, Neale JJ (Ed) The practice of SCM. Kluwer Academic Publishers, Boston
Ivanov D (2009) DIMA – A research methodology for comprehensive multi-disciplinary modelling of production and logistics networks. Int J Prod Res 47(5):1133–1155
Jahns Ch (2008) Einkauf, Logistik und SCM. Gabler, Wiesbaden
Ketikidis PH, Koh SCL, Dimitriadis N, Gunasekaran A, Kehajova M (2008) The use of information systems for logistics and SCM in South East Europe: current status and future direction. Omega 36(4):592–599
Koechel P, Nielander U (2005) Simulation-based optimisation of multi–echelon inventory systems. Int J Prod Econ 93-94:505–513
Kok de AG, Graves SC (2004) Supply Chain Management: design, coordination and operation. Amsterdam, Elsevier
Kuehnle H (2008) A system of models contribution to production network (PN) theory. J Intell Manuf 18(5):543–551
Mangan J, Lalwani C, Butcher T (2007) Global logistics and supply chain management. John Wiley, New York
McLaren TS, Vuong DCH (2008) A "genomic" classification scheme for supply chain management information systems. J Enterpise Inform Manag 21(4):409–423
Mertins K, Schallock B (2009) Restructuring of value chains. In: Ivanov D, Meinber U (Eds) Logistics and SCM: Modern trends in Germany and Russia. Cuvillier Verlag, Goettingen
Oliver RK, Webber MD (1982) Supply chain management: logistics catches up with strategy. In: Christopher M (Ed) Logistics: the strategic issues. Chapman & Hall, London
Olle W (2008) Production in partnership. Logistics and supply chain management 2(25):32–38, (in Russian)
Ross A (2004) Creating agile supply chain. IEEE Manufacturing Engineer 82(6):18–21
Rudolph T, Drenth R (2007) Kompetenzen fuer Supply Chain Manager. Springer, Berlin
Sarimveis H, Patrinos P, Tarantilis CD, Kiranoudis CT (2008) Dynamic modeling and control of supply chain systems: a review. Comput Oper Res 35(11):3530–3561
Shen W, Norrie DH, Barthes JP (2001) Multi–agent systems for concurrent intelligent design and manufacturing. Taylor & Francis Group, London
Simchi-Levi D, Wu SD and Zuo-Yun S (2004) Handbook of quantitative supply chain analysis. Springer, New York
Subramanian GH, Iyigungor AC (2006) Information systems in supply chain management: a comparative case study of three organisations. Int J Bus Inform Syst 1(4):370–386

Surana A, Kumara S, Greaves M, Raghavan UN (2005) Supply chain networks: a complex adaptive systems perspective. Int J Prod Res 43(20):4235–4265
Swaminathan JM, Smith SF and Sadeh NM (1998) Modeling supply chain dynamics: a multiagent approach. Decis Sci 29(3):607–632
Tayur S, Ganeshan R, Magazine M (1999) Quantitative models for supply chain management. Kluwer Academic Publishers, Boston
Wannenwetsch H (2005) Vernetztes supply chain management. Springer, Berlin
Werner H (2008) Supply Chain Management: Grundlagen, Strategien, Instrumente und Controlling, 3. Auflage. Gabler Verlag, Wiesbaden
Yusuf MA, Gunasekaran A, Adeleye EO, Sivayoganathan K (2004) Agile supply chain capabilities. Determinants of competitive objectives. Eur J Oper Res 159(2):379–392

Chapter 2
Conceptual Frameworks for Supply Chain Management

No great discovery was ever made without a bold guess.
Isaac Newton

2.1 Agile, Flexible and Responsive Supply Chains

To ensure long-term competitiveness and survival, companies implement new strategies, based on collaboration with business partners and an advanced utilization of IT and Web services (Geunes *et al.* 2002). Various *competitive strategies* of agile, responsive and flexible SCs have been developed over the last decade.

In many branches, hierarchical SCs with long-term predetermined suppliers' structures and product programmes evolve into *flexible dynamic SC structuring* (Sarkis *et al.* 2007). Nowadays, agile organizations with heterogeneous structures, core competences, buyer-focused cells and extensive application of Web services are being increasingly introduced in practice (van Donk and van der Vaart 2007). Collin and Lorenzin (2006) emphasize that "an agile SC is a basic competitive requirement in the industry and building agility into operations requires a continuous planning process together with customers".

According to Vonderembse *et al.* (2006), "an agile SC profits by responding to rapidly changing, continually fragmenting global markets by being dynamic and context specific, aggressively changing, and growth oriented. They are driven by customer designed products and services".

Chandra and Grabis (2007) identified the key triggers for designing and implementing SC with regard to agility, flexibility and responsiveness. They are as follows:

- introduction of new product(s), or upgrade for existing product(s);
- introduction of new, or improvement in existing, process(es);
- allocation of new, or re-allocation of existing, resource(s);
- selection of new supplier(s), or deselection of existing ones;
- changes in demand patterns for product(s) manufactured;
- changes in lead times for product and/or process life cycles; and
- changes in commitments within or between SC members.

Within the strategy of agility, different concepts like VE, agile SCs and responsive SCs exist (Christopher and Towill 2001, Camarinha-Matos and Afsarmanesh

2004, Ross 2004, Yusuf *et al.* 2004). Lee (2004) specifies that the main objectives of SC agility are to respond to short-term changes in demand or supply quickly and to handle external disruptions smoothly. The most distinguished cases of agile SC applications are those of DELL, Benetton, AT&T, Nissan, BMW, Nokia etc. Bustelo *et al.* (2007), Collin and Lorenzin (2006) and Gunasekaran *et al.* (2008) ground the practical need and efficiency of agile SCs on the basis of empirical tests.

The advantages of agility, responsiveness and flexibility lie in customer-oriented networking and flexible configurable SCs conditioned by an enlargement of alternatives to search for suitable partners for a cooperation enabled by enterprise resource and advanced planning systems (APS) and Internet technologies. The agility, responsiveness, and flexibility ensure the following:

- flexibility and adaptation to market changes;
- building integrated business processes to unify customer relationships, forecasting, planning, replenishment, distribution, and manufacturing;
- systematic information coordination; and
- supply chain event management.

To narrow the literature analysis to the objectives of this research, we will concentrate on the problems of (1) modularization/postponement and agility, (2) virtualization and agility, and (3) coordination and agility.

2.1.1 Postponement, Modularization and Agility

Van Hoek (2001) defines *postponement* as "an organizational concept whereby some of the activities in the SC are not performed until customer orders are received". Recent quantitative models have evaluated the cost and benefits of applying postponement to a large variety of stochastic and deterministic settings (Li *et al.* 2008).

Ernst and Kamrad (2000) introduced a conceptual framework for evaluating different SC structures in the context of modularization and postponement. In the analysis, modularization is linked to postponement. The paper introduces taxonomy and develops a corresponding framework for the characterization of four SC structures, defined according to the combined levels of modularization and postponement: rigid, postponed, modularized and flexible. The study provides examples of efficient postponement and modularization combining by HP, Suzuki, and Benetton. Additional case examples include Dell Computers, Nike, IBM and General Motors and are given by Tully (1993) and Gunasekaran *et al.* (2008).

Reichhart and Holweg (2007) synthesize the existing contributions to manufacturing and SC flexibility and responsiveness, and draw on various related bodies of literature that affect a SC's responsiveness, such as the discussion of product architecture and modularization. Picot *et al.* (2001), Warnecke and Braun (1999) and Wirth and Baumann (2001) elaborated concepts and models of value-adding chain organization based on the integrated, customer-oriented networking of small autonomous elements (module, fractals, competence-cells and segments). Coordination between these autonomous elements usually leads to non-hierarchical organizational forms. The ideas of integrating the product and process modularity have also been extensively investigated in the mass customization approach (Chandra and Kamrani 2004).

A crucial issue in postponement and modularization is the determination of an *order penetration point* (OPP) (see Fig. 2.1). Towill and Mason-Jones (1999) have demonstrated that there are actually two decoupling points in SCs – the "material" decoupling point, or OPP, where strategic inventory is held in as generic a form as possible (this would correspond to the β-line in Fig. 2.1), and the "information" decoupling point (this would correspond to the α-line in Fig. 2.1).

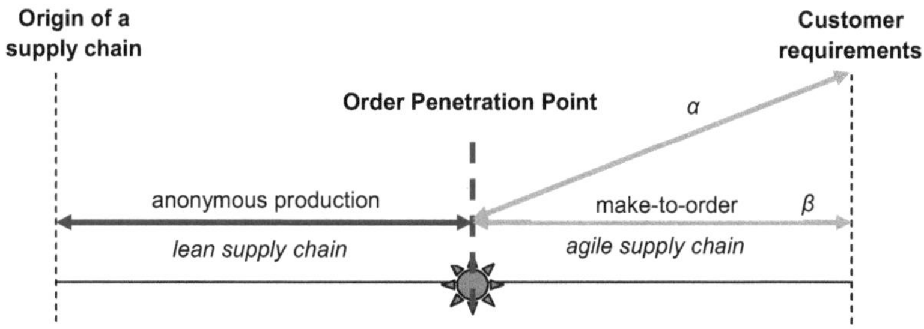

Fig. 2.1 Order penetration point

By efficient coordination in relation to these two decoupling points, a powerful opportunity for agile response can be created (Christopher and Towill 2000). The integration of lean (upstream of the OPP) and agile (downstream of the OPP) SCs was extensively discussed in Mason-Jones *et al.* (2000) and Christopher and Towill (2000). Recent study by Wang *et al.* (2009) reports on a three-dimensional concept based on the integration of product, engineering and production activities to define customer order decoupling points.

2.1.2 Virtualization and Agility

In SC agility, aspects of virtualization play a significant role. The main objective of a VE is to allow a number of organizations to develop a common working environment or *virtual breeding environment* with the goal of maximizing flexibility and adaptability to environmental changes and developing a pool of competencies and resources (Camarinha-Matos and Afsarmanesh 2004, Gunasekaran *et al.* 2008). VEs focus on speed and flexibility. A virtual enterprise is enabled by building a united information space with extensive usage of Web services.

VE structures are highly dynamic and their life cycles can be very short. The existence of a number of alternatives for SC configuration is remarkable. This is a great advantage because of a possibility to react quickly to customers' requirements. This also builds a structural-functional reserve for SC running. Unfortunately, VEs are considered mostly from the information perspective without dealing properly with managerial and organizational aspects. Besides, our practical experiences show that there are only a few (if any) organizations that have managed to apply the main idea of the VE, collaborate for a short time and then disperse, perhaps to form new networks with other enterprises. There are two main obstacles: trust and technical project documentation.

2.1.3 Coordination and Agility

The agility and coordination of SCs have strong links to manufacturing and logistics postponement strategies. A recent AMR Research study shows the great importance of demand-oriented SC coordination: demand forecast accuracy creates high responsiveness and cuts costs right through the SC (Friscia *et al.* 2004). According to the study, the companies that are best at demand forecasting maintain on average 15% less inventory, 17% stronger perfect-order fulfilment and 35% shorter cash-to-cash life cycles.

Different concepts of *coordination* have been developed over the last two decades, such as efficient consumer response (ECR), collaborative planning, forecasting, and replenishment (CPFR) in retail as well as JIT and VMI in industries. Enablers of the coordination are IT, such as enterprise resource planning (ERP), APS, electronic data interchange (EDI), and RFID.

Collin and Lorenzin (2006) emphasize that, in practice, coordination determines the postponement strategy and the position of the OPP or decoupling point in the SC. The coordination has become a key factor in mitigating the *bullwhip effect* and in overcoming information asymmetry (Lee *et al.* 1997, Chen *et al.* 2000,). Moreover, due to Internet technologies, it has become possible to integrate customers into SC considerations, resulting in the development of the build-to-order (BTO) SCM (Gunasekaran and Ngai 2005, Sharif *et al.* 2007).

2.1.4 Flexibility, Coordination and Agility

Unlike the well-grounded manufacturing flexibility, SC *flexibility* is still an under-investigated area. Swafford *et al.* (2008) showed that achieving SC agility is a function of other abilities within the organization, specifically SC flexibility and IT integration. Using empirical data, this study grounded a domino effect among IT integration, SC flexibility, SC agility and competitive business performance. The results from this study indicate that IT integration enables a firm to tap into its SC's flexibility, which in turn results in higher SC agility and ultimately higher competitive business performance.

Tachizawa and Thomsen (2007) empirically investigated the aspects of flexibility related to the upstream SC. The results show that firms need supply flexibility for a number of important reasons (manufacturing schedule fluctuations, JIT purchasing, manufacturer slack capacity, a low level of parts commonality, demand volatility, demand seasonality and forecast accuracy), and that companies increase this type of flexibility by implementing two main strategies: "improved supplier responsiveness" and "flexible sourcing". The results also suggest that the supply flexibility strategy selected depends on the type of uncertainty (mix, volume or delivery).

Coronado and Lyons (2007) investigated the implications of operation flexibility in industrial SCs and the effect it has on supporting initiatives designed for BTO manufacturing. The analysis has revealed the close relationship between operation flexibility and the SC flexibility dimensions of people and information systems. Wadhwa *et al.* (2008) presented a study on the role of different flexibility options (i.e., no flexibility, partial flexibility and full flexibility) in a dynamic SC model based on some key parameters and performance measures. Fotopoulos *et al.* (2008) analysed flexible supply contracts under price uncertainty. Ozbayrak *et al.* (2006) showed that flexibility is interrelated with adaptation. The study considered a number of performance metrics such as work-in-progress (WIP), tardiness, responsiveness, and mean flow time with regard to three localized control policies.

The main observation from literature analysis is that the collaborative organization with heterogeneous structures, core competences, buyer-focused cells and extensive application of Web services makes it possible to form SCs based on a project-oriented networking of core competences through a partner selection from a pool of available suppliers in a virtual environment according to customer requirements. Such SCs are expected to be more flexible and reactive, and capable of rapid evolution and surviving competition. SC agility reserves are usually considered in relation to postponement, product modularization and different inventory redundancies on the cooperation side as well as demand-driven roll-out planning and collaborative forecasting on the coordination side. Another agility reserve is a temporary customer-oriented dynamical SC structuring with operative outsourcing alternatives (Ivanov and Teich 2009).

2.2 Vision of the Adaptive Supply Chain Management (A-SCM) Conceptual Framework

2.2.1 Adaptive Supply Chains: State of the Art

The first use of the term "*adaptive supply chain management*" (A-SCM) is regarded as being in 2001–2002 and in the area of information technologies. The SAP's (SAP 2002) initiative on adaptive SC networks can be considered as the first step in automating the SC networks using new technologies including agent-based, RFID, and Web services. Subsequently, a number of white papers on *A-SCM* appeared from different consulting houses.

During 2000–2009 a number of concepts were developed to meet the requirements for speed, agility, responsiveness and flexibility (Goranson 1999, Christopher and Towill 2001, Ross 2004, Yusuf *et al.* 2004, Camarinha-Matos and Afsarmanesh 2007, Gunasekaran *et al.* 2008). In these conceptual business models, SCs with heterogeneous structures and an extensive application of IT are expected to be more flexible and reactive, and capable of rapid evolution and surviving competition.

The *mathematical research* on adaptive SCs is rooted in CAS and control theory. Choi *et al.* (2001) claimed that emergent patterns in a supply network can be managed much better through positive feedback than through negative feedback from control loops. They conclude that imposing too much control detracts from innovation and flexibility; conversely, allowing too much emergence can undermine managerial predictability and work routines. Therefore, managers must appropriately balance the control and the emergence areas.

Surana *et al.* (2005) investigated how various concepts, tools and techniques used in the study of CAS can be exploited to characterize and model SC networks. These tools and techniques are based on the fields of non-linear dynamics, statistical physics, and information theory. In the study by Pathak *et al.* (2007), a theory-based framework is developed that combines aspects of CAS theory, industrial growth theory, network theory, market structure, and game theory. This framework specifies categories of rules that may evoke different behaviours in the two fundamental components of any adaptive supply networks, i.e., the environment and the firms in that environment. The framework is implemented as a multi-paradigm simulation utilizing software agents and it joins discrete-time with discrete-event simulation formalisms. Another agent-based model has been elaborated by Kaihara and Fujii (2008) to reflect SCs' abilities to adapt. The study considered the VE environment and developed a computer simulation model, clarifying the SC formulation dynamism on the negotiation process with adaptive behaviour. Many other papers have also dealt with agent-based modelling and SC adaptivity; e.g., Ahn *et al.* (2003) suggested a flexible agent system for SCs that can adapt to the changes in transactions introduced by new products or new trading partners.

Another research stream has been dealing with control policies and algorithms to adapt SCs by means of different techniques. Shervais *et al.* (2003) employed a set of neural networks to develop control policies that are better than fixed, theoretically optimal policies with regard to a combined physical inventory and distribution system in a non-stationary demand environment. The study analysed the control policies embodied by the trained neural networks and fixed policies (found by either linear programming or genetic algorithms) in a high-penalty cost environment with time-varying demand.

Scholz-Reiter *et al.* (2004) presented an adaptive control (AC) concept for production networks. This study also employed an agent-based method concerning the adaptive coordination of customer orders along the SC to handle flexibly disturbances in relation to the reallocation of alternative suppliers to ensure a timely and accurate fulfilment of customer orders. Kim *et al.* (2005) proposed centralized and decentralized adaptive inventory-control models for a SC consisting of one supplier and multiple retailers. The objective of the two models is to satisfy a target service level predefined for each retailer using a reinforcement learning technique called the action-value method, in which the control parameters are designed to change adaptively as customer-demand patterns change.

Jang (2006) developed a new control architecture originating from modern political systems that are designed to mediate conflicts among people and to accommodate a nation's global benefits. Similarly, the proposed model should also resolve conflicts among controllers and maximize the shop floor's overall benefits. Pandey *et al.* (2007) distributed a feedback control algorithm, called the adaptive logistics controller (ALC), which simultaneously decides the order quantities for each stage of the SC subject to minimizing the total WIP (work-in-progress) in the entire SC for a given demand. Cai *et al.* (2008) presented a fuzzy adaptive model with an adaptive proportional–integral–derivative (PID) controller. A further discussion on control theory application to the SCM domain will be provided in Chap. 9.

In this book we will consider *A-SCM as a new research direction* that requires comprehensiveness with regard to interrelations and consistency of conceptual business models, engineering frameworks, mathematical models and IT. We propose to name this particular concept as an *A-SCM* (and not as, e.g., a flexible or agile SCM). Namely, the *adaptation* is the most comprehensive category defined in systems and control science that covers the system's ability to change its behaviour regarding changes in the execution environment and with regard to the system's goals. Even the ability to change is the most important driver of competitiveness in modern and feature markets.

Moreover, in particular, the adaptation is the category that corresponds to the modern stage of state of the art in management and information systems. Theoretical discussions on self-configuring and self-learning SCs cannot be properly perceived and implemented in practice with existing management systems and because of the lack of standard "mass" software solutions. However, in future, adaptive SCs should evolve into self-organizing and self-learning SCs. The difference between *adaptive* and *self-organizing* SCs is that in the adaptation approach the system's shape and goals are fixed while in self-organization both the system and its goals evolve. The system's borders become fuzzy, the system can broaden by "acquiring" a space from the environment, or the system can narrow in the reverse way.

As a new research direction, *A-SCM* requires comprehensiveness with regard to the interrelations and consistency of conceptual business models, engineering frameworks, mathematical models and IT. Recent research shows a gap regarding the engineering frameworks and mathematical models. Gaining advancements in this direction is a critical and timely issue because of the critical role of this level with regard to the practical applicability of business concepts and the development of IT that would be adequate for the business concepts.

In the further course of this chapter, we will consider the vision of the conceptual framework of *A-SCM*. In the subsequent chapters, the engineering and mathematical frameworks will be presented. These frameworks extend the narrow understanding of adaptive SCs as mobile IT or agent systems to a comprehensive new research direction that is composed of conceptual business research as well as model-based and IT-based advanced decision-making techniques in SCM.

2.2.2 Basic Terms and Definitions

In this section, the conceptual basics of the A-SCM approach are considered. We start with the main definitions, and then we consider the A-SCM framework. Based on the frameworks of the control and systems theory, let us introduce some basic definitions.

Definitions

The SC *adaptability* is the ability of a SC to change its behaviour for the prevention, improvement or acquisition of new characteristics for the achievement of SC goals in environmental conditions that vary in time and the aprioristic information about which dynamics is incomplete.

Adaptive management is a management method of a SC with varying unknown environmental characteristics, in which for the final time defined (satisfactory, wished for, or optimum) goals of SCM are reached by means of a change of the SC parameters, processes, and structures or characteristics of control influences on the feedback loop driven basis.

Adaptive planning is a method of planning in which the plan of a SC is modified periodically by a change of parameters of the SC or characteristics of control influences on the basis of information feedback about a current condition of the SC, the past and the updated forecasts of the future.

An adaptive SC is a networked organization wherein a number of various enterprises:

- collaborate (cooperate and coordinate) along the entire value-adding chain and product life cycle to acquire raw materials, convert these raw materials into specified final products, deliver these final products to retailers, design new products, and ensure post-production services;
- apply all modern concepts and technologies to make SCs stable, effective, responsive, flexible, robust, sustainable, cost-efficient and competitive in order to increase SC stability, customer satisfaction and decrease costs, resulting in increasing SC profitability.

A-SCM studies the resources of enterprises and human decisions with regard to stability, adaptability and profitability of cross-enterprise collaboration processes to transform and use these resources in the most rational way along the entire value-adding chain and product life cycle, from customers up to raw material suppliers, based on cooperation, coordination, agility and sustainability throughout.

2.2.3 A-SCM Framework

As discussed in Sect. 2.1, various strategies of integrated production and logistics in industrial organizations – from SCM, VE, agile/responsive SCs up to sustainable SCs – have been developed over the last two decades. Although the strategies appear to differ in targets, presumptions, application areas, enabling technologies, and research methodologies, each compliments the others, endeavouring to improve competitiveness. Considering the significance of all the strategies for organizations, the developed approach integrates the elements of these strategies to develop a framework of *A-SCM* (see Fig. 2.2).

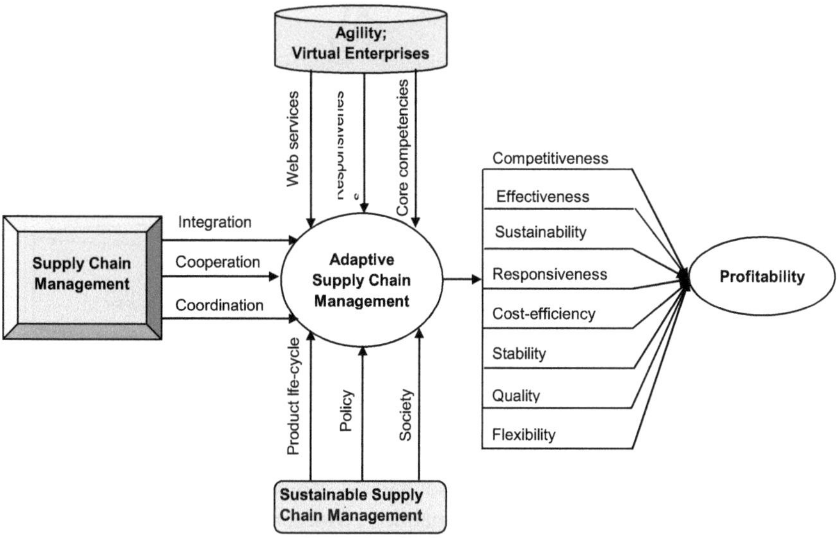

Fig. 2.2 Framework of *A-SCM*

In the *A-SCM* framework, we do not set off different value chain strategies with each other, but consider them as an integrated framework. The encapsulation of the advantages of SCM, agility, and sustainability enables *A-SCM*.

SCM serves as a basis for integration (organizational: suppliers and customers; functional: collaborative business processes; managerial: strategic, tactical, and operative decision-making levels), cooperation, and coordination. The strategies of *agility* enrich SCM by means of a general information space with the help of Web services and higher flexibility/responsiveness through concentration on core competencies and building virtual alliances/environments. *Sustainable SCM* integrates the consideration of the product development, utilization, product end-of-life, and recovery processes. On the other hand, sustainable SCM brings into consideration policy and society issues, which may affect the SCs and which may be affected by SCs.

2.2.4 A-SCM Drivers and Organization

Fig. 2.3 depicts the A-SCM strategy as drawn from elements of SCM, agility, and sustainability.

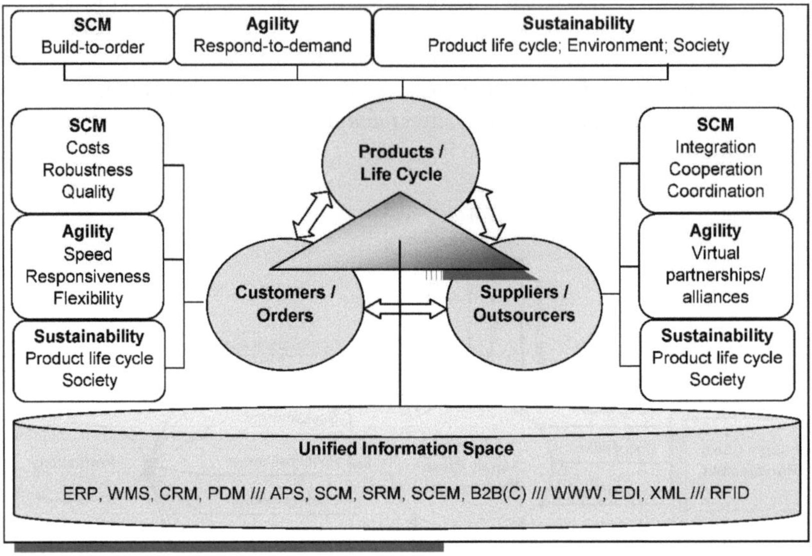

Fig. 2.3 The A-SCM drivers

In *A-SCM*, all three value chain drivers – products and their life cycles, customers and their orders, and suppliers/outsourcers – are enhanced by combining the elements from SCM, agility and sustainability. Moreover, these drivers are interlinked within a unified information space.

A-SCM unites an SC owner (an original equipment manufacturer (OEM) or a logistics service provider), customers and suppliers. The organizational structure consists of a real SC environment and a virtual alliance/partnership environment (see Fig. 2.4).

In the real SC environment, the SC owner collaborates with its customers and suppliers in regard to the existing products and product lines in all the stages of the product life cycle. The virtual alliance/partnership environment is an adaptation structural–functional reserve of the real SC environment. In the case of market changes, new products, or an impact of operational inefficiencies due to a variety of disruptive factors (machine failures, human decision errors, information systems failure, cash-flow disruption or simply catastrophic events), these *structural–functional reserves* are activated to adapt the SC. Second, in the virtual alliance/partnership environment, new products are designed (with the integration of potential customers and suppliers).

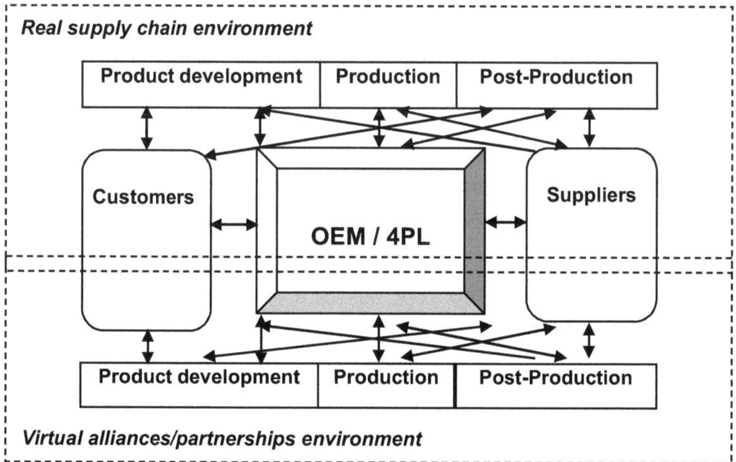

Fig. 2.4 The A-SCM organization

In traditional SCs, decisions about a customer's order acceptance or rejection are made on a stable long-term predetermined supplier structure. In A-SCM, it is possible to build *new order-oriented structures*, taking into account technological product individualization, demand volume fluctuations, or operative disruptions in SCs (see Fig. 2.5).

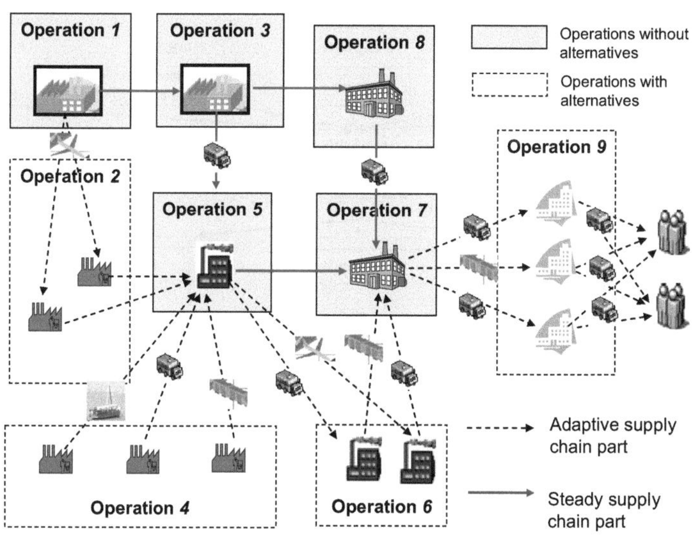

Fig. 2.5 Case description of adaptive SC organization

Figure 2.6 depicts a case example from special machinery building. Similar cases can be found in textile and or electronics industries. SCs are formed dynamically based on the offer parameters of the enterprises, customers' requirements and so-called soft factors (e.g., reputation, trust, etc.). Remarkable is the existence of alternative suppliers for various project operations, which differ from each other by operations parameters. The problem consists of an evaluation of alternative SCs to select the best one for the following scenario:

- new products (customer individualized products or new product lines);
- technological disruptions (machines, IT);
- collaboration problems (errors or IT failure); and
- demand fluctuations.

The special feature of this concept lies in a customer-oriented networking of core competences and flexible configurable SCs conditioned by an enlargement of alternatives to search for suitable partners for a cooperation enabled by ERP and APS systems and Internet technologies (EDI and business-to-business).

Finally, let us consider the goal tree of A-SCM (see Fig. 2.6).

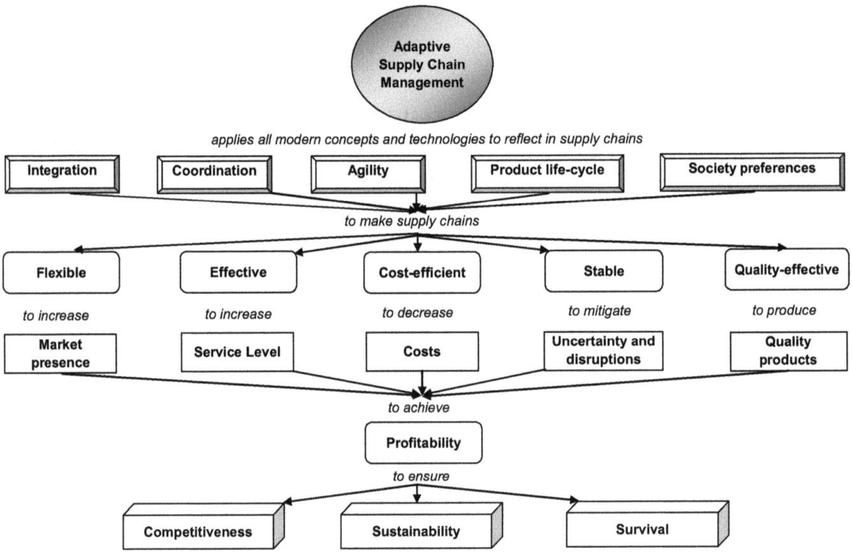

Fig. 2.6 Goal tree of the A-SCM

Figure 2.6 depicts the goal tree of A-SCM. The goal tree shows the drivers of A-SCM: integration, coordination, agility and sustainability. By reflecting these drivers, SCs can be made flexible, responsive, cost-effective, stable and quality-effective to achieve maximum profitability, which ensures long-term competitiveness, sustainability and survival.

2.2.5 Application Issues

The main aspects on which a particular emphasis should be set by practical implementation of the proposed framework are the following:

- identifying core competencies, making them describable and analysable;
- establishing trust and long-term collaboration in partnerships;
- elaborating product and process documentation throughout, especially for products and their life cycle (i.e., on the basis of CALS – continuous acquisition and life cycle support – standards);
- unifying product data within an electronic catalogue connected to an ERP system;
- integrating into the ERP/APS landscape SC event management systems;
- keeping the information architecture as simple as possible in relation to linking different information systems as well as operating by users; and
- thinking about the duality of IT's impact: on one hand, IT is an infrastructure to enable efficient coordination in SCs; on the other, IT enables new organizational methods.

Finally, let us discuss the limitations of the proposed framework. Generally, the proposed approach can be implemented in all branches. However, some particular aspects of the approach show limitations regarding the branch independence, i.e., with regard to the flexible suppliers' structuring, the approach can be applied especially to the cases where there is the possibility to attach alternative suppliers quickly to a number of operations in the value-adding process. The proposed concept can be applied in two main cases: (1) for unique products or (2) for products without strict technical quality policy).

Another very important point is the *trust and collaboration* in the network. Before automation, a huge amount of organizational work should be carried out to convince the OEMs and suppliers to collaborate within a common informational space, share the data, actualize the data and ensure financial trust. While automating, it is important to elaborate and to maintain throughout product and process technological documentation and classification. Last, but not least, the firms themselves should perceive the necessity for such collaboration.

References

Ahn HJ, Lee H, Park SJ (2003) A flexible agent system for change adaptation in supply chains. Expert Syst Appl 25:603–618

Bustelo D, Avella L, Fernández E (2007) Agility drivers, enablers and outcomes: empirical test of an integrated agile manufacturing model. Int J Oper Prod Manag 27(12):1303–1332

Cai Z-Y, Xiao R-B, Tan Y, Gong F-M (2008) Fuzzy adaptive production plan dispatching of cycle supply chain under uncertainty conditions. Control and Decision/ Kong Zhi Yu Jue Ce 23(5):525–529

Camarinha-Matos LM, Afsarmanesh H (2004) The emerging discipline of collaborative networks. In: Camarihna-Matos L (Ed) Virtual enterprises and collaborative networks. Kluwer Academic Publishers, Boston
Camarinha-Matos LM, Afsarmanesh H (2007) A comprehensive modelling framework for collaborative networked organizations. J Intell Manuf 18(5):527–615
Chandra C, Grabis J (2007) Supply chain configuration. New York, Springer
Chandra C, Kamrani A K (2004) Mass customization: a supply chain approach. Kluwer Academic, New York
Chen F, Drezner Z, Ryan JK, Simchi-Levi D (2000) Quantifying the bullwhip effect in a simple supply chain: the impact of forecasting, lead times, and information. Manag Sci 46(3):436–443
Choi TY, Dooley KJ, Rungtusanatham M (2001) Supply networks and complex adaptive systems: control versus emergence. J Oper Manag 19(3):351–366
Christopher M, Towill DR (2000) Supply chain migration from lean and functional to agile and customized. Supply Chain Manag: Int J 5(4):206–13
Christopher M, Towill, DR (2001) An integrated model for the design of agile supply chains. Int J Phys Distrib Oper Manag 31:235–244
Collin J, Lorenzin D (2006) Plan for supply chain agility at Nokia. Int J Phys Distrib Logist Manag 36(6):418–430
Coronado M, Lyons AE (2007) Evaluating operations flexibility in industrial supply chains to support build-to-order initiatives. Bus Process Manag J 13(4):572–587
Ernst R, Kamrad B (2000) Evaluation of supply chain structures through modularization and postponement. Eur J Oper Res 124:495–510
Fotopoulos SB, Hu X, Munson CL (2008) Flexible supply contracts under price uncertainty. Eur J Oper Res 191(1):251–261
Friscia T, O'Marah K and Souza J (2004) The AMR research supply chain top 25 and the new trillion-dollar opportunity. AMR Research Report, November
Geunes J, Pardalos P, Romeijn HE (2002) Supply chain management: models, applications, and research directions. Kluwer academic publishers. Dordrecht
Goranson HT (1999) The agile virtual enterprise – cases, metric, tools. Quorum Books, London
Gunasekaran A, Ngai NWT (2005) Build-to-order supply chain management: literature review and framework for development. J Oper Manag 23(5):423–51
Gunasekaran A, Kee-hung Laib and TCE Cheng (2008) Responsive supply chain: a competitive strategy in a networked economy. Omega 36(4):549–564
Ivanov D, Teich T (2009) Management concepts and tools of competence-based agile supply chains. Proceedings of the 13 IFAC Symposium INCOM. Moskau. Preprints: 868–873
Jang PY (2006) A flexible and adaptive control architecture for the efficient supply chain management (SCM). WSEAS Transactions on Communications 5(6):1015–1025
Kaihara T, Fujii S (2008) Supply chain management for virtual enterprises with adaptive multi-agent mechanism. Int J Manuf Tech Manag 14(3-4):299–310
Kim CO, Jun J, Baek JK, Smith RL, Kim YD (2005) Adaptive inventory control models for supply chain management. Int J Adv Manuf Tech 26(9-10):1184–1192
Lee H (2004) The triple-A supply chain. Harv Bus Rev October:102–12
Lee HL, Padmanabhan V and Whang S (1997) Information distortion in a supply chain: the bullwhip effect. Manag Sci 43(4):546–558
Li J, Wang S, Cheng TCE (2008) Analysis of postponement strategy by EPQ-based models with planned backorders. Omega 36(5):777–788
Mason-Jones R, Naylor B, Towill DR (2000) Engineering the leagile supply chain. International Journal of Agile Management Systems 2(1):54–61
Ozbayrak M, Papadopoulou TC, Samaras EA (2006) Flexible and adaptable planning and control system for an MTO supply chain system. Robot Comput-Integrated Manuf 22(5-6):557–565
Pandey A, Masin M, Prabhu V (2007) Adaptive logistic controller for integrated design of distributed supply chains. J Manuf Syst 26(2):108–115

Pathak SD, Day JM, Nair A, Sawaya WJ, Kristal MM (2007) Complexity and adaptivity in supply networks: building supply network theory using a complex adaptive systems perspective. Decis Sci 38(4):547–580

Picot A, Reichwald R and Wigand RT (2001) Die grenzenlose Unternehmung. Gabler Verlag, Wiesbaden.

Reichhart A, Holweg M (2007) Creating the customer-responsive supply chains: a reconciliation of concepts. Int J Oper Prod Manag 27(12):1144–1172

Ross A (2004) Creating agile supply chain. IEEE Manufacturing Engineer 82(6):18–21

SAP (2000, 2002) Adaptive Supply Chain Networks. White Paper

Sarkis J, Talluri S and Gunasekaran A (2007) A strategic model for agile virtual enterprise partner selection. Int J Oper Prod Manag 27(12):1213–1234

Scholz-Reiter B, Hoehns H, Hamann T (2004) Adaptive control of supply chains: building blocks and tools of an agent-based simulation framework. CIRP Annals Manufacturing Technology 53(1):353–356

Sharif A, Irani Z and Lloyd D (2007) Information technology and performance management for build-to-order supply chains. Int J Oper Prod Manag 27(12):1235–1253

Shervais S, Shannon TT, Lendaris G G (2003) Intelligent supply chain management using adaptive critic learning. IEEE Trans Syst Man Cybern Syst Hum 33(2):235–244

Surana A, Kumara S, Greaves M, Raghavan UN (2005) Supply-chain networks: a complex adaptive systems perspective. Int J of Prod Res 43(20):4235–4265

Swafford PM, Ghosh S, Murthy N (2008) Achieving supply chain agility through IT integration and flexibility. Int J Prod Econ 116: 288–297

Tachizawa EM, Thomsen CG (2007) Drivers and sources of supply flexibility: an exploratory study. Int J Oper Prod Manag 27(10):1115–1136

Towill D and Mason-Jones R (1999) Using the information decoupling point to improve supply chain performance. Int J Logist Manag 10(2):13–26

Tully S (1993) The modular corporation. Fortune 08.02.1993:52–56

Van Donk D, Van der Vaart T (2007) Responsiveness through buyer-focused cells: exploring a new supply strategy. Int J Oper Prod Manag 27(12):1362–1379

Van Hoek RI (2001) The rediscovery of postponement a literature review and directions for research. J Oper Manag 19:161–84

Vonderembse MA, Uppal M, Huang SH, Dismukes JP (2006) Designing supply chains: towards theory develeopment. Int J Prod Econ 100:2, 223–238

Wadhwa S, Saxena A, Chan FTS (2008) Framework for flexibility in dynamic supply chain management. Int J Prod Res 46 (6):1373–1404

Wang F, Lin E, Liu H (2009) Three-dimensional model of customer order decoupling point position in mass customisation. Int J Prod Res, DOI: 10.1080/00207540902865746, in press

Warnecke HJ, Braun J (1999) Vom Fraktal zum Produktionsnetzwerk: Unternehmenskooperation erfolgreich gestalten. Springer-Verlag, Berlin, Heidelberg, New York

Wirth S, Baumann A (2001) Wertschoepfung durch vernetzte Kompetenz. Huss Verlag, Munich

Yusuf MA, Gunasekaran A, Adeleye EO, Sivayoganathan K (2004) Agile supply chain capabilities. Determinants of competitive objectives. Eur J Oper Res 159(2):379–392

Chapter 3
Decision-making Support for Supply Chain Management

Science may set limits to knowledge,
but should not set limits to imagination.
Bertrand Russell

3.1 Model-based Decision-making Support

3.1.1 Models: Basic Terms and Classifications

The concept of a model is widely applicable in natural human languages and is a general scientific term. It is characterized by polysemy, that is, widely expressed and reflecting different meanings of this concept depending on the applications and contexts. At present, there are several hundred definitions of the concept of a model and modelling. In summoning up different definitions, the following views of models and modelling can be presented.

A *model* is:

- a system whose investigation is a tool for obtaining information about another system;
- a method of knowledge existence; and
- a multiple system map of the original object that, together with absolutely true content, contains conditionally true and false content, which reveals itself in the process of its creation and practical use.

Modelling is one of the stages of cognitive activity of a subject, involving the development (choice) of a model, conducting investigations with its help, obtaining and analysing the results, the production of recommendations on the further activity of the subject and the estimation of the quality of the model itself as applied to the solved problem and taking into account specific conditions.

Because of the *finiteness* of the designed (applied) model (a limited number of elements and relations that describe objects belonging to infinitely diverse reality) and the *limited* resources (temporal, financial and material) supplied for modelling, the model always reflects the original object in a simplified and approximate way. However, the human experience testifies that these specific features of a model are admissible and do not oppose the solution of problems that are faced by subjects.

In the course of modelling, it is advisable to distinguish the following basic elements and relations: *first*, a subject or subjects, an original object, the model object and an environment in which the modelling is performed; and, *second*, binary relations between the listed elements. Subjects of modelling mean the following classes of social subjects: decision makers, persons who substantiate the decisions, experts, persons who use the models and persons who design the models.

It is worth noting that one of the main specific features of original objects (real or abstract) is their exceptional complexity, which reveals itself in the form of structural complexity, complexity of functioning, complexity of the choice of behaviour and complexity of development. As shown by Sokolov and Yusupov (2004), to describe such objects we should use several models rather than a unique model. In other words, we should perform system modelling (a polymodel description of the application domain).

Another specific feature of the tools of abstract modelling consists of considerable intensification of works in the automation of this process and, first of all, the phase connected with the design of a computer model. Moreover, within the framework of new IT based on the concepts of knowledge bases, the concept of a "model" has considerably extended the limits of its application – from the field of passive informational resources to the field of *active* ones. Under these conditions, algorithms that are elements of procedural knowledge turn into operating environments that provide the solution to problems with a subject in the language of models.

The main properties of good models are the following:

- *Adequacy* (from Latin *adaequatus*, which means equated, completely suitable, or comparable). The model should possess the specified property relative to certain aspects of the original object.
- *Simplicity* and *optimality* of the model. The property of adequacy is directly associated with the properties of simplicity and optimality. Indeed, sometimes, to achieve the required degree of adequacy, we should essentially complicate the model. On the other hand, if we can choose different models that have approximately the same adequacy, it is advisable to use the simplest model.
- The *flexibility* (*adaptability*) of models assumes that parameters and structures that can vary in given ranges are introduced into the composition of models in order to achieve the goals of modelling.
- *Universality* and *task orientation* of models. It is advisable to design models specialized relative to an admissible class of modelled objects and universal with respect to a list of supported functions.

Among other properties of models, we should distinguish reliability, unification, openness and accessibility, intelligence, the efficiency of computer implementation, complexity, identifiably, stability, sensibility, observability of models, their invariance, self-organization and self-learning. On the whole, each variant of implementation of the system modelling techniques is characterized by its own time consumption, the expenditure of resources and the final results).

Finally, let us provide a general classification of different kinds of models. There are different options for classifying models. The first and the most common option is to use the morphologic analysis and to distinguish two or three possible states for each feature. An example of such a classification is shown in Table 3.1.

Table 3.1. Morphological model classification (from Sokolov and Yusupov 2004)

Classification feature	Kinds of mathematical models I	II	III
Scientific basis and building logic	Axiomatic	Empirical	Semi-empirical
	Hypothesis deductive (phenomenon logic)	Deductive–asymptotic	Inductive
	Deductive	Inductive	Deductive–inductive
Exactness	Analytical	Simulation	Analytical–simulation
Data and scope	Qualitative	Quantitative	Hybrid
Main model's function	Descriptive	Predictive	Hybrid
Alternativity of model	Satisfaction	Optimization	Non-alternative
Self-learning and self-organization	Self-learning	Self-organizing	Strictly predetermi-ned
Time	Static	Kinematic	Dynamic
	Continuous	Discrete	Hybrid
Certainty	Deterministic	Stochastic	Uncertainty

Another approach to model classification is the determination of interconnections and mutual associations between different types and kinds of information fusion models, the detection of models' generalized specifics, and the structuring of the investigated objects' space as well (Kalinin and Reznikov 1987). In this approach, the elements and interrelations between them build a new original object – a *developing situation*. This developing situation becomes the modelling subject. An important feature of this taxonomy is that the elements in the particular model classes are not strictly predefined. Hence, depending on the developing situations, new classes may be formed. Figures 3.1 and 3.2 show different variants of classification of developing-situation models.

With regard to SC models, different frameworks have been constituted so far. Beamon (1998) distinguished between deterministic analytical models, stochastic analytical models, economic models and simulation models. A very similar classification with a more detailed division between types of analytical models was proposed by Dong (2001). Riddalls and Bennett (2002) and Disney *et al.* (2006) considered the applications of continuous differential equation-based models in modelling SC dynamics. Min and Zhou (2002) dealt with hybrid and IT-driven models. Kim and Rogers (2005) extended the importance of IT-driven models by adding a category of business process engineering models.

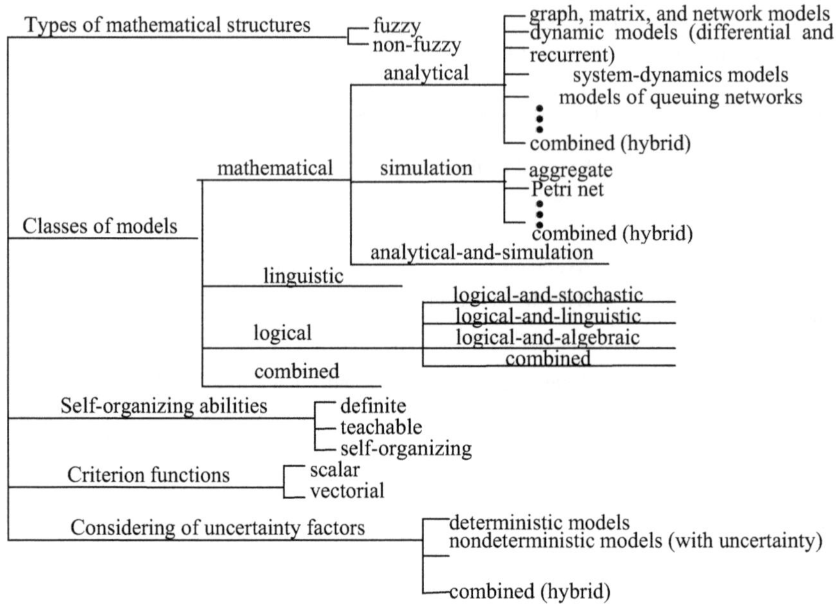

Fig. 3.1 Classification of models for developing-situation analysis

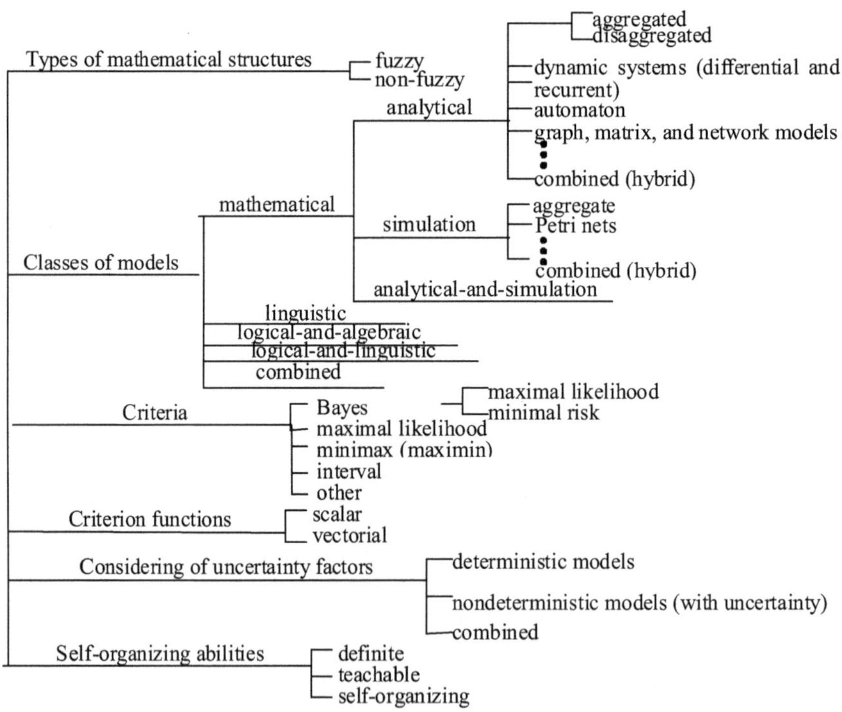

Fig. 3.2 Classification of models for the evaluation of states in developing situations

Chandra and Grabis (2008) demonstrated the use of modern IT techniques and methods for the *integration of SC decision-making models*. In their early work (Chandra *et al.* 2007), they classified SC models into information models, analytical models, simulation models and hybrid models (in the wide interpretation, not necessarily confined to a combination of optimization and simulation models). Gunasekaran and Ngai (2009) classified the models for BTO SC models based on the nature of the decision-making areas and sub-classified to focus on solving problems with modelling and analysis. According to Tolujev (2008), the most widespread modelling approaches in practice can be divided into the graphical (qualitative) and mathematical (quantitative) approaches. In the graphical (qualitative) model class are included topological models and business process information models. Mathematical (quantitative) models are statistical descriptive models, static optimization models and dynamic simulation models.

3.1.2 Mathematical Models

In the class of mathematical models, a number of decision-making support research paradigms and solution methods can be distinguished. The main research paradigms are OR, control theory and agent-based approaches. The main solution techniques are optimization, simulation, statistics, heuristics, and hybrid models (Beamon 1998, Swaminathan *et al.* 1998, Tayur *et al.* 1999, Fox *et al.* 2000, Shen *et al.* 2001, de Kok and Graves 2004, Simchi-Levi *et al.* 2004, Koechel and Nielandaer 2005, Surana *et al.* 2005, Chopra and Meindl 2007, Chandra *et al.* 2007, Sarimveis *et al.* 2008).

OR on the SCM can be divided into three primary approaches to conducting SC modelling. These are optimization, simulation and heuristics. Hybrid models (e.g., optimization-based simulation models) also exist. Optimization is an analysis method that determines the best possible method of designing a particular SC. Optimization has been a very visible and influential topic in the field of OR. (Tayur *et al.* 1999, de Kok and Graves 2004, Simchi-Levi *et al.* 2004). The drawback of using optimization is difficulty in developing a model that is sufficient detailed and accurate in representing complexity and uncertainty of SCM, while keeping the model simple enough to be solved (Harrison 2005).

Simulation is imitating the behaviour of one system with another. By making changes to the simulated SC, one expects to gain understanding of the dynamics of the physical SC. Simulation is an ideal tool for further analysing the performance of a proposed design derived from an optimization model. Simulation models can be classified in macroscopic (system dynamics models), microscopis (discrete-event models) and mesoscopic models. The philosophy behind the mesoscopic approach can be described by the phrase "discrete time/continuous quantity" (Schenk *et al.* 2009).

Heuristics are intelligent rules that often lead to good, but not necessarily the best, solutions. Heuristic approaches typically are easier to implement and require fewer data. However, the quality of the solution is usually unknown. Unless there is a reason not to use the optimization, heuristics is an inferior approach. Heuristics do not guarantee the optimal solution but allow a permissible result to be found within an acceptable period of time. The quality of this solution with regard to the potential optimum, however, remains unknown. Second, the multiple-criteria problems are still a "bottleneck" of the heuristics. An option to estimate the quality of heuristic algorithms and the directions and scope of their development may be the usage of optimization as a tool to gain "ideal" solutions to problems.

The most popular mathematical modelling approaches to SCM in OR are static optimization models (usually formulated as the mixed-integer linear programming – MILP), dynamic programming and dynamic simulation models. With regard to the problems of high dimensionality, heuristics are usually applied.

To take into account both the problem dynamics and high-dimensionality, control theory can be used (Disney and Towill 2002, Braun *et al.* 2003, Anderson *et al.* 2006, Disney *et al.* 2006, Lalwani *et al.* 2006, Ivanov *et al.* 2007, van Houtum *et al.* 2007, Wang *et al.* 2007, Ivanov 2009). Control theory is a multi-disciplinary scientific discipline that contains powerful conceptual and constructive tools to conduct research into the dynamic problems of the flexible (re)distribution of a variable set of jobs to a variable set of resources (Kalinin and Reznikov 1987, Stefani *et al.* 2002, Sarimveis *et al.* 2008). Besides, CAS (Surana *et al.* 2005) and system dynamics (Sterman 2000) can also be applied to SC dynamics modelling at different precision, space, and time levels. The detailed analysis of mathematical modelling will be undertaken in Chap. 8.

3.1.3 Information Models

Information models describe the problems from an information processing perspective (Fiala 2005, Chandra and Grabis 2008, Chatfield *et al.* 2009). Information modelling can be referred to as descriptive modelling and serves as an interface for information systems development. Information modelling of SCs is subject to business process reengineering models, IT-driven models and the use of modern IT techniques and methods for the integration of SC decision-making models.

A *process* is a content and logic sequence of functions that are needed to create an object in a specified state. Processes have input and output parameters. Processes do not exist autonomously but are tightly interrelated with other parallel, following, past, underordinated and superordinated processes.

Processes can be optimized subject to finding their best state with regard to costs, quality, service level, reliability, flexibility and assets. The analysis of process results may be characterized by effectiveness (the achievement of process goals) and efficiency (performing the process with minimum costs). Processes that are both effective and efficient are named optimal. *Optimal* processes are characterized by effectiveness, efficiency, controllability, stability, flexibility, analysability, observability, reliability, documentability and permanent improvement capabilities (Becker 2008).

In SCs, processes may be optimized subject to different indicators and application areas. These are:

- SC design and configuration (i.e., new market acquisition, too long time-to-market, too complex SC configurations, weak consistency within SC configurations);
- SC planning (weak flow capacities, too long a supply cycle, excessive replenishment);
- SC operations (false priorities of customers' orders, imbalance of capacities and order volumes, too frequent disruptions and high costs for their recovery);
- SC performance evaluation (performance of different departments such as logistics, transport and production is evaluated local for each department without any general links from the SCM perspectives); and
- SC execution (different levels of managers' qualifications, false or incomplete process documentation, weak consistency in process performance evaluation).

At each of these levels, more detailed optimization reserves exist. These can be the usage of Kanban, direct supply to assembly lines, technological converting optimization, Six Sigma, etc. For SC process cost estimation, a number of methods, such as TCO – total cost of ownership, SC total costs, and material flow costs, exist.

In the literature, the increase in sales and reduction in costs in the value-adding chain due to SC management amount to 15–30%. Partial effects, such as inventory reductions, an increase in service level, SC reliability, and flexibility, a reduction in transaction costs, etc., amount to 10–60%.

As SCs rarely emerge as a "green-field" concept (Harrison 2005), the primary focus is usually directed to rationalizing the existing structures and processes from the SC management perspective. The key questions here are the business process identification, analysis and improvement.

For *business process modelling*, a number of techniques and tools can be used (Kamath *et al.* 2003). The most popular of these are SCOR (SC Operations Reference), ARIS (Architecture of Information Systems), UML (Unified Modelling Language) and IDEF (Integration Definition for Function Modelling). These approaches can also be used to model the workflow of decision–making processes. The process modelling serves for (1) describing processes and structures in SCs and (2) for clear illustrating those entities. For these purposes, different solutions have been developed, e.g., activity diagrams in UML, event-process chains (EPC) in ARIS, etc. Among these techniques, SCOR possesses a special place.

SCOR has been widely presented in the literature (Bolstorff and Rosenbaum 2007). Without repeating these materials, let us analyse the SCOR advantages and shortcomings. The main value of SCOR from the business process modelling point of view is the standardized business process models that are interlinked at three levels. Besides, a coherent system of performance indicators is correlated with the process models. Finally, the data origins to calculate the performance indicators are explicitly provided.

To the shortcomings of SCOR belong the orientation of an enterprise (as a rule, a focal enterprise) but not of the SC as a whole. The process models are constructed as an "ideal" process and do not consider adjustment actions in the case of possible disruptions. SCOR mostly considers the transport–manufacturing part of the value-adding chain and does not take into account interrelations with other product life cycle phases. Finally, the level of SC execution is outside the process models.

With regard to the use of modern IT techniques and methods for the *integration of SC decision-making models*, Lau and Lee (2000) used the distributed objects approach to elaborate on an infrastructure of integrated component-based SC information systems. Kobayashi *et al.* (2003) conceptually discussed the workflow-based integration of planning and transaction processing applications, which allows for an effective integrated deployment of heterogeneous systems. Verwijmeren (2004) developed the architecture of component-based SC information systems. The author identified the key components and their role throughout the supply network. Themistocleous *et al.* (2004) described the application of enterprise application integration technologies to achieve the physical integration of SC information systems. However, approaches and technologies for logical integration at the decision-modelling level, where a common understanding of managerial problems is required, are still insufficiently developed (Delen and Benjamin 2003).

3.2 Information Systems-based Decision-making Support

3.2.1 Coordination and Information Technologies

Coordination in SCM plays a fundamental role in mitigating uncertainty with the help of synchronizing information flows from a point-of-sale up to the raw material suppliers and material flows in the reverse direction (Holweg and Pil 2008). One key problem in nearly all SCs for which the coordination is especially needed is the so-called *bullwhip effect* (Lee *et al.* 1997).

To run their business successfully, enterprises should not just coordinate their internal activities but also the links to their suppliers and customers. Coordination in SCM is based on building in a SC an information and communication environment to ensure complete, timely, correct, and full information about demand and supply along the entire value-adding chain. Coordination is tightly interlinked with the *integration*. Figure 3.3 depicts these ideas.

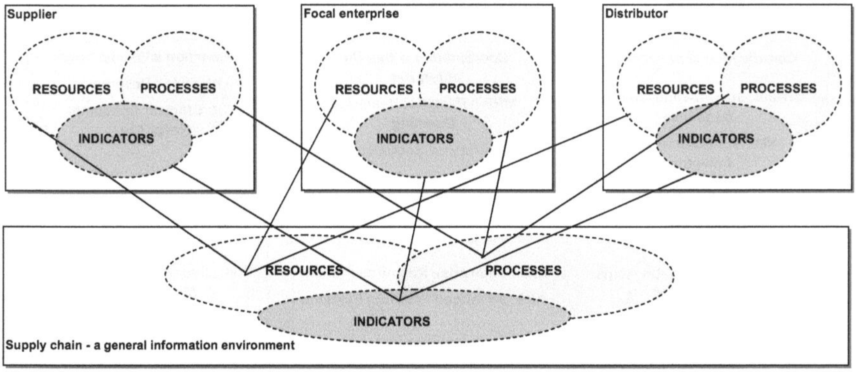

Fig. 3.3 Components of SC coordination and integration

Insufficient coordination in a SC can be caused by three main problems (Chopra and Meindl 2007):

- different conflicting goals in different SC parts;
- information distortion and incompleteness in the SC links; and
- insufficient degree of automation of SC participants.

These problems may result in situations when each SC participant aims at maximizing only its own profitability or not providing information about capacities and inventories on commercial grounds. Besides these organizational problems, technical problems may inhibit coordination in SCs. Nowadays, the data transferring volumes are becoming even more complex and dynamic.

In considering SC coordination, coordination strategies and IT for coordination should be distinguished (see Fig. 3.4). The main coordination strategies are as follows:

- JIT (Just-In-Time);
- JIS (Just-In-Sequence);
- VMI (Vendor-Managed Inventory);
- Kanban with suppliers' responsibility;
- ECR (Efficient Consumer Response);
- QR (Quick Response); and
- CPFR (Collaborative Planning, Forecasting, and Replenishment).

To implement these strategies, IT are needed. These can be either customized standard constructs or self-developed constructs (as made by a number of national companies). Efficient information handling in a SC is of the most significant importance. The information environment is the backbone of SCs. Information flows connect SC participants, SC functions both vertically and horizontally and management decision levels.

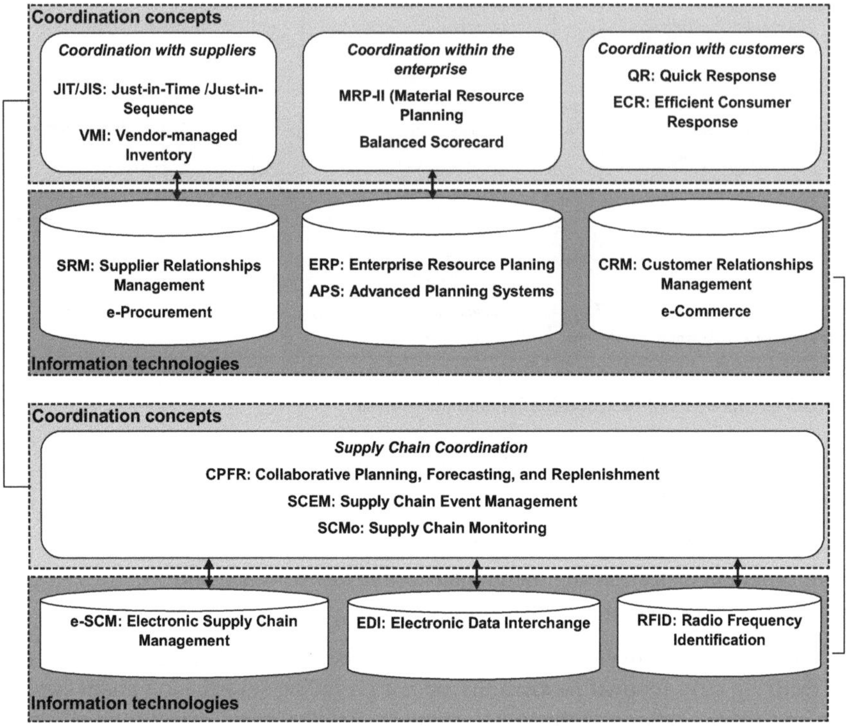

Fig. 3.4 SC coordination concepts and IT

Some examples follow. Master production planning requires information about demands. Inventory management policies affect supply cycle time. Operative changes in demand affect tactical plans in production, sales and replenishment as well as budgeting. Warehouse information is crucial to inventory management with regard to collaboration with suppliers and customers.

3.2.2 Classification of Information Technologies in SCM

There is a wealth of literature on IT in SCM (i.e. Gunasekaran and Ngai 2004, Subramanian and Iyigungor 2006, Chandra and Grabis 2008, Ketikidis *et al.* 2008). For example, McLaren and Vuong (2008) reported on the taxonomy that describes 83 major functional attributes that form five top-level categories: primary SC processes, data management, decision support, relationship management and performance improvement. The codes representing SC processes agree with the widely used SCOR process model.

Here, we regard it as sensible to limit the classification to a general classification of IT in SCM without describing their particularities in detail. The variety of IT can be distinguished in the following four groups.

Planning at the enterprise level

- ERP (Enterprise Resource Planing)
- MES (Manufacturing Execution Systems)
- WMS (Warehouse Management Systems)
- Budgeting systems

Planning and control at the SC level

- APS (Advanced Planning Systems)
- SCEM (SC Event Management)
- SCMo (SC Monitoring)

Transportation control

- RFID (Radio Frequency Identification)
- Trace and Tracking

Business intelligence

- OLAP (On Line Analytical Processing)
- DSS (Decision Support Systems)
- Data miming

Communication and data interchange

- EDI (Electronic Data Interchange)
- XML (Extensible Markup Language)
- Mobile technologies, WAP (Wireless Application Protocol), web browsers

Electronic payment systems with security services

- SSL (Secure Sockets Layer)
- SET (Secure Electronic Transfer)
- PCI (Peripheral Component Interconnect)

Finally, it must be said that the IT provide a new level of coordination capabilities in SCs and enable a breakthrough in SC responsiveness and flexibility. IT, on one hand, serve as an environment to support SC management. On the other hand, they are in turn the enabler of much advancement in SC management. Modern IT can potentially enable almost any coordination concept. More important problems for efficient coordination lie in the organization sphere, collaboration culture and trust.

3.2.3 Impact of Information Technology on Management

Currently, computer science and IT are becoming one of the basic factors (catalysts) for the evolution of civilization and scientific and technological progress. In line with these general trends, IT influence development of SCM (Rai *et al.* 2006, Sarker *et al.* 2006, Kahli and Grover 2008). Exactly this reason gave rise to the emergence and widespread circulation of such fundamental notions as informatization, the information society and the information economy. In the present section, we attempt to draw the attention of specialists to the influence of informatics and IT on the progress in management processes of various kinds (Yusupov and Sokolov 2009).

A universal property of the management process, irrespective of the problem domain, is that it has a notably *informational nature*, i.e. is connected, first of all, with the collection, processing, analysis and usage of data, information, and knowledge. Currently, general topics related to the collection, processing, representation, transmission and protection of information are studied in informatics. The results of these investigations are realized as IT. This term means a family of methods for realizing informational processes in various fields of human activity aimed at the creating an informational product, including those in management systems. It is obvious that the effectiveness and efficiency of management depends on the progress in informatics and IT.

The creation and wide practical realization of the theory and technologies of intellectual control are a striking example of the influence of computer science and IT on management theory and systems. These technologies greatly expand the possibilities of previously developed methods and self-tuning algorithms, adaptation and self-organization with application to the solution of complex problems of automatic and automated control under the conditions of substantial structural and parametric uncertainty of the management object and environment models.

Today, the influence of computer science and IT on the progress in management theory and systems is of a global nature. Specialists note that, recently, the second stage of convergence between general control theory (cybernetics) and computer science has taken place; we are observing revolutionary progress in management and control systems caused by the influence of IT. In these settings, the problem under discussion deals with the formation of a new interdisciplinary direction towards cybernetics, telecommunication theory, and general systems theory. This direction can be called *neocybernetics*.

3.3 Integrated Frameworks of Decision-making Support

3.3.1 Information View of Integrated Modelling Frameworks

Recent studies have provided significant advancement in integrating information models and decision-making support (Chandra *et al.* 2007, Chandra and Grabis 2008, Chatfield *et al.* 2009). Shapiro (2000, 2001) emphasized the requirement for tight integration between decision making and the IT support tool. The components of the proposed framework include the database management system (central component), corporate database, SC decision database, model generator with an advanced optimizer and analytical tools. Chandra and Grabis (2008) propose the architecture of the SC configuration DSS. Their basic components are as follows: an SCM information system, data warehouse, information modelling system, knowledge base and decision-modelling database, groupware, decision modelling system and decision modelling components (optimizer, simulator, and statistical data analysis). This framework is presented in Fig. 3.5.

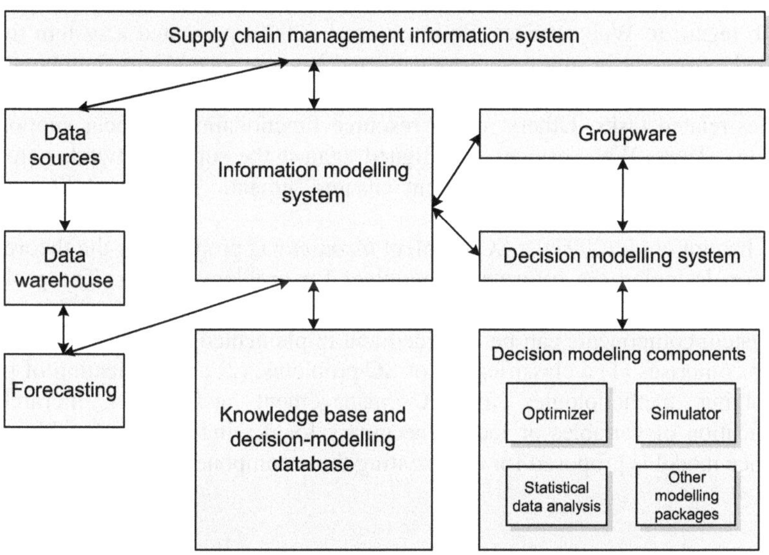

Fig. 3.5 Architecture of the SC configuration DSS (from Chandra and Grabis 2007)

Dolk (2000) reported on structuring SC decision making around data warehousing using OLAP tools. Dotoli *et al.* (2003) proposed a SC configuration framework that includes data analysis, network design and solution evaluation modules. With regard to dynamical aspects and SC reconfiguration, Piramuthu (2005) proposed a knowledge-based approach and using various intelligent decision-making algorithms. Kim and Rogers (2005) considered a SC's process from four views – function, structure, process, and behaviour – using UML syntax. Chandra and

Grabis (2008) reported on an overall approach to using IT at various stages of model development. Data and process modelling techniques are used to develop a semi-formalized representation of integrated models. These models support the integration of decision-making components with other parts of the SC information system.

Smirnov *et al.* (2006) proposed an agent-based technological framework for dynamic SC configuration applying different techniques of computational intelligence: (1) the theory of games with fuzzy coalitions, (2) genetic algorithms, and (3) constraint satisfaction problem solving. Ontology formalized as object-oriented constraint networks is used for task description and breakdown.

Another research stream is exploring various ways to utilize XML-based standard document types to assist simulation modelling and execution. Bradley (2003) reviewed some of the recent XML initiatives related to OR. Kim (2001) described an open architecture for the definition, storage and exchange of decision models and developed the Object-Oriented Structured Modelling Language (OOSML). Chatfield *et al.* (2009) developed an open information standard to assist SC modelling, analysis and decision support. The SC Modelling Language (SCML) is a platform- and methodology-independent XML-based mark-up language for storing SC structural and managerial information.

With regard to Web services, Smirnov *et al.* (2009) presented a system that is intended to operate in smart environments and has a service-oriented architecture. Some of the Web services making up the architecture are intended to model the logistics-related tasks. Others model resource functionalities or bear supporting functions. These Web services are aligned against the supply network ontology (application ontology). This alignment ensures semantic interoperability of the heterogeneous resources.

In Chandra *et al.* (2007), a SC problem *taxonomy* is proposed as the theoretical basis for designing the information required for problem solving. The problem taxonomy provides the overall framework under which problem-oriented information system components can be designed and implemented. The SC problem taxonomy comprises (1) a classification of SC problems, (2) a classification of problem-solving methodologies for SC management and (3) a hierarchical classification of variables or factors necessary for dealing with the problems. A reference model is proposed for representing these components formally.

3.3.2 Mathematical View of Integrated Modelling Frameworks

In the mathematical modelling of SCs, a number of particular features in the SCM domain should be taken into account. The processes of SC execution are non-stationary and non-linear. It is difficult to formalize various aspects of SC functioning. Besides, the SC models have high dimensionality. There are no strict criteria of decision making for SCM and no *a priori* information about many SC parameters.

The SC execution is always accompanied by perturbation impacts. The perturbation impacts initiate the SC structure dynamics and predetermine a sequence of control inputs compensating for the perturbation. Unlike the automatic systems, adjustment control decisions in SCs are taken by managers and not by automats. This results in individual interests, risk perceptions, and time delays (from minutes up to months) between disruption identification and taking adjustment measures. Hence, SC modelling problems are more complex than they appear in some studies. Most SC modelling issues are multi-disciplinary and cross-linked.

In the class of mathematical models, a number of decision-making support research paradigms and solution methods can be distinguished. The main research paradigms are OR, control theory, and agent-based approaches. The main solution techniques are optimization, simulation, statistics, heuristics and hybrid models.

The works on integrating optimization and simulation models (Koechel and Nielaender 2005) can be considered as the first step to the integrated mathematical modelling framework. Recent research (Ivanov 2009) has extended these contributions and emphasizes that various modelling approaches are not to be set off with each other, but should be considered as a united modelling framework. For example, multi-agent ideology is considered as a basis for the modelling of active elements. Control theory serves as a theoretical background to dynamic systems analysis and synthesis. OR provides fundamentals for optimization and analytical models. The theoretical fundamentals of such models' combinations were proposed by Sokolov and Yusupov (2004) within the concept of model qualimetry. These aspects will be considered in detail in Chap. 9.

3.3.3 Main Requirements for Integrated Decision-Support Systems

The leading role in decision-making assurance in SCM belongs to the integrated decision-support systems (IDSS) and to their core, special software and mathematical tools for decision support. The requirements for IDSS could be divided into two groups:

- general requirements for IDSS that define its suitability for various existing and perspective problems in the area of SC design, planning, operation, and evolution; and
- detailed requirements for particular IDSS elements and subsystems; these requirements are mostly concerned with the processes of SC elements' design and use.

IDSS general requirements could be set as follows.

The validity of IDSS-assisted decisions at various SC life cycle phases
The decision validity can be improved, first of all, on the basis of more precise and operative methods and algorithms for optimal choice and information processing. It is supposed that the algorithms can manage the amount of information, being operated by appropriate control bodies without mathematical models. In the

second place, the decision validity can also be raised by means of a wider application of various mathematical models for the estimation of multiple quantitative decision alternatives.

Harmonious interaction between a decision maker and the computing environment (intellectual user interface creation)
The experience of various IDSS operations demonstrates that formal models (first of all, mathematical models) could not take into account the whole variety of the real SC dynamics. Therefore, the interconnection of IDSS formal procedures of analysis, choice and the decision maker's creative ability becomes particularly significant.

IDSS openness and its ability to adapt, self-organize and evolve
The analysis of processes of SC design, planning, operation and evolution confirms that the "external environment" the SC is interacting with, as well as the SC itself, permanently changes. The changes produce variations in the structures and parameters of all the interacting objects. In these settings, the models, methods and algorithms developed at different phases with various purposes and embodied in special software and mathematical tools for decision support are able to produce only an approximate reflection of the necessary characteristics of SCs. It is rather difficult to create universal models and algorithms for the implementation of all the main functions in space and time within the problem domain under consideration. Therefore, substantively, in practice, the most suitable type of model and appropriate algorithm (from IDSS) for management problem solving should be chosen or constructed in accordance with its properties and actual situation.

SCs and their IDSS will be able to produce effective planning and regulative inputs under the influence of a non-stationary external environment as soon as they obtain special mechanisms (procedures) of adaptation and, in perspective, self-organization. The procedures will enable the directed modification of model parameters as well as the alteration of IDSS models and algorithms as a whole subject to feature possible control inputs. Finally, this will adjust to the future evolution of the control objects and the external environment too.

The choice or construction of an IDSS particular model or algorithm should be regarded as a function of a special IDSS subsystem (adapter). The adapter fulfils parametric and structural adaptation (and self-organization, in perspective) and adjusts the characteristics of SCs and their models in accordance with the actual environment and so ensures the reduction to a minimum value of the number of situations at various phases of the SC life cycle when IDSS is unable to produce management recommendations for a decision maker.

The latter opportunity is particularly significant for SC reconfiguration in the case of disruptions in information, material, and financial flows or in the SC as a whole. The adaptation should include the adaptation to the "past" and to the "future". To implement the types of adaptation mentioned, IDSS should be equipped with procedures that are able to accumulate and keep the unique experience of control bodies, to discover objective laws of control processes, to fix the experience and laws in a formalized form: as the SC states' information-processing algorithms, as control law parameters, or in the form of records in a database or a knowledge base.

Timeliness of control inputs
This requirement is especially related to the stage of SC execution. The significance of these requirements may be explained as follows. The execution of programmes, realizing IDSS methods and algorithms, is always concerned with time consumption and computational burdens that are necessary for information-processing completeness and quality of decision validity. When information processing and input generation ends later than established in accordance with SC real-time functioning specificity, it becomes necessary to fulfil reengineering of the appropriate IDSS models, methods and algorithms to increase the productivity and economic characteristics of SCs.

A necessary degree of SC model adequacy
This requirement is obligatory for all cases of model construction and simulation system (SS) creation within IDSS. Hence, it is obvious that not the complete adequacy, but adequacy in a certain sense (the required degree of adequacy) can be established in practice. A single model of a SC as a complex system can reflect only some aspects of the original, and so the notion of adequacy "in general" does not exist for it; only the adequacy of the aspect reflection can be considered. The model adequacy degree estimation should be evaluated in accordance with the possible extent of goal achievement in a specific model-based problem investigation.

The main *detailed requirements* for particular IDSS elements and subsystems are as follows.

Simplicity and optimality of particular IDSS models and system of models
This requirement is directly bound up with the requirement of the necessary degree of model adequacy. Really, one should essentially complicate a model or even replace it with a system of models to provide the required level of adequacy. However, when there is an opportunity for a choice between different classes of models, or combinations of these models, providing approximately equal degrees of modelling adequacy, the simplest model should evidently be selected. The above considerations explain the meaning of the optimality of model construction or choice.

Efficiency of the computer implementation of systems of models
The accomplishment of this requirement means high efficiency of the computational process organized subject to specific characteristics of constructed models and algorithms (the degree of the relationship of algorithms, the possibility of multi-sequencing and overlays in problem solving).

Other requirements

- the possibility of modelling with different time scales;
- universality and problem orientation of IDSS;

- unification of IDSS (meaning the use of standard application packages, modelling languages, intelligent systems design tools within software environments);
- the joining of formal and informal procedures for the purposes of modelling;
- easiness and availability; and
- reliability of IDSS functioning (the following types of reliability are distinguished: algorithmic, programme, information, and computing).

The fulfilled analysis of the enumerated requirements demonstrates that IDSS construction on the basis of models belonging to one class (mathematical, logic-linguistic, logic-algebraic, etc.) leads to doubtful, sometimes erroneous, results because of the low degrees of adequacy and openness and because of the absence of the necessary programme and information facilities ensuring the adaptability of the single-model DSS.

The concrete composition and the form of interaction of the IDSS sub-systems should be determined for each SCM hierarchy level, for each phase and with reference to each management function in accordance with the functioning specificity of the appropriate SC elements. Besides, when determining the IDSS elements and structure, the following relationship between the SCM hierarchy level and the characteristics of decision-making procedures should be taken into account. While moving from the lowest hierarchy level to the highest one, the importance and the costs of decisions (from the point of view of the SCM mission) increase, the required levels of accuracy and thoroughness of information presentation decrease and the duration of decision realization increases.

The coordination problem of various decision-making automation tools remains the central problem of IDSS design. To find the solution to this problem, one of the mathematical structures used in most of the models should be selected as the basic structure.

In conclusion, it can be mentioned that every enumerated automation tool should answer the defined requirements of adaptation and evolution. For that, every tool should have, first, the property of redundancy (functional, structural, etc.) and, second, should contain special subsystems (internal and external adapters) implementing mechanisms (procedures) of adaptation and of the adjustment IDSS element.

References

Anderson Jr EG, Morrice DJ, Lundeen G (2006) Stochastic optimal control for staffing and backlog policies in a two-stage customized service supply chain. Prod Oper Manag 15(2):262–278

Beamon BM (1998) Supply chain design and analysis: models and methods. Int J Prod Econ 55(3):281–294

Becker T (2008) Prozesse in Produktion und Supply Chain optimieren. Springer, Berlin Heidelberg

Bolstorff P, Rosenbaum R (2007) Supply chain excellence – a handbook for dramatic improvment using the SCOR model, 2 Edition. McGraw-Hill, New York

Bradley G (2003) Introduction to extensible markup language (XML) with operations research examples. INFORMS Computing Society Newsletter 24(1):1–14
Braun MW, Rivera DE, Flores ME, Carlyle WM, Kempf KG (2003) A model predictive control framework for robust management of multi-product, multi-echelon demand networks. Annu Rev Control 27:229–245
Chandra C, Grabis J (2008) Information technology support for integrated supply chain modeling. Hum Syst Manag 27(1):3–13
Chandra C, Grabis J (2007) Supply chain configuration. Springer, New York
Chandra C, Grabis J, Tumanyan A (2007) Problem taxonomy: a step towards effective information sharing in supply chain management. Int J Prod Res 45(11):2507–2544
Chatfield DC, Harrison TP, Hayya JC (2009) SCML: an information framework to support supply chain modeling. Eur J Oper Res 196 (2):651–660
Chopra S, Meindl P (2007) Supply chain management. Strategy, planning and operation. Pearson Prentice Hall, New Jersey
Delen D, Benjamin PC (2003) Towards a truly integrated enterprise modelling and analysis environment. Comput Ind 51:257–268
Disney SM, Towill DR (2002) A discrete linear control theory model to determine the dynamic stability of vendor managed inventory supply chains. Int J Prod Res 40 (1):179–204
Disney SM, Towill DR, Warburton RDH (2006) On the equivalence of control theoretic, differential, and difference equation approaches to modeling supply chains. Int J Prod Econ 101:194–208
Dolk DR (2000) Integrated model management in the data warehouse era. Eur J Oper Res 122:199–218
Dong M (2001) Process modelling, performance analysis and configuration simulation in integrated supply network design. PhD Dissertation Virginia Polytechnic Institute and State University, Blacksburg, VA
Dotoli M, Fanti MP, Meloni C, Zhou MC (2003) A decision support system for the supply chain configuration. Proc IEEE Int Conf Syst Man Cybern. Washington, DC:1867–1872
Fiala P (2005) Information sharing in supply chains. Omega 33:419–423
Fox MS, Barbuceanu M, Teigen R (2000) Agent-oriented supply chain management system. Int J Flex Manuf Syst 12:165–188
Gunasekaran A, Ngai EWT (2004) Information systems in supply chain integration and management. Eur J Oper Res 159:269–295
Gunasekaran A, Ngai NWT (2009) Modeling and analysis of build-to-order supply chains. Eur J Oper Res 195(2):319–334
Harrison TP (2005) Principles for the strategic design of supply chains. In: Harrison TP, Lee HL, Neale JJ (Ed) The practice of supply chain management. Springer
Holweg M, Pil FK (2008) Theoretical perspectives on the coordination of supply chains. J Oper Manag 26(3):389–406
Ivanov D (2009) DIMA – a research methodology for comprehensive multi-disciplinary modelling of production and logistics networks. Int J Prod Res 47(5):1133–1155
Ivanov D, Arkhipov A, Sokolov B (2007) Intelligent planning and control of manufacturing supply chains in virtual enterprises. Int J Manuf Tech Manag, Special issue on modelling and optimisation of supplier-based manufacturing and management 11(2):209–227
Kahli R, Grover V (2008) Business value of IT: an essay on expanding research directions to keep up with the times. J Assoc Inform Syst 9(1):23–39
Kalinin V, Reznikov B (1987) Systems and control theory: structural-mathematical approach. VIKI, Leningrad (in Russian)
Kamath M, Dalal N, Chaugule A, Sivaraman E, Kolarik W (2003) A review of enterprise process modeling techniques. In: Prabhu V, Kumara S, Kamath M (Eds) Scalable enterprise systems: an introduction to recent advances. Kluwer academic publishers, Dordrecht

Ketikidis PH, Koh, SCL, Dimitriadis N, Gunasekaran A, Kehajova M (2008) The use of information systems for logistics and supply chain management in South East Europe: current status and future direction. Omega 36(4):592–599

Kim H (2001) An XML-based modeling language for the open interchange of decision models. Decis Support Syst 31:429–441

Kim J, Rogers KJ (2005) An object-orineted approach for building a flexible supply chain model. Int J Phys Distib Logist Manag 35:481–502

Kobayashi T, Tamaki M, Komoda N (2003) Business process integration as a solution to the implementation of supply chain management systems. Inform Manag 40:769–780

Koechel P, Nielaender U (2005) Simulation-based optimisation of multi-echelon inventory systems. Int J Prod Econ 93-94:505–513

Kok de AG and Graves SC (2004) Supply chain management: design, coordination and operation. Elsevier, Amsterdam

Lalwani CS, Disney S, Towill DR (2006) Controllable, observable and stable state space representations of a generalized order-up-to policy. Int J Prod Econ 101:172–184

Lau HCW, Lee WB (2000) On a responsive supply chain information system. Int J Phys Distrib Logist Manag 30:598–610

Lee H L, Padmanabhan V and Whang S (1997) Information distortion in a supply chain: the bullwhip effect. Manag Sci 43(4):546–558

McLaren TS, Vuong DCH (2008) A "genomic" classification scheme for supply chain management information systems. J Enterpise Inform Manag 21(4):409–423

Min H, Zhou G (2002) Supply chain modelling: past, present and future. Comput Ind Eng 43(1-2):231–249

Piramuthu S (2005) Knowledge-based framework for automated dynamic supply chain configuration. Eur J Oper Res 165(1):219–230

Rai A, Patnayakuni R, Seth N (2006) Firm performance impacts of digitally enabled supply chain integration capabilities. MIS Quarterly 30(2):225–246

Riddalls CE, Bennett S (2002) The stability of supply chains. Int J Prod Res 40(2):459–475

Sarimveis H, Patrinos P, Tarantilis CD, Kiranoudis C T (2008) Dynamic modeling and control of supply chain systems: a review. Comput Oper Res 35(11):3530–3561

Sarker S, Sarker S, Sidorova A (2006) Understanding business process change failure: an actor-network perspective. J Manag Inform Syst 23(1):51–86

Schenk M, Tokujew J, Reggelin T (2009) Comparison of three methods of implementation of mesoscopic flow models. In: Ivanov D, Meinberg U (Ed) Logistics and supply chain management. Trends in Germany and Russia. Cuvillier Verlag, Goettingen

Shapiro JF (2000) Modeling the supply chain. Duxbury Press, New York

Shapiro JF (2001) Modeling an IT perspectives on supply chain integration. Inform Syst Front 3:455–464

Shen W, Norrie DH, Barthes JP (2001) Multi-agent systems for concurrent intelligent design and manufacturing. Taylor & Francis Group, London

Simchi-Levi D, Wu SD and Zuo-Yun S (2004) Handbook of quantitative supply chain analysis. Springer, New York

Smirnov AV, Sheremetov LB, Chilov N, Sanchez-Sanchez C (2006) Agent-based technological framework for dynamic configuration of a cooperative supply chain. Comput Intell 28:217–246

Smirnov AV, Levashova T, Kashevnik A, Shilov N (2009) Self-configuring logistics networks: architecture and major technologies. In Ivanov D, Meinberg U (Ed) Logistics and supply chain nanagement: trends in Germany and Russia. Civillier-Verlag, Goettingen

Sokolov BV, Yusupov RM (2004) Conceptual foundations of quality estimation and analysis for models and multiple-model systems. J Comput Syst Sci 6:5–16

Stefani RT, Shahian B, Savant CJ, Hostetter GH (2002) Design of feedback control systems. Oxford University Press, Oxford

Sterman JD (2000) Business dynamics: systems thinking and modeling for complex world. McGraw-Hill/Irwin, New Jersey

Subramanian GH, Iyigungor AC (2006) Information systems in supply chain management: a comparative case study of three organisations. Int J Bus Inform Syst 1(4):370–386

Surana A, Kumara S, Greaves M, Raghavan UN (2005) Supply-chain networks: a complex adaptive systems perspective. Int J Prod Res 43(20):4235–4265

Swaminathan JM, Smith SF and Sadeh NM (1998) Modeling supply chain dynamics: a multiagent approach. Decis Sci 29(3):607–632

Tayur S, Ganeshan R and Magazine M (1999) Quantitative models for supply chain management. Kluwer Academic Publishers, Boston

Themistocleous M, Iran Z, Love PED (2004) Evaluating the integration of supply chain information systems: a case study. Eur J Oper Res 159:93-405

Tolujev J (2008) Practical issues in simulation modeling logistics systems. Logistics and Supply Chain Management 25(2):53–63 (in Russian)

Van Houtum GY, Scheller-Wolf A, Yi J (2007) Optimal control of serial inventory systems with fixed replenishment intervals. Oper Res 55(4): 674–687

Verwijmeren M (2004) Software component architecture in supply chain management. Comput Ind 53:165–178

Wang W, Rivera D, Kempf KG (2007) Model predictive control strategies for supply chain management in semiconductor industry. Int J Prod Econ 107:56–77

Yusupov R, Sokolov B (2009) Influence of Computer Science and Information Technologies on Progress in Theory and Control Systems for Complex Plants. In: Proceedings of the 13 IFAC Symposium INCOM: Key Note Papers. Moskau.

Chapter 4
Challenges in Research on Modern and Future Supply Chains

Science may set limits to knowledge,
but should not set limits to imagination.
Bertrand Russell

The problems that exist in the world today
cannot be solved by the level of thinking that created them.
Albert Einstein

4.1 Potential Supply Chain Performance and Stability

During the last few years, research on global production and logistics systems has usually been concentrated on the creation of executable predictive baseline (optimal) plans, not considering, however, that during execution, a plan may be subject to numerous unplanned disruptions. After long-lasting research on SC optimality from the service level's and costs' points of view, the research community has begun to shift to a paradigm that the performance of SCs is to consider *adaptable, stable and crisis-resistant processes* to compete in a real perturbed execution environment (Kleindorfer and Saad 2005, Sheffy 2005, Van de Vonder *et al.* 2007).

The real SC performance is based on maintaining the planned execution and a quick cost-efficient recovery once disturbed. The profit losses through non-purposeful (e.g. demand fluctuations) and purposeful (e.g. terrorism or thefts) perturbation impacts can amount to up to 30% of the annual turnover. For example, in 2000, the material damage to the European retail trade amounted to 13.4 billion euros, and the material damage to the European manufacturers reached 4.6 billion euros (Beck *et al.* 2003). With regard to empirical data of international insurance, companies lose up to 15% of their turnover as a result of threats alone. The discrepancies between demand and supply caused by coordination failures or demand fluctuations can influence up to 30% of added value.

Mulani and Lee (2002) showed that SC managers spend about 40-60% of their working time to handle the disruptions. Due to the economic crisis for the last few months, these figures have become even worse. That is why the issue of the composite objective of maximizing both the SC stability and the SC economic performance can be considered as a timely and crucial topic in modern SCM. In these setting, it becomes necessary to consider the criterion of SC *stability* as a *primary* SC *planning and performance criterion* (see Fig. 4.1).

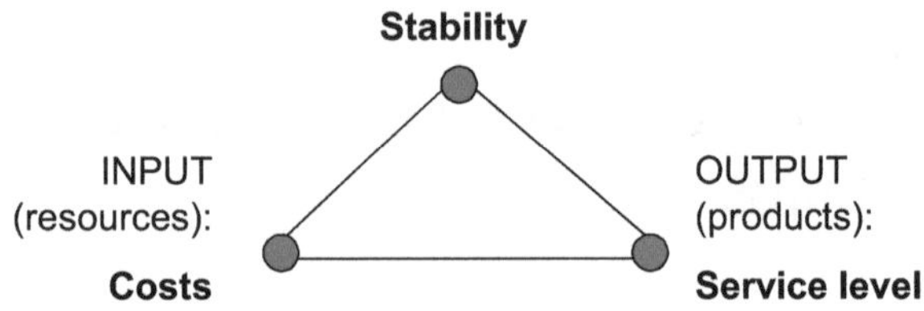

Fig. 4.1 Triangle of SCM goals

Figure 4.1 depicts the considerations of SC economic performance (service level and costs) that should be brought together with the SC stability. The duality of the main goals of SCM – maximizing the service level and minimizing costs – should be enhanced by the third component – maintaining SC stability.

The current economic decrease and its impacts on SCs confirm the necessity of new viewpoints on the SC optimization vision. Striving for maximal profitability in the hope of an unperturbed environment and unlimited economic growth has led to tremendous collapses and losses in SCs. The crisis even provided the ultimate evidence that the main task of SCM is to balance profitability and stability in order to remain competitive in the perturbed economic environment. Besides, the stability criterion meets the SCM nature to a greater extent. Increases in sales and cost reductions may be related to operational logistics improvements at local knots of SC. But the stability of the whole SC is even the direct performance criterion of SCM.

4.2 Uncertainty and Dynamics

Planning and scheduling in the SC environment should be considered not as a static jobs appointment to behaviourally passive machines but as dynamic scheduling in accordance with current demand fluctuations, resource availability and active behaviour of SC elements (own interests, risks, etc.). The subjective multi-criteria of the schedules and the uncertainty of the SC execution dynamics challenge the modern theory and practice of SC planning and scheduling (Proth 2006, Chauchan *et al.* 2007, Pfund *et al.* 2008, Van de Vonder *et al.* 2007, Dolgui and Proth 2009). In the most tactical–operational problems, SC dynamics and uncertainty considerations are mandatory.

Although the feedback loops in SCs have been extensively investigated in system dynamics (Sterman 2000), these models have been successfully applied only for strategic issues of SC configuration and showed many limitations with regard to the tactical and operation control levels. With regard to these two levels, recent literature has indicated an increasing renewed interest in the theoretical background of control theory (Disney *et al.* 2006, Ivanov *et al.* 2007, van Houtum *et al.* 2007, Ivanov and Ivanova 2008, Ivanov 2009b, Ivanov *et al.* 2010).

Control theory is a multi-disciplinary scientific discipline that contains powerful conceptual and constructive tools to conduct research on dynamic problems of flexible (re)distribution of a variable set of jobs to a variable set of resources. The closed-loop control systems are of particular interest in these settings. The current challenge is to transit from simple open time slots incremental planning to dynamic, feedback-based, closed-loop adaptive SC planning and scheduling frameworks and dynamic models to implement adaptability, stability and crisis-resistance throughout the value chain.

A promising area is to combine research on control theory and CAS. Choi *et al.* (2001) claimed that emergent patterns in a supply network can be managed much better through positive feedback than through negative feedback from control loops. The ideas of CAS can be applied to the SCM domain; however, the formal aspects of models are subject to a special analysis. CAS reflects a dynamic network of many agents and decentralized control. However, the tools and techniques of CAS that are based on the fields of non-linear dynamics, statistical physics and information theory are very specific and require specific mathematical background. Besides, the tactical-operational dynamic models should be brought in to correspondence with strategic system dynamics models (Sterman 2000).

4.3 Interrelations and Optimality of Decisions

Over the last decade, a wealth of valuable approaches to SC strategic, tactical and operational planning has been extensively developed (Simchi-Levi *et al.* 2004, de Kok and Graves 2004, Chopra and Meindl 2007). However, conventionally, the planning decisions at different management levels have been considered as being isolated from the other levels. In practice, the interrelation of these three management levels is very important. Not only a problem solution in a fixed environment (the system under control) but also a simultaneous consideration of system formation and solutions to management problems in this system should be the focus of investigations. This aspect is of significant practical importance.

This problem is related to the compatibility of decision theory and managerial tendencies as highlighted by Rittel and Webber (1973) and Peck (2007). Scientists and engineers commonly deal with clear identifiable problems with a known desirable outcome. Such problem localizations frequently lead to unrealistic simplifications and the connection of the model to reality fails. Real problems involve multiple decision makers and different interests and value sets (e.g. individual risk perception). Hence, a danger that an optimal solution may negatively influence the processes of another management level or structure is evident.

The efforts to optimize each process level in isolation can potentially result in the danger that the optimal solutions at the process level (or, more precisely, the solution that is held for the optimal solutions) can cause damage to strategic management, assets management or technological infrastructures, e.g. optimizing the lead time of a set of customers' orders at the operational level may cause a decrease in the service level or an increase in costs at the tactical level. Moreover,

SC optimization as a whole may potentially negatively affect multi-corporative relations or lead to disruptions in ecological systems due to an enormous increase in transportation.

Let us provide a short example. In practice, the challenge is not to calculate optimal schedules to optimize local order fulfilment parameters but to schedule SCs subject to the real dynamics (Dolgui and Proth 2009) for the achievement of SC goals with regard to performance, service level, and stability. That is why the efforts should be directed not (only) to improving algorithms for a benchmarking problem with regard to their speed, but also to scheduling SCs in dynamics with regard to the goals of a superordinated planning level (e.g. service level). To achieve this, it can be necessary to employ more resources or to plan fewer customers' orders, etc.

This evidence challenges the SC models to provide not (only) an output value but a number of alternative solutions with respect to diverse management styles. The other challenge is to conduct research not only into artificial localized problems, but to consider the modelling level with a higher degree of abstraction and to develop generic methodical constructs, which can be localized in concrete environments with the help of methodical guidelines.

4.4 Multi-structural Nature of Supply Chains

Conventionally, the investigations into SC planning (SCP) are performed on the structure of physical distribution, manufacturing and procurement. However, SCs can be considered not only from this organizational point of view, but also from a process point of view (Beamon 1998). It should be emphasized that SCs consist of different structures: business processes and technological, organizational, technical, topological, informational, and financial structures. All of these structures are interrelated and change in their dynamics (Ivanov *et al.* 2010).

The literature on SCM indicates various multi-structural frameworks that received managerial attention when designing SCs (Lambert and Cooper 2000, Bowersox *et al.* 2002). The issue of how to avoid structural incoherency and inconsistency by designing SCs is very important, first for SC design (SCD) itself and second for designing robust SCs (Van Landeghem and Vanmaele 2002) and re-designing SCs for new products (Graves and Willems 2005), new OPP and a variety of disruptive factors.

Especially in adaptive SCs with high dynamics, the issue of how to achieve structural comprehensiveness, responsiveness and flexibility as well as to avoid structural incoherency and non-consistency by SC planning and operations is very important. The adaptation of one structure causes changes in the other related structures. To ensure a high responsiveness level, the SC plans must be formed extremely quickly, but must also be robust. That is why it becomes very important to plan and run SC plans in relation to all the structures. This can be realized if (1) different SC structures are considered simultaneously and (2) the execution dynamics in all the structures can be reflected to establish bilateral feedback between SC plans and operations.

Some examples of the structural interrelations follow. Business processes are designed in accordance with SC goals and are executed by organizational units. These units fulfil management operations and use certain technical facilities and information systems for planning and coordination. Business processes are supported by information systems. Organizational units have a geographical (topological) distribution that also may affect the planning decisions. Collaboration and trust (the so-called "soft facts") in the organizational structure do affect other structures, especially the functional and informational structures. Managerial, business processes (distribution, production, replenishment, etc.), technical and technological activities incur SC costs, which also correspond to different SC structures. So the SC can be interpreted as a complex multi-structural system.

The SC planning and operations decisions are dispersed over different structures. Furthermore, the SC execution is accomplished by permanent changes of internal network properties and external environment. In practice, structure dynamics is frequently encountered. Decisions in all the structures are interrelated. Changes in one structure affect the other structures. Furthermore, the structures and decisions on different stages of SC execution change in dynamics. Output results of one operation are interlinked with other operations (the output of one model is at the same time the input of another model). This necessitates structure dynamics considerations. In the case of disruptions, changes in one structure will cause changes in other relevant structures. Structure dynamics considerations may allow the establishing of feedback between SC design and operations (Ivanov *et al.* 2010, Ivanov 2009c).

4.5 Multi-Disciplinary Modelling

As emphasized in Simchi-Levi *et al.* (2004), Chandra and Grabis (2007), Kuehnle (2008), Chatfield *et al.* (2009), Ivanov (2009a), SC problems are tightly interlinked with each other and have multi-dimensional characteristics that require the application of different integrated frameworks of decision-making support. Beamon (1998) emphasize that SC systems are inherently complex. Thus, the models and methods used to study these systems accurately are, expectedly, also complex. The activity and autonomy of SC elements should also be considered (i.e., simulation of suppliers' selections are connected not only to optimizing certain criteria, but also to their interactions, taking their goal-oriented behaviour into account).

Cross-linked SC planning and operations control problems require combined application of various modelling techniques (optimization, statistics, heuristics and simulation). At different stages of the SC life cycle, a particular problem can be solved by means of different modelling techniques due to changeability of data nature, structure and values, as well as requirements for output representation. Selection of a solution method depends on data fullness, problem scale, one or multiple criteria, requirements on output representation and inter-connection of a problem with other problems. Different approaches from the OR, control theory, systems analysis and agent-based modelling have a certain application area and a certain solution procedure. Isolated application of only one solution method leads to a narrowing in problem formulation, overdue constraints and sometimes unrealistic or impracticable goals.

4.6 Establishing Links to Product Life Cycle, Related Enterprise Management Functions and the Environment

All the enterprise management drivers are tightly interlinked for maximum performance. PLM can be interlinked with SCM, i.e., through the suppliers' and customers' participation in new product development and engineering. A better synchronization of material, informational and financial flows will have a positive impact on all the three flows. The simultaneous optimization of manufacturing and logistics processes and the links between these processes also brings positive effects with regard to shareholders' satisfaction. Marketing can have a profound impact on adjusting SC imbalances with regard to over-inventories.

Enhancing SCs through sustainability with the help of establishing interactions between operations, environment and the product life cycle is the critical next step in SCM as driven from recent examinations in relation to operations and sustainability (Kleindorfer *et al.* 2005) and operations and the environment (Corbett and Kleindorfer 2003). In doing so, the focus on environmental management and operations is moved from local optimization of environmental factors to consideration of the entire SC during the production, consumption, customer service and post-disposal disposition of products. Linton *et al.* (2007) emphasize that sustainable development is a rich area for academic research that is still in its infancy and has the potential to affect future government policy and current production operations, and to identify new business models. It is critical to move forward towards holistic conceptual frameworks and mathematical formulations of the sustainable SCM.

With regard to the above-mentioned issues, the concepts of sustainable SCM, reverse logistics and closed-loop SCs have been developing over the last few years (Guide and Wassenhove 2009). With increasing transportation, the minimizing of negative impacts on ecology may become one of the primary objectives in SCM and logistics. Actually, this is already stated by global automotive companies. Hence, we are moving towards a triangle of goals: profitability, stability and ecological goals.

4.7 Information Technologies and Organizational Aspects

The successful application of SCM depends to a very large extent on intra-organizational changes. Even the collaborative processes with an extended information systems application are managed by people who work in different departments: marketing, procurement, sales, production, etc. The interests of these departments are usually in conflict with each other. Hence, not only outbound synchronizations but also internal organizational synchronization and an integrated performance controlling from the SCM viewpoints is included in the context of SC organization.

Some levels of SC organization can be distinguished. These are communication, cooperation, integration, coordination and collaboration. Actually, only a few SCs in the world have achieved the highest collaboration level. Between 15 and 20% of SCs are at the stage of advanced coordination, and about 50% can be placed between integration and simple coordination.

Before automation, a huge amount of organizational work should be carried out to convince suppliers to collaborate within a common informational space, share the data, actualize the data and ensure financial trust. Modern IT can potentially enable almost any coordination concept. More important problems for efficient coordination lie in the organization sphere, collaboration culture and trust. Last, but not least, the firms themselves should perceive the necessity for such collaboration.

IT provide a new level of coordination capabilities in SCs and enable a breakthrough in SC responsiveness and flexibility. IT, on one hand, serve as an environment to support SCM. On the other hand, they are in turn the enabler of much advancement in SCM. In practice, the building of an IT infrastructure is often based on the "rules of thumb" without properly analysing the business processes. This may result in IT infrastructures that are too complex and too expensive. There are many enterprises that really use only 20–25% of the bought IT systems. Another aspect is the operability and compatibility of IT systems. There are operators in enterprises that must operate several IT systems in parallel with the same functionalities, while the supplier is in several SCs with different OEMs.

4.8 Twelve Main Misunderstandings of SCM

Let us summarize 12 main misunderstandings of SCM that we have experienced in our teaching and consulting practice so far.

Misunderstanding 1. SCs "start from scratch"
SCs rarely emerge as a "green-field" concept. The primary focus is usually directed to rationalizing the existing structures and processes from the SCM perspective. The key questions here are the business process identification, analysis and improvement.

Misunderstanding 2. SCM is just a "hot spot" for several years and will disappear in the near future
Time will tell. The development of SCM has been driven by objective market factors such as customer orientation, market globalization and establishing an information society. These trends have caused changes in enterprise competitive strategies and required new adequate value chain management concepts. The practice of SCM has provided enough evidence that intra-organizational integration and inter-organizational coordination along the entire value-adding chain have a profound influence on profitability and competitiveness, rather than local optimization of intra-organizational functions.

Misunderstanding 3. SCM replaces logistics
In analysing the existing research literature and empirical case studies, the following can be concluded: logistics deals mostly with local functions for implementing the physical transition of material flows and SCM deals with the value-adding chain as a whole and concentrates on the links between the local functions for implementing the physical transition of inbound and outbound material flows. Logistics is attracted to optimizing the realization of physical transitions; SCM is attracted to the management level. In other words, logistics takes care of providing the right goods, in the right place, at the right time, in the right volume, in the right package, in the right quality, with the right costs, and SCM takes care of balancing supplies and demand along the entire value-adding chain subject to the full customer satisfaction. Both the logistics and SCM will exist in future. Logistics and SCM are tightly interlinked with each other. Operational logistics performance influences the increase in sales and decrease in costs, while SCM ensures SC stability.

Misunderstanding 4. The main goal of SCM is to maximize profitability
Yes and no. The basis of entrepreneurship is the creation and maximization of added value. However, this potential performance may be achieved only if the processes are fulfilled in accordance with a plan that in turn may be subject to different disruptions that may inhibit the achievement of SCM goals in a real execution environment. Hence, the important goal of SCM is to ensure SC stability with regard to possible disruptions as the potential performance may be achieved only through stability.

Misunderstanding 5. A SC is a linear sequence of enterprises
This can seem amazing but many people have such an understanding of SCs. This may partially result from figures in study books that depict a SC in this way. Of course, SCs are networks with lots of branching points and parallel operations.

Misunderstanding 6. SCM is just the optimization of customer and supplier relationships
Indeed, SCM primarily implies the inter-organizational level. However, the collaborative processes with an extended information system application are managed by people in your own enterprise who work in different departments: marketing, procurement, sales, production, etc. Successful implementation of SCM also requires balancing the interests of diverse departments that are usually in conflict

with each other. Hence, not only outbound synchronizations but also internal organizational synchronization is in the context of SC organization.

Misunderstanding 7. SCM is just of a strategic nature
This is the next false assertion. SCM is subject to strategic, tactical and operational management levels as well as the execution level. Only the level of physical product transformation and transition is primarily in the competence of logistics and manufacturing.

Misunderstanding 8. SCM is all about IT
Indeed, initially, SCM has been primarily considered from the IT point of view. Nowadays, IT is the backbone of SCM realization. Its primary contribution is speed and transparency in SC processes. On the other hand, IT is just an "enabler" of the business concepts and organizational schemes.

Misunderstanding 9. SCM is a number of functions (procurement, production and distribution)
SCs should always be considered from both the object and the process points of view. Even the balancing of different SC structures (functional, organizational, informational, technological, and financial) challenge the modern SCM.

Misunderstanding 10. Optimizing SCs consists of optimizing local SC problems (replenishment, production, etc.)
SC optimization should be subject to key performance indicators (KPI) and not (only) to lead time, batches and inventory optimization. Second, the practice does not need the optimal plans that will fail in a real execution environment but adaptive and stable plans that may be executed under perturbation impacts.

Misunderstanding 11. Integrated management of the entire SC
In practice, the OEMs manage their collaboration with the first tier level, and in some cases with the second tier level. Indeed, the integrated management of the entire SC is of an abstract nature and emerges through iterative procedures of coordination in partial SC links.

Misunderstanding 12. Modern OEMs aim at a low production depth (at about 30%) and outsourcing of most of their competencies
Indeed, the official data report 30% production depth as a benchmarking value. However, if considering the OEMs, e.g. in the automotive sector, in detail, it can be seen that a large part of these "external" suppliers are in close legal relations with the OEMs and cannot be considered as really autonomous and independent.

References

Beamon BM (1998) Supply chain design and analysis: models and methods. Int J Prod Econ 55(3):281–294

Beck A, Chapman P, C Peacock Shrinkage (2003) A collaborative approach to reducing stock loss in the supply chain. ECR Europe, Brussels

Bowersox D, Closs D and Cooper M (2002) Supply chain logistics management. McGraw-Hill/Irwin, New York
Chandra C, Grabis J (2007) Supply chain configuration. Springer, New York
Chatfield DC, Harrison TP, Hayya JC (2009) SCML: an information framework to support supply chain modeling. Eur J Oper Res 196 (2):651–660
Chauhan SS, Gordon V and Proth J-M (2007) Scheduling in supply chain environment. Eur J Oper Res 183(3):961–970
Choi TY, Dooley KJ, Rungtusanatham M (2001) Supply networks and complex adaptive systems: control versus emergence. J Oper Manag 19(3):351–366
Chopra S, Meindl P (2007) Supply chain management. Strategy, planning and operation. Pearson Prentice Hall, New Jersey
Corbett CJ, Kleindorfer PR (2003) Environmental management and operations management: introduction to the third special issue. Prod Oper Manag 12(3):287–289
Disney SM, Towill DR, Warburton RDH (2006) On the equivalence of control theoretic, differential, and difference equation approaches to modeling supply chains. Int J Prod Econ 101:194–208
Dolgui A, Proth JM (2009) Supply Chains Engineering: useful methods and techniques. Springer, Berlin *et al.*
Guide VDR, Wassenhove LNV (2009) The evolution of closed-loop supply chain research. Oper Res 57(1):10-18
Graves SC and Willems SP (2005) Optimizing the supply chain configuration for new products. Manag Sci 51(8):1165–1180
Guide VDR, Wassenhove LNV (2009) The evolution of closed-loop supply chain research. Oper Res 57(1):10–18
Ivanov D (2009a) DIMA – a research methodology for comprehensive multi-disciplinary modelling of production and logistics networks. Int J Prod Res 47(5):1133–1155
Ivanov D (2009b) Supply chain multi-structural (re)design. Int J Integrated Supply Manag 5(1):19–37
Ivanov D (2009c) An adaptive framework for aligning (re)planning decisions on supply chain strategy, design, tactics, and operations. Int J Prod Res, DOI: 10.1080/00207540902893417, in press
Ivanov D, Ivanova M (2008) A framework of adaptive control for production and logistics networks. In: Haasis HD, Scholz-Reiter B, Kreowski J (Eds) Dynamics in logistics. Springer, Berlin
Ivanov D, Kaeschel J, Sokolov B (2007) Integrated modeling of agile enterprise networks. International Journal of Agile Systems and Management 2(1):23–49
Ivanov D, Sokolov B, Kaeschel J (2010) A multi-structural framework for adaptive supply chain planning and operations control with structure dynamics considerations. Eur J Oper Res 200: 409–420
Kleindorfer PR, Saad GH (2005) Managing disruption risks in supply chains. Prod Oper Manag 14(1):53–68
Kleindorfer PR, Singhal K, van Wassenhove LN (2005) Sustainable operations management. Prod Oper Manag 14(4):482–492
Kok de AG and Graves SC (2004) Supply chain management: design, coordination and operation. Elsevier, Amsterdam
Kuehnle H (2008) A system of models contribution to production network (PN) theory. J Intell Manuf 18(5):543–551
Lambert DM and Cooper MC (2000) Issues in Supply Chain Management. International Marketing Management 29(1):65–83
Linton J, Klassen R, Jayaraman V (2007) Sustainable supply chains: an introduction. J Oper Manag 25:1075–1082

Mulani NP, Lee HL (2002) New Business Models for Supply Chain Excellence. In N. Mulani (ed.), Achieving Supply Chain Excellence Through Technology (Vol. 4), Montgomery Research, Inc.

Peck H (2007) Supply chain vulnerability, risk, robustness, and resilience. In: Mangan J, Lalwani C, Butcher T (Ed) Global logistics and supply chain management. John Wiley, New York

Pfund ME, Balasubramanian H, Fowler JW, Mason JS and Rose O (2008) A multi-criteria approach for scheduling semiconductor wafer fabrication facilities. J Sched 11(1): 29–47

Proth JM (2006) Scheduling: new trends in industrial environment. Information control problems in manufacturing. Annu Rev Control 31(1):157–166

Rittel HWJ, Webber MM (1973) Dilemmas in a general theory of planning. Pol Sci 4:155-169

Sheffy J (2005) The resilient enterprise. MIT Press, Massachusetts

Simchi-Levi D, Wu SD and Zuo-Yun S (2004) Handbook of quantitative supply chain analysis. Springer, New York

Sterman JD (2000) Business dynamics: systems thinking and modeling for complex world. McGraw-Hill/Irwin, New Jersey

Van de Vonder S, Demeulemeester E and Herroelen W (2007) A classification of predictive-reactive project scheduling procedures. J Sched 10(3):195–207

Van Houtum GY, Scheller-Wolf A, Yi J (2007) Optimal control of serial inventory systems with fixed replenishment intervals. Oper Res 55(4):674–687

Van Landeghem H, Vanmaele H (2002) Robust planning: a new paradigm for demand chain network design problems. Eur J Oper Res 94:362–376

Chapter 5
Uncertainty, Risk and Complexity

Lots of folks confuse bad management with destiny.
Kin Hubbard

5.1 Origins and Variety of Uncertainty

5.1.1 Uncertainty and Perturbation Influences

Uncertainty is a system property characterizing the incompleteness of our knowledge about the system and the conditions of its development. Uncertainty is a polysemic term (poly – many, sema – a sign). Historically, the first terms related to uncertainty were accident, probability and possibility, which we relate to Aristotle. Up to the twentieth century, the mathematical basics of uncertainty factor description were founded on probability–frequency interpretation and are related to Pascal, Ferma, Bernoulli and Laplace. Modern probability theory is based on the research of Kolmogorov, who introduced an axiomatic definition of probability as a measure related to a system of axioms of a so-called probability space.

In contrast to risk, uncertainty is a more comprehensive term, considering situations that cause both positive (chance) and negative (threats) deviations from an expected outcome. Modern system theory defines uncertainty as "a gradual assessment of the truth content of a proposition, e.g. in relation to the occurrence of the event" (Möller and Beer 2004).

One of the main dangers of uncertainty is the perturbation influences, leading to a change in a planned course of events in the SC functioning and (or) a threat of goal default. There are different external and internal, objective and subjective perturbation influences altering the execution conditions of a SC. Let us analyse the main types of perturbation influences that can be divided into two groups:

- purposeful perturbation influences; and
- non-purposeful perturbation influences.

The *purposeful perturbation influences* can be antagonistic (impeding SC functioning) or non-antagonistic (promoting SC functioning). Examples of purposeful perturbation impacts are thefts, terrorism, piracy and financial misdeeds.

The *non-purposeful perturbation influences* can be natural, economic or technological. The former can be caused by phenomena of the geo-, hydro- or biosphere. An example of an economic non-purposeful perturbation impact is demand fluctuations and the bullwhip-effect.

Hence, there are two types of uncertainty affecting SCs: (1) risks arising from the problems of coordinating supply and demand and (2) risks arising from purposeful disruptions to normal activities (Kleindorfer and Saad 2005).

5.1.2 Sources of Uncertainty

In Fig. 5.1, a classification of uncertainty origins is undertaken.

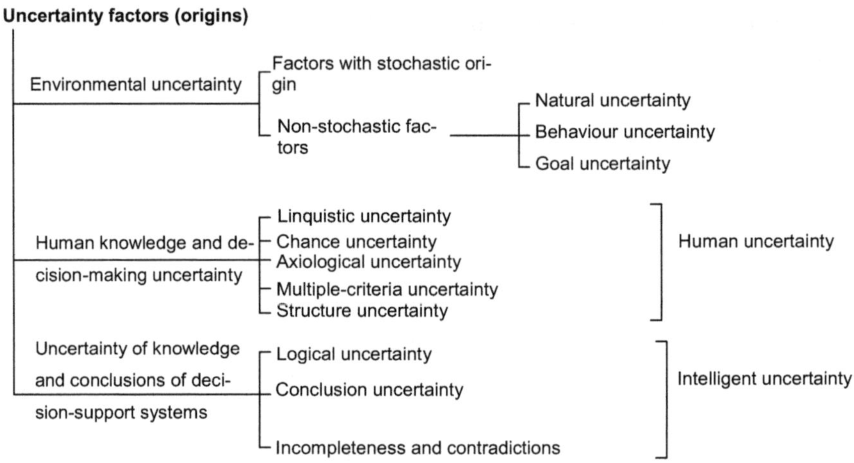

Fig. 5.1 Classification of uncertainty factors

The uncertainty factors are usually divided into two groups: stochastic factors and non-stochastic factors. The first group can be described via probability models. The factors described as aleatory variables (functions, fields) with known distributions are statistically defined. Aleatory variables with unknown distributions can be of two types: those with known or unknown characteristics. The following factors produce non-stochastic uncertainty:

- Purposeful opposition of a rival system, while its actions are unknown. This type of uncertainty is called behavioural.
- Phenomena interrelated with SC operation and insufficiently studied. This type of uncertainty is called uncertainty of nature.
- Uncertainty of human thinking. This kind of uncertainty arises when the system is being managed or investigated. It can be called personnel uncertainty.
- Uncertainty of knowledge in the system of artificial intellect.

For the formal description of non-stochastic uncertainty, fuzzy description with known membership functions, subjective probabilities for the uncertainty factors, interval description, and combined description of the uncertainty factors are used.

In analysing uncertainty, four aspects are usually encountered. The one is uncertainty itself, the second is risks, the third is perturbation influence (disturbances), and the last is the perturbation impact influences (deviations). In the further course of this and the following chapters, we will frequently encounter this constellation (see Fig. 5.2).

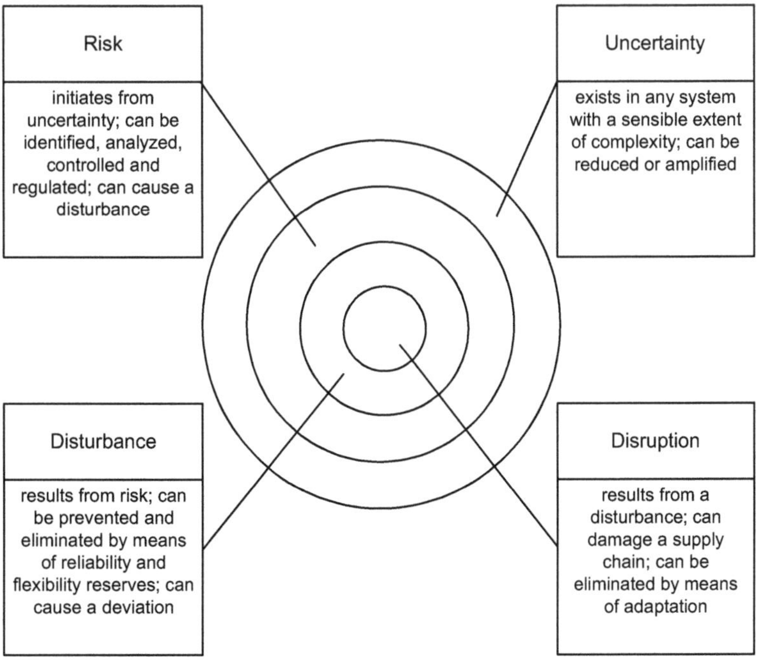

Fig. 5.2 Interrelations of uncertainty, risk, disturbance and disruption

Uncertainty is the general property of a system environment that exists independent of us for any system of a sensible complexity degree. As shown in Fig. 5.4, we can broaden and narrow the uncertainty space.

Risk initiates from uncertainty (see also Sect. 5.2). Risks can be identified, analysed, controlled and regulated. If we return to the discussions in Sect. 5.1.1, we consciously talked about uncertainty factors and risks arising (e.g. risk of demand fluctuation as a result of the environmental uncertainty).

A *disturbance* (perturbation influence) is the consequence of risks. These may be purposeful (i.e. thefts) and non-purposeful (i.e. demand fluctuations or the occurrence of some events that may necessitate adapting the SC). They may cause a deviation (disruption) in the SC or not (e.g. a SC can be robust and adaptive enough to overcome the disturbance).

Deviations (disruptions) are the result of perturbation influences. They may affect operations, processes, plans, goals or strategies. To adjust the SC in the case of deviations, adaptation measures need to be taken.

For the SCM domain, uncertainty factors and measures for their handling can be distinguished as follows (see Table 5.1).

Table 5.1 Uncertainty factors and measures for their handling in SCs

Decision-making level	Uncertainty factors	Handling measures
Strategic	Multiple management goals	Multi-criteria analysis techniques
	Terrorism, piracy	SC security management
	Financial and political crises	Liquid assets reserves
	Natural disasters	Strategic material inventories
		Market diversification and outsourcing
		Product lines' flexibility and modularity
Tactical and operational	Weak coordination	Safety stocks and time buffers
	Stockless processes	Reserves of SC capacities
	Weak control of cargo security	SC coordination, monitoring, and event management
	Technological breaks	
	Human errors	

In Table 5.2, some examples of disturbances and disruptions in SCs are presented.

Table 5.2 Examples of disturbances and disruptions in SCs (Beck *et al.* 2003, Zeller 2005)

Factor	Example	Impacts
Thefts and damages of goods	Retail	Losses in Europe 13.4 billion euros a year
	Production	Losses in Europe 4.6 billion euros a year
		Losses up to 15% of annual turnover
Terrorism	September 11	Five Ford plants have been closed for a long time
Piracy	Somali, 2008	Breaks in many SCs
Natural disasters	Earthquake in Thailand, 1999	Apple computers' production in Asia has been paralysed
	Flood in Saxony, 2002	Significant production decrease at VW, Dresden
	Earthquake in Japan, 2007	Production breakdown in Toyota's SCs amounted to 55.000 cars
Political crises	"Gas" crisis 2009	Breaks in gas supply from Russia to Europe, billions of losses to GAZPROM and customers
Financial crises	Autumn 2008	Production decrease or closing; breaks in SCs throughout
Coordination problems	Demand fluctuations,	Losses through not receiving customers' orders; penalties and forfeits of up to 15% of annual turnover
	Internal unbalancing	

In sum, the profit losses through non-purposeful (e.g. demand fluctuations) and purposeful (e.g. terrorism or thefts) perturbation impacts can amount to up to 30% of the annual turnover. That is why the issue of uncertainty handling can be considered as a timely and crucial topic in modern SCM.

5.1.3 Uncertainty Within the Complexity Management

Complexity has been one of the most challenging phenomena in business and science over the last 60 years (Ashby 1956, Simon 1962, Casti 1979, Holland 1995, Anderson 1999, Lissak and Letiche 2002, Richardson 2004, 2005, 2007, Pathak *et al.* 2007). Complexity is a multi-spectral category and one of the basic properties of systems of any nature (see Sect. 5.3).

The fulfilled analysis confirmed that a well founded concept for the uncertainty analysis in the SC models is a system-cybernetics one (see Fig. 5.3).

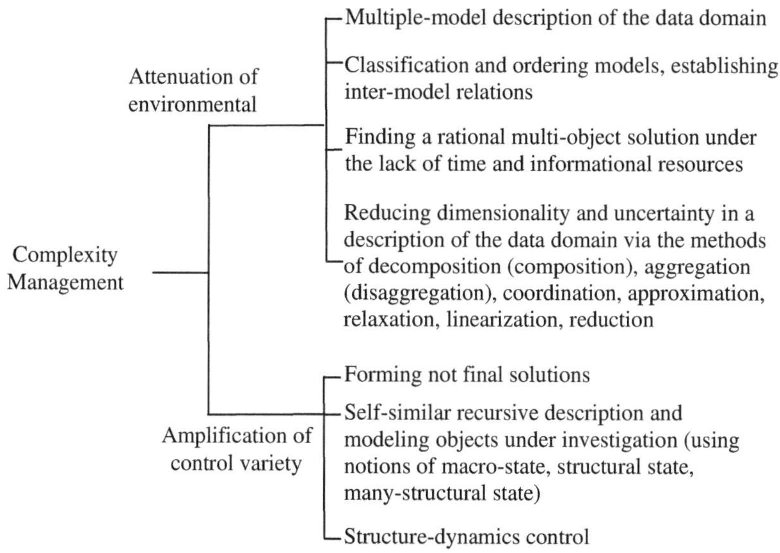

Fig. 5.3 Directions for realizing the law of requisite variety

This concept presumes that all the input signals of a dynamic system (SC, in our case) can be divided into two classes: control inputs and perturbation inputs. Moreover, it is assumed that the control inputs are known, and thus the SC and its control processes can be regarded as deterministic mathematical constructions. All the perturbation factors are called factors of uncertainty. They belong to the environment into which the deterministic object is "plunged".

Complexity management and system modelling can be considered as a theoretical basis for handling uncertainty in SCs. From the perspective of complexity management, the problem of a system under control and uncertainty is related to an area under control and an area under uncertainty. This idea is based on Ashby's principle (Ashby 1956) of requisite variety. By broadening the control area (Fig. 5.4b) and narrowing the uncertainty area or reverse (Fig. 5.4a), the system control can be adapted. Hence, the mutual relations between the system and environment spaces fall into the categories of amplification of a *control variety* and attenuation of an *environmental variety (*see Fig. 5.4).

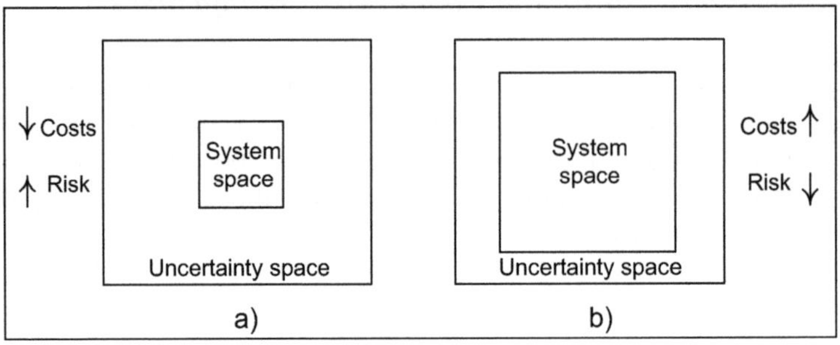

Fig. 5.4 System space and uncertainty space

Thus, amplifying the variety of our control area and reducing the area of uncertainty, (1) a balance of control and perturbed impacts as well as (2) the maintenance of the planned execution processes and a quick cost-efficient process recovery once disturbed can be reached (see also Sect. 6.3).

5.2 Risk Management in Supply Chains

Uncertainty initiates risk. *Risk management* is a methodological approach to the management of outcome uncertainty. The concept of risk is subject to various definitions. (Knight 1921) classified under 'risk' the 'measurable' uncertainty. From the financial perspective of Markowitz (1952), risk is the variance of return. From the project management perspective, risk is a measure of the probability and consequence of not achieving a defined project goal. According to March and Shapira (1987), risk is a product of the probability of occurrence of a negative event and the resulting amount of damage.

Generally, in decision theory, risk is a measure of the set of possible (negative) outcomes from a single rational decision and their probabilistic values. In the literature on SCM, the term "risk" is also replaced with "vulnerability", which means "at risk" (Kersten and Blecker 2006, Peck 2007). Pfohl (2008) provided an overview of strategic and constructive methods for SC risk and security management. Risk management usually comprises the stages of *identifying* possible risks, an *analysis* of these, elaborating *control* actions and risk *controlling* (Götze and Bloech 2002, Hallikas *et al.* 2004, Khan and Burnes 2007, Pfohl 2008, Rao and Goldsby 2009).

A particular feature of risk management in SCs (unlike in technical systems) is that people do not strive for a *100% guarantee* of the result: they consciously tend to take risks. Some literature (e.g. Sokolov and Yusupov 2006, Peck 2007) points out the problem of contradiction between *objective risk* (determined by experts, applying quantitative scientific means) and *perceived risk* (perception of managers).

Actually, the objective risk treatment is rooted in technical science where 100% reliability is mandatory. In socio-economic systems, like SCs, a value of 95% as an orientation for SCs is empirically suggested (e.g. Sheffy 2005). Different managers perceive risk to different extents, and these perceptions can change in the same manager due to changes in his environment. That is why the models for SCs should not strive for a unique optimal solution but allow the formation of a number of alternative solutions with different degrees of potential economic performance and risk. Summarizing, we will note that the risk can be considered from three basic positions:

1. the risk is a likelihood estimation of a negative outcome of the event leading to losses/losses (the technological approach);
2. the risk is an individual estimation by the person of the danger of a negative outcome of the event leading to losses/losses; risk is ultimately a property of any entrepreneurship (the psychological approach);
3. the risk is an integral property of any process or system, the management of which is a key problem in economic performance and stability maintenance (the organizational approach).

Let us describe the proposed concept of risk handling. In order to analyse risks, the following main categories are introduced: the risk factor, the risk source, the risk situation, and the dangerous situation. The risk factor is a global category that characterizes a system at the goal-orientation level (e.g. upsetting of the production plan, delivery breakdown, etc.). Risk sources consider certain events that may cause risk factors. The dangerous situation characterizes the state of a system when a probability of risk sources' appearance and their direct influence on this system is high. A risk situation means a condition when the active influences of risk sources cause disturbances and deviations in system functioning (see Fig. 5.5).

The problem of SC functioning in terms of risk consists of the following main phases: risk factors' identification → risk sources and dangerous situations' identification → identification of interdependences between risk situation appearance and changes of system functioning parameters → decision-making about compromise while SC configuration by aggravation of some goal criteria (e.g. cost increase while keeping the planned production volume and deadline; production volume reduction while keeping the same cost level and deadline; change of deadline while keeping the same costs and production volume, etc.) → control decision development in order to compensate for possible disturbances in system functioning caused by risk situations → development of a managed object monitoring system.

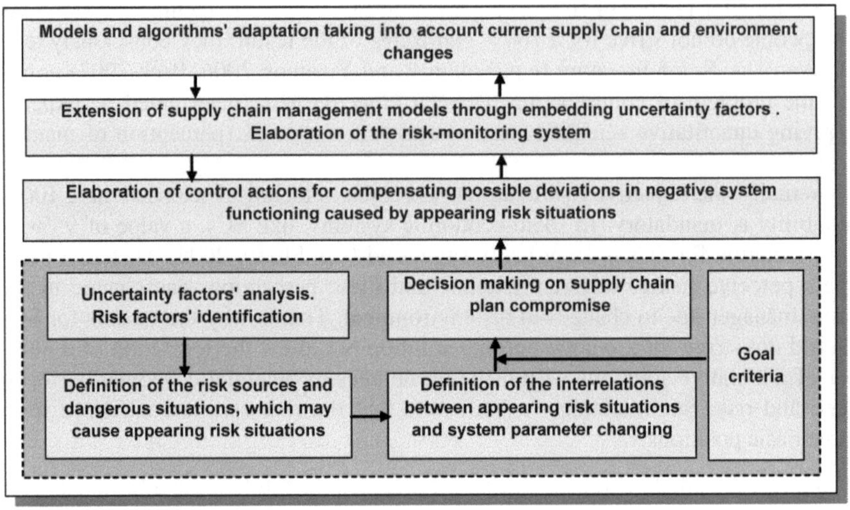

Fig. 5.5 Handling risks in SCs

5.3 Supply Chains as Complex Systems

This section is a logical follow-up to the Sect. 5.1.3. The distinction of this section comes from the evidence that, although SCs are often related as complex systems, a discussion on why they are those is rarely given. The problem of complexity has various aspects and applications (Simon 1962, Bertalanffy 1968, Mesarovic and Takahara 1975, Casti 1979). The literature on complexity shows that the viewpoints regarding the concept of 'complexity' tend to be as richly varied as complexity itself. Although no unified definition of a complex system exists, a number of complexity views may be distinguished.

The first group of complexity factors is related to *structural* complexity. This consists of a number of elements in a system and a number of interrelations between these elements. Moreover, the variety of the elements and the interrelations is under consideration.

The second group of complexity factors is related to *functional* complexity. This includes the dynamics of the change in the elements, their variety and interrelations between the elements. Another aspect is the consideration of system complexity at certain instants of time. A system can be composed of a great number and variety of elements and interrelations, but in a snap-shot at an instant of time, the system may appear to be very simple. Last but not least in the functional complexity is the uncertainty of the change in the elements, their variety and interrelations between the elements. This point is one of the most critical while considering system complexity.

The third group of complexity factors is related to *modelling complexity*. To this group is related the well-known calculation complexity, resulting, e.g. in NP (nondeterministic polynomial)-hard problems and decision-making complexity, resulting from conflicting goals that are in turn difficult to formalize (Casti 1979). The problems in systems are tightly interrelated. However, different methods and data are needed for solving different tasks. Usually, investigations into a complex system are performed by means of the combined application of different methods and involve specialists in economy, mathematics and computer science.

Fig. 5.6 depicts the above-described complexity and uncertainty factors and proposes approaches to handle these factors.

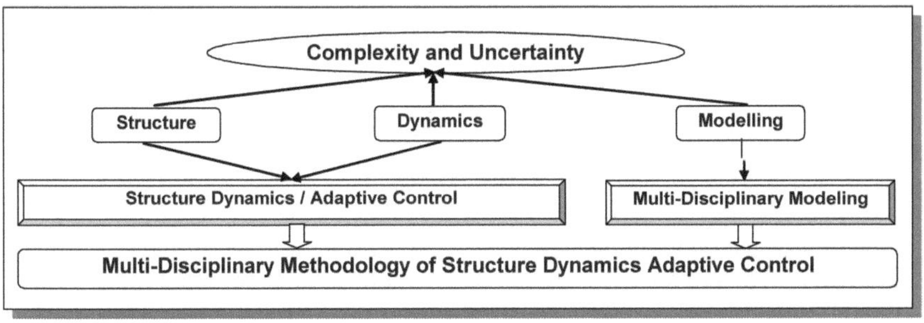

Fig. 5.6 SC complexity and necessary research approaches

SCs are characterized by a great number and variety of elements and the interrelations between them. Moreover, decisions in SCs are dispersed over different structures and management levels. The SC structures change in dynamics, so structure dynamics is frequently encountered. SC dynamics is characterized by uncertainty. Besides, SCs are described by means of different modelling approaches and model classes. Moreover, elements of SCs are active. They act as self-goal-oriented, are autonomous but collaborative and have conflicting goals. These goals are also subject to multiple criteria and are difficult to formalize. Here, the tight interlinking of the complexity and uncertainty can be perceived.

As SCs continuously interact with the environment and evolve through these interactions, they may be also considered as open systems. To remain manageable, SCs should maintain certain steady states for a specific period of time (Chandra and Grabis 2007). These steady states result from the continuous balancing of inflow and outflow from and to the SC environment. This balancing is based on managerial control inputs of both a planned and a regulative nature regarding perturbation influences from the environment. We understand the environment as being everything that is not in our system (SC). As discussed in Sect. 5.2, this connection "system–environment" is subject to changes by narrowing and broadening the system's borders. In actually open systems, these narrowing and broadening processes are based on self-organizing and self-learning principles. Modern SCs are still far from this stage but they are following the path to self-organization.

Hence, SCs may be justifiably called *complex dynamic multi-structural systems with active elements of free-will behaviour*. Research on such systems requires the application of different methods and disciplines. Actually, the research on SCs as complex systems should impart much more universality than is really considered in today's social and business systems.

In this book, we consider SCs from the above-mentioned point of view. The research approaches of structure dynamics control, adaptive control, and multi-disciplinary modelling will be applied to approach handling complexity, uncertainty, and operations dynamics in SCs. Special consideration in the follow-up frameworks and models will be given to the aspects of SC dynamics at the tactical and operational decision levels.

5.4 Practical Issues of Uncertainty in Supply Chains

Uncertainty inevitably exists in SC. Uncertainty can be scattered over different SC parts but it cannot be eliminated completely. In considering the uncertainty problem, two main aspects can be highlighted:

- uncertainty can be mitigated and;
- SCM is always connected with risk that should be taken by somebody.

In Chaps. 6 and 7, we will discuss the ways to handle uncertainty in SCs. But even if a good balance of a system space and an uncertainty space can be found, certain risks of failure and disturbances will exist. For these cases, different measures in the SC contracting and insurance should be taken. As we have already shown, a balance of an uncertainty space (risk area) and a system space (control area within which deviations can be eliminated with the planned reliability and flexibility reserves) should be found with regard to individual risk perceptions of SC managers (see Chap. 14 and Sect. 15.3).

In practice, one of the most important challenges of uncertainty and risk analysis is the identification and strengthening of so-called bottlenecks. These are not all perturbation impacts that will affect the SC. As practice shows, the robustness of the bottlenecks and their flow capacity capabilities determine the economic performance and stability of SCs to a very high degree.

To the SC bottlenecks belong:

- a SC part that is permanently subject to disturbances (these parts should be eliminated if possible);
- a SC part that is critical to the supply flow capacity (this may be handled, e.g. by the DBR (Drum–Buffer–Rope) technique);
- a SC part in which small deviations cause large deviations in performance indicators (this may be handled, e.g., by sensitivity analysis); and
- a SC part whose adjustment after being disturbed requires significant financial and (or) time consumption (no explicit methodologies for this issue have been identified so far).

There is no doubt that the identification and strengthening of bottlenecks requires an individual procedure for almost every concrete SC. Any methodical approaches should be considered in these settings as methodical frameworks that will be fulfilled with concrete contents within an actual SC environment. Nevertheless, the elaboration of the methodical orientations for the identification and strengthening of bottlenecks in SCs, enhanced by the specificity of the SC under analysis, may potentially contribute to reducing the costs of disruption elimination, reducing the frequency of the disruptions, and making SC managers' and operators' work more comfortable.

In summarizing the practical advancements in uncertainty handling within a SC environment, the following can be concluded:

1. Uncertainty space may be reduced by means of, e.g.:

- introducing excessiveness in SC structures (e.g. time buffers, safety stocks, additional resources, capacity reserves, etc.);
- improving coordination and information flows to make better quality, timeliness, and accessibility;
- introducing SC monitoring and event management systems to react quickly to disturbances and disruptions; and
- forming a set of not final decisions, i.e. postponement and rolling/adaptive planning.

2. It is impossible to avoid uncertainty.

In the further course of this book, we will consider SCs from the point of view of both economic performance and stability. We will analyse how uncertainty may affect SCs, how it can be handled and how to balance the system space of uncertainty and the system space under control to achieve maximum SC economic performance in a real perturbed execution environment.

References

Anderson P (1999) Complexity theory and organization science. Organ Sci 10:216–232

Ashby WR (1956) An introduction to cybernetics. Chapman & Hall, London

Beck A, Chapman P, Shrinkage PC (2003) A collaborative approach to reducing stock loss in the supply chain. ECR Europe, Brussels

Bertalanffy L (1968) General system theory: foundations, development, applications. George Braziller, New York

Casti JL (1979) Connectivity, complexity and catastrophe in large-scale systems. Wiley-Interscience, New York and London

Chandra C, Grabis J (2007) Supply chain configuration. Springer, New York

Götze U, Bloech J (2002) Investitionsrechnung, Modelle und Analysen zur Beurteilung von Investitionsvorhaben, 3. Auflage. Springer, Berlin

Hallikas J, Virolainen V, Tuominen M (2004) Risk analysis and assessment in network environments: a dyadic case study. J Prod Econ 78:45–55

Holland JH (1995) Hidden order: how adaptation builds complexity. Helix Books, Cambridge
Kersten W, Blecker T (2006) Managing risks in supply chains. How to build reliable collaboration in logistics. Erich Schmidt Verlag, Berlin
Khan O, Burnes B (2007) Risk and supply chain management: creating a research agenda. Int J Logist Manag 18(2):197–216
Kleindorfer PR, Saad GH (2005) Managing disruption risks in supply chains. Prod Oper Manag 14(1):53–68
Knight F (1921) Risk, Uncertainty, and Profit. Hart, Schaffner, and Marx Prize Essays 31. Houghton Mifflin, Boston and New York
Lissak MR, Letiche H (2002) Complexity, emergence, resilence, and coherence: gaining perspective on organizations and their study. Emergence 4(3):72–94
March JG, Shapira Z (1987) Managerial perspectives on risk and risk taking. Manag Sci 33:1404–1418
Markowitz H (1952) Portfolio selection. J Finance VII(1):77–91
Mesarovic MD and Takahara Y (1975) General systems theory: mathematical foundations. Academic Press, New York, Can Francisco, London.
Möller B, Beer M (2004) Fuzzy randomness. Uncertainty in civil engineering and computational mechanics. Springer, Berlin
Pathak SD, Day JM, Nair A, Sawaya WJ, Kristal MM (2007) Complexity and adaptivity in supply networks: building supply network theory using a complex adaptive systems perspective. Decis Sci 38(4):547-580
Peck H (2007) Supply chain vulnerability, risk, robustness, and resilience. In: J Mangan, C Lalwani, T Butcher (Ed) Global logistics and supply chain management. John Wiley and Sons, Chippenham
Pfohl HC (2008) Sicherheit und Risikomanagement in der Supply Chain. DVZ-Verlag, Hamburg
Rao S, Goldsby TJ (2009) Supply chain risks: a review and typology. Int J Logist Manag 20(1):97–123
Richardson KA (2004) Systems theory and complexity: Part 1. E:CO 6(3):75–79
Richardson KA (2005) Systems theory and complexity: Part 3. E:CO 7(2):102–114
Richardson KA (2007) Systems theory and complexity: Part 4. The evolution of systems thinking. E:CO 9(1):166
Sheffy J (2005) The resilient enterprise. MIT Press, Massachusetts
Simon HA (1962) The architecture of complexity. Proc Am Phil Soc 106(6):467–482
Sokolov B, Yusupov R (2006) Risk integrated modelling for control in complex organizational-technical systems. Journal of Control and Informatics 1:1–22
Zeller AJ (2005) Automatisierung von Diagnose und Therapie beim Controlling von Liefernetzen. Shaker Verlag, Aachen

Chapter 6
Handling Uncertainty in Supply Chains

Once men are caught up in an event, they cease to be afraid.
Only the unknown frightens men.
Antoine de Saint-Exupery

What is outside is harder to change than what is inside.
Paulo Coelho

6.1 Supply Chain Security

6.1.1 Purposeful Perturbation Influences on Supply Chains

A purposeful perturbation influence is an external input influence on a SC to harm or damage a SC (for example, plundering of cargoes, terrorism or piracy). Modern SCM is facing the challenge of designing secure supply networks with high economic performance. The recent crashes in financial markets, disruptions in global SCs like gas supply, terrorist attacks, piracy and numerous natural disasters provide the evidence that integrated considerations of SC performance and security is a crucial and timely topic in SCM research and practice. That is why the concept of SC security has been developed over the last few years (Rice and Caniato 2003, Sheffy 2005, Sheffy and Rice 2005, Kleindorfer and Saad 2005, Closs *et al.* 2008, Williams *et al.* 2008). This evidence is also reflected in the newly published international standards ISO 28000 on SC security management.

SC *security* is a general system property characterising uninterrupted performance of a SC functioning to achieve its goals under protection against external purposeful threats.

Lets us bring some practical data. In a Stanford University study of 11 manufacturers and 3 "innovator" logistic providers, samples of the commercial benefits of SC security were a 26% reduction in customer attrition and a 20% increase in the number of new customers, a 38% reduction in theft/loss/pilferage and a 37% reduction in tampering, and a 30% reduction in problem identification and problem resolution time (ISO 28000, 2007). For example, in 2000 the material damage to the European retail trade amounted to 13.4 billion euros, and the material damage to the European manufacturers reached 4.6 billion euros (Beck *et al.* 2003).

The literature on SC security is just at the beginning of its development. Sheffy (2005) argued that a company's survival and prosperity depend more on what it does before disruptions occur than on the actions it takes as the event unfolds. Sheffy (2005) provided a number of case studies from international companies. Sheffy and Rice (2005) provided evidence of terrorism influence on SCs on the examples of Toyota and Ford SC disruptions after September, 11, 2001. They concluded that building redundancy and flexibility in SCs to recover from disruption quickly is mandatory. Kleindorfer and Saad (2005) provided a conceptual framework that reflects the joint activities of risk assessment and risk mitigation that are fundamental to disruption risk management in SCs. Knight (2003) distinguished the following *key elements* of SC security: risk analysis, physical security, access control, personnel security, education and training awareness, procedural security, information security, incident reporting and investigations, documentation processing security, trading partner security, conveyance security, crisis management and disaster recovery.

Closs *et al.* (2008) underlined that "recent terrorist threats and security incidents have heightened awareness regarding SC security. But many managers still underestimate SC vulnerability and struggle with where to focus their security efforts". The paper also proposed a conceptual framework for SC security. Within the framework, 10 security competencies are distinguished within and across each firm in the SC. Security competencies are created through the development of security capabilities such as infrastructure, processes, assets and resources that achieve and maintain SC security.

A very important topic in SC security is the *insurance*. Lodree and Taskin (2008) emphasized that disaster relief assumes a major role in the logistics activities associated with responding to disasters. They provided an insurance risk management framework for disaster relief and SC disruption inventory planning. Knemeyer *et al.* (2009) considered a process to plan proactively for catastrophic risk events through an integration of diverse research streams related to the risk management. In particular, the proposed process builds upon an existing risk analysis framework by incorporating an innovative approach used by the insurance industry to quantify the risk of multiple types of catastrophic events on SCM.

The essential conclusion from the existing literature is that SC security is understood (in the narrow interpretation) as a protection against purposeful threats such as terrorism, piracy, financial misdoings and thefts. These threats influence SC drastically. With regard to empirical data of international insurance, companies lose up to 15% of the turnover only by threats. Alone the damage and spoiling of goods are amount to up to 3-4% of inventories.

6.1.2 The Concept and Tools of Supply Chain Security

While analysing the issues in SC planning and execution under uncertainty practically and theoretically, we classified five main categories of integrated SCM and security aspects. These are:

- regulations level;
- risk management level;
- event and process management level;
- informational level; and
- physical security level.

These levels build the framework which endeavours to provide a holistic approach to the integrated SC security management. Within the framework, appropriate management levels, methods and tools at these levels are positioned and presented in their interrelations. The general framework of integrated SCM and security is presented in Fig. 6.1.

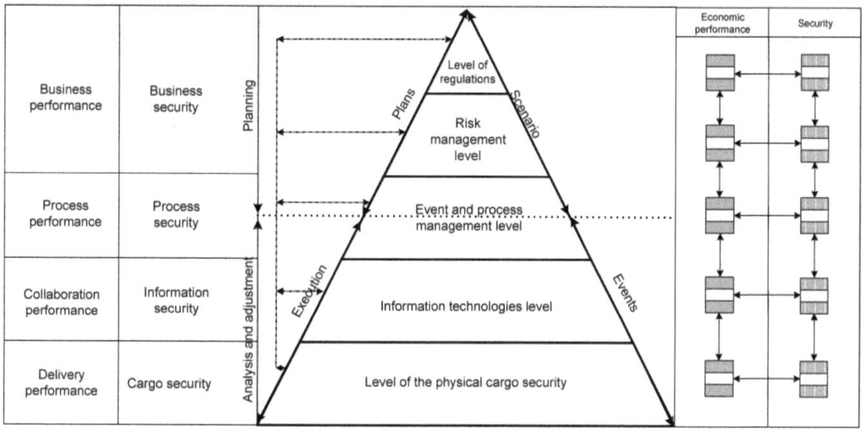

Fig. 6.1 The general framework of integrated SC security management

The evidence of a significant increase in the importance of SC security regulations is reflected in the practical area, i.e. in regulatory compliance (for example, ISO 28000:2007 and "voluntary" initiatives such as COSO Enterprise Risk Management, TAPA – Transported Asset Protection Association, RILA (Retailer Initiative Leadership Association), C-TPAT, and Authorised Economic Operator). ISO 28000 considers risk management to be a fundamental corporate activity. ISO 28000:2007 takes a pragmatic and business-centric approach to security management. The standard promotes security management as a central component of effective management.

The standards of the ISO 28000 family extend the treatment of the security management as a key category of risk management coined in ISO/IEC 15288:2002 "System engineering. System life cycle processes" the regulations level and the risk management level refer to the planning stage. The application of international standards and guidelines by SC partners aims at ensuring management process compatibility. Companies adopting standards can also gain brand equity through the clear demonstration of their commitment to security. Companies adopting ISO 28000 make an organizational commitment not only to security, but also to efficient management and continual improvement.

The level of risk management contains the primary methods of risk management. Essential process-related procedures take place at the interface between the risk management level and the process management level. Here, comprehensive uncertainty analysis, the identification of possible risks and their analysis take place. Then the identified risks will be mapped with the process plans, and the degree the risks will affect the processes (i.e. by means of the event probability index, (Ijioui *et al.* 2007) and the scenario of managers' handling in the case of disturbances (i.e. by means of the event management plan) will be determined.

The key level of the proposed framework is the event and process management level. As shown in Fig. 6.1, this level is explicitly divided into the planning (process management level) and execution stages (event management level). At the planning stage, different reserves to mitigate uncertainty and to ensure SC security are built. This results in a number of alternative SC plans with different values of economic performance and security indices. The planning ends with the simulation of different execution scenarios for different SC con-figurations and plans with a subsequent evaluation of these alternatives by managers according to their individual risk perceptions.

After planning, the stage of SC operations execution follows. At the physical security level, cargo movement control takes place. The data from primary control devices (enabled by RFID, GPSS) are transmitted, accumulated and evaluated within the information systems level. At the interface between the information systems level and the event management level, based on SCEM tools, SCMo and reinstating (adaptation) take place. This results in decisions on SC processes, plans or goal correcting, amending or replacing on the basis of the disturbances that occurred and the control actions that existed.

Summarizing this section, it is necessary to say that the elaborated composition contributes to linking conceptual SC security as a strategic goal and "engineering" quantitative models and methods to ensure SC performance and security at the tactical and operations level. A number of important interrelations and details can be observed while considering the framework levels and the methods positioned within the levels. The framework also contributes to linking issues of SC configuration, planning and execution.

6.2 Supply Chain Vulnerability

6.2.1 The Prevention of Non-Purposeful Perturbation Impacts on Supply Chains

A non-purposeful perturbation influences an external input influence on a SC of a casual nature (for example, demand fluctuations and resource failure). The literature on non-purposeful perturbation influences on SCs is widespread. At the stage of the SC synthesis (configuration and planning), uncertainty is a category that is mostly used in relation to risk management. Risk management is a methodological approach to the management of outcome uncertainty (Peck 2007, Ritchie and Brindley 2007).

In operations management, the research on how to cope with non-purposeful disturbances has mostly been concentrated on the *environmental uncertainty* (i.e. demand fluctuations and the so-called bullwhip-effect) (Chen *et al.* 2000, Lee *et al.* 1997) by means of stochastic or robust optimization (i.e., Santoso et al. 2005) and SC coordination (Holweg and Pil 2008)). Sha and Che (2006) analysed SC network design under uncertainty with regards to partner selection and production/distribution planning. Fisher *et al.* (1997) emphasized the necessity to consider costs of uncertainty reduced while configuring SCs. Kleindorfer and Saad (2005) provided a conceptual framework of risk assessment and risk mitigation that are fundamental to disruption risk management in SCs.

In analysing SC uncertainty, the property of SC robustness has been investigated (Van Landeghem and Vanmaele 1996, Meepetchdee and Shah 2007). Stability is another general system property that has been widely investigated in related literature (i.e., Daganzo 2004). Disney and Towill (2002) provided an extensive overview of control theory based approaches for the stability analysis. Van de Vonder *et al.* (2007) evaluated several predictive-reactive resource-constrained project-related scheduling procedures under the composite objective of maximizing both the schedule stability and the timely project completion probability. Cheng and Wu (2006) elaborated a multi-product, multi-criterion supply-demand network equilibrium model. Ostrovsky (2008) studied matching in vertical networks, generalizing the theory of matching in two-sided markets.

The other research stream deals with the uncertainty caused by *human decisions* and goals. Sterman (1989) sees wrong decisions made by human decision makers as the major cause of the bullwhip effect. Hallikas *et al.* (2004) consider organizational risks and propose an approach to reduce uncertainty by means of increasing entire network transparency. Sokolov and Yusupov (2006) distinguish seven psychological types of managers and consider this criterion in the model of risk management.

6.2.2 Supply Chain Adjustment in The Case of Non-purposeful Perturbation Impacts

Unless the SC is designed and planned to be stable and robust with respect to the uncertain control conditions, the impact of operational inefficiencies and disruptions can change the actual SC execution from the planned scenarios. For the SC execution analysis and replanning, a number of concepts and models have been developed, such as ATP/CTP (available-to-promise/capable-of-promising) (Zschorn 2006), information-update (Sethi *et al.* 2005), SCMo, and SCEM (Ijioui *et al.* 2007). Decision-making in the case of deviations is one of the main challenges in SC execution.

Proth (2006) and Dolgui and Proth (2009) referred to dynamic real-time scheduling. Kopfer and Schoenberger (2006) considered the online-optimization of problems with several multi-layered objectives with preventive and reactive SC adaptation. Ahn *et al.* (2003) suggested a flexible agent system for SCs that can adapt to the changes in transactions introduced by new products or new trading partners. Ivanov *et al.* (2009) considered SC reconfiguration based on the structure dynamics control framework.

The challenge of SC execution is faced in the SCEM approach (Otto 2003). SCEM has been increasingly developed over the last few years as a link between SC planning and execution as well as a strategy to recover SCs in the case of disturbances (Wildemann 2007, Ijiouni *et al.* 2007). SCEM aims at a timely identification of deviations or the dangerousness of deviations in SCs, the analysis of deviations and alerting about what disruptions that have occurred or may occur, and elaborating control actions to recover SC operability. A basis for the alerting and disruption recovery is a tolerance area of execution parameters' admissible deviations. If a parameter value is out of this area's borders, the alerting takes place. Two important questions still remain open: (1) how to determine the borders of the tolerance area and (2) what adjustment steps should be taken to overcome a particular disruption. In practice, these decisions are made on the basis of weak-grounded heuristics or just an expert analysis. SCEM is currently understood from the information systems' point of view. The methodical aspects of SCEM are still underdeveloped.

6.3 Managerial Impacts to Handle Uncertainty in Supply Chains

Recent literature has also dealt extensively with methods to strengthen SCs to mitigate uncertainty impacts. First, different *reserves* (material inventory, capacities) can be referred to. For this issue, valuable approaches and models for SC design and planning under uncertainty were elaborated, widely presented in Tayur *et al.* (1999) and de Kok and Graves (2004). Second, new *strategies* such as leagile, agile and responsive SCs as well as structural-functional reserves (like a pool of alternative suppliers from the VE concept) can be applied to make SCs more flexible in a wider sense of the word (Christopher and Towill 2001, Gunasekaran and Ngai 2009, Ivanov *et al.* 2007, Ivanov *et al.* 2010). The third one is related to better *coordination* in SCs and refers to the concepts like CPFR and ECR. Fourth, a set of *postponed decisions* (product postponement, rolling/adaptive planning) can be used. All these approaches can be called as SC *excessiveness.*

The above-mentioned redundancies generally serve for two problem areas (Fisher 1997). First, they are intended to protect the SC against perturbation impacts based on certain reserves (de Kok and Graves 2004). This issue is related to the SC *reliability*. Second, redundancies are created to amplify the fork variety of SC paths to react quickly and flexibly to changes of a real execution environment. This issue is related to SC *flexibility* (Vickery *et al.* 1999, Swafford *et al.* 2008).

6.3.1 Supply Chain Reliability

The reliability of SCs is a complex characteristic of a non-failure operation, durability, recoverability, and the maintaining of SC processes and an SC as a whole; this is connected with the creation of a reserves system (the introduction of resource excessiveness) for the prevention of failures and deviations in SC processes.

Recent literature has identified different methods to strengthen SCs to mitigate uncertainty impacts and ensure SC reliability. Different reliability reserves (material inventory, capacities buffers, etc.) can be referred to. For this issue, valuable approaches and models for SC design and planning under uncertainty were elaborated, widely presented in Tayur *et al.* (1999) and de Kok and Graves (2004).

The formulation of strategic production–distribution models SCD has been widely investigated. Most of these formulations are introduced in the form of MILP models. Beamon (1998), Tayur *et al.* (1999), Goetschalckx *et al.* (2002), de Kok and Graves (2004), Simchi-Levi *et al.* (2004), Harrison (2005), Chopra and Meindl (2007), Shen (2007) provide a systematic summary of OR on quantitative models for SCD.

Yan *et al.* (2003) propose a strategic production–distribution model for SC design with consideration of bills of materials (BOM) formulated as logical constraints in a mixed integer programming (MIP) model.Graves and Willems (2005) developed a dynamic programme with two state variables to solve the SC configuration problem for SCs that are modelled as spanning trees and applied it to optimizing the SC con-figuration for new products. Kleindorfer and Saad (2005) provided a conceptual framework that reflects the joint activities of risk assessment and risk mitigation that are fundamental to disruption risk management in SCs. Meepetchdee and Shah (2007) develop a framework of logistical network design with robustness and complexity considerations and used an MILP model for concept implementation.

The main elements of the *reliability reserves* are time buffers, safety stocks, and additional facilities, reservation of capacities, and IT-based coordination and monitoring. These elements cause certain *costs* for the creation of the reliability reserves, their maintaining, and recovery handlings in the case of disruptions and application of these reserves to recover the SC processes, multi-variant and modular production. However, in the case of disruptions, these reserves may also be an *income* origin because of uninterrupted SC processes.

6.3.2 Supply Chain Flexibility and Adaptation

The flexibility of SC*s* is a property of a SC concerning its ability to change itself quickly, structurally and functionally depending on the current execution state and reaching SCM goals by a change in SC structures and behaviour. This is connected with the creation of an *adaptation system* (with regard to operations and resources) for the prevention, improvement, or acquisition of new characteristics for the achievement of goals under the current environmental conditions varying in time.

Tachizawa and Thomsen (2007) empirically investigated the aspects of flexibility related to the upstream SC. Coronado and Lyons (2007) investigated the implications of operations flexibility in industrial SCs and the effect it has on supporting initiatives designed for BTO manufacturing. Wadhwa *et al.* (2008) presented a study on the role of different flexibility options (i.e. no flexibility, partial flexibility and full flexibility) in a dynamic SC model based on some key parameters and performance measures. (Swafford *et al.* 2008) showed that that IT integration enables a firm to tap its SC flexibility which in turn results in higher SC agility and ultimately higher competitive business performance. Ozbayrak *et al.* (2006) and Jang (2006) showed that flexibility is interrelated with adaptation.

Adaptation is a changing of functioning and the abilities to function in unsettled conditions by a goal-oriented change of the SC parameters and/or structures. The main elements of the *adaptation reserves* are unification of management functions between different SC decision making points, "rolling" or adaptive planning, not final decisions (e.g. postponement), virtual reserves (e.g. alternative suppliers' pool), dynamic pricing and flexible contracting.

As with reliability, flexibility is enabled by the introduction of certain excessiveness (redundancy) in an SC. This also implies additional "unproductive" costs. However, in contrast to reliability, the application domain which stabilizes SCs by means of assessing the excessive resources in the case of disturbances, flexibility contributes to the flexible use of these excessive resources and even to the adaptation of the excessiveness amount and structure to changes in the execution environment (see Fig. 6 2).

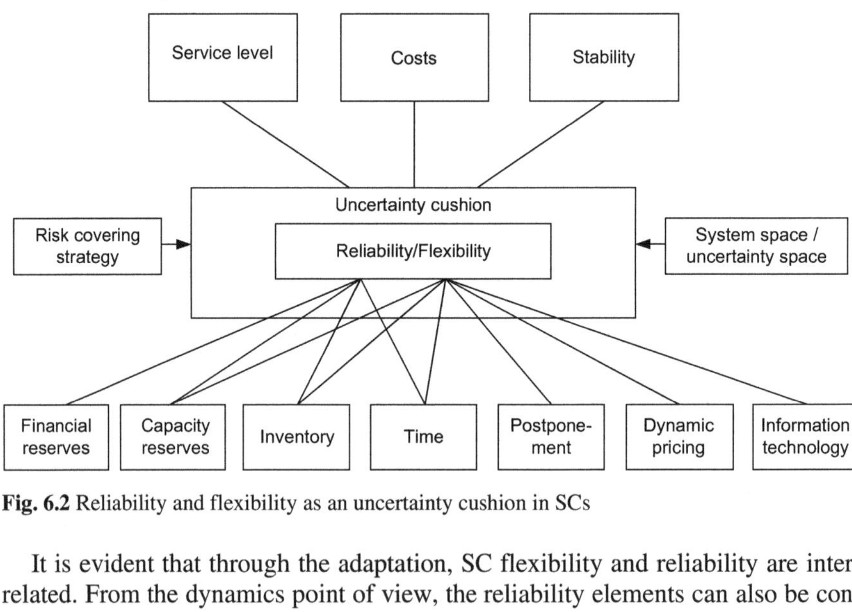

Fig. 6.2 Reliability and flexibility as an uncertainty cushion in SCs

It is evident that through the adaptation, SC flexibility and reliability are interrelated. From the dynamics point of view, the reliability elements can also be considered as flexibility elements and the flexibility elements can also be considered as reliability drivers. This is quite natural because both the reliability and flexibility serve as an "uncertainty cushion" of a SC.Balancing the elements of flexibility and reliability, different constellations of service level, costs and stability can be analysed methodically well-founded and with regard to a risk covering strategy and an SCM strategy (see also Chap. 14 and Sect 15.3).

From the perspective of the complexity management the problem of a system under control and uncertainty is related to an area under control (SC system space) and an area under uncertainty (environment space) (see Chap. 5). There are also a number of other interrelations between different verbal and formal SC properties with regard to uncertainty. These interrelations will be considered further in the next chapter.

References

Ahn HJ, Lee H, Park SJ (2003) A flexible agent system for change adaptation in supply chains. Expert Syst Appl 25:603–618

Beamon BM (1998) Supply chain design and analysis: models and methods. Int J Prod Econ 55(3):281–294

Beck A, Chapman P, C Peacock Shrinkage (2003) A collaborative approach to reducing stock loss in the supply chain. ECR Europe, Brussels

Chen F, Drezner Z, Ryan JK and Simchi-Levi D (2000) Quantifying the bullwhip effect in a simple supply chain: the impact of forecasting, lead times, and information. Manag Sci 46(3):436–443

Cheng TCE, Wu YN (2006) A multiproduct, multicriterion supply-demand network equilibrium model. Oper Res 54(3):544–554

Chopra S, Meindl P (2007) Supply chain management. Strategy, planning and operation. Pearson Prentice Hall, New Jersey

Christopher M, Towill DR (2001) An integrated model for the design of agile supply chains. Int J Phys Distrib Oper Manag 31:235–244

Closs D, Speier C, Whipple J, Voss MD (2008) Supply chain security: a framework for protecting your supply chain. Logist Manag 47(9):45–48

Coronado M, Lyons AE (2007) Evaluating operations flexibility in industrial supply chains to support build-to-order initiatives. Bus Process Manag J 13(4):572–587

Daganzo CF (2004) On the stability of supply chains. Oper Res 52(6):909–921

Disney SM, Towill DR (2002) A discrete linear control theory model to determine the dynamic stability of vendor managed inventory supply chains. Int J Prod Res 40(1):179–204

Dolgui A, Proth JM (2009) Supply Chains Engineering: useful methods and techniques. Springer, Berlin *et al.*

Fisher ML (1997) What is the right supply chain for your product? Harv Bus Rev march-april: 83–93

Goetschalckx M, Vidal CJ and Dogan K (2002) Modeling and design of global logistics systems: a review of integrated strategic and tactical model and design algorithms. Eur J Oper Res 143(1):1–18

Graves SC and Willems SP (2005) Optimizing the supply chain configuration for new products. Manag Sci 51(8):1165–1180

Gunasekaran A, Ngai NWT (2009) Modeling and analysis of build-to-order supply chains. Eur J Oper Res 195(2):319–334

Hallikas J, Karvonen I, Pulkkinen U, Virolainen V-M and M Tuominen (2004) Risk management processes in supplier networks. Int J Prod Econ 90(1):47–58
Harrison TP (2005) Principles for the strategic design of supply chains. In: Harrison TP, Lee HL, Neale JJ (Ed) The practice of supply chain management. Kluwer academic publishers, Boston
Holweg M, Pil FK (2008) Theoretical perspectives on the coordination of supply chains. J Oper Manag 26(3):389–406
Ijioui R, Emmerich H, Ceyp M (2007) Supply Chain Event Management Konzepte, Prozesse, Erfolgsfaktoren und Praxisbeispiele. Springer, Berlin
Ivanov D, Kaeschel J, Sokolov B (2007) Integrated modeling of agile enterprise networks. International Journal of Agile Systems and Management 2(1):23–49
Ivanov D, Kaeschel J, Sokolov B (2009) Structure dynamics control-based framework for adaptive reconfiguration of collaborative enterprise networks. Int J Manuf Techn Manag 17(1-2):23–41
Ivanov D, Sokolov B, Kaeschel J (2010) A multi-structural framework for adaptive supply chain planning and operations control with structure dynamics considerations. Eur J Oper Res 200: 409–420
Jang PY (2006) A flexible and adaptive control architecture for the efficient supply chain management (SCM). WSEAS Transactions on Communications 5(6):1015–1025
Kleindorfer PR, Saad GH (2005) Managing disruption risks in supply chains. Prod Oper Manag 14(1):53–68
Knemeyer AM, Zinn W and Eroglu C (2009) Proactive planning for catastrophic events in supply chains. J Oper Manag 27(2):141–153
Knight P (2003) Supply chain security guidelines. IBM, White Paper
Kok de AG and Graves SC (2004) Supply chain management: design, coordination and operation. Elsevier, Amsterdam
Kopfer H, Schoenberger J (2006) Die Lösung von Optimierungsproblemen mit mehrschichtigen Zielsetzungen durch die Adaption von selbststeuernden Planungsagenten. In: Ivanov D (Ed) Logistics, supply chain management, and information systems. SPbGPU, Saint Petersburg.
Lee HL, Padmanabhan V and Whang S (1997) Information distortion in a supply chain: the bullwhip effect. Manag Sci 43(4):546–558
Lodree EJ Jr, Taskin S (2008) An insurance risk management framework for disaster relief and supply chain disruption inventory planning. J Oper Res Soc 59(5):674–684
Meepetchdee Y and Shah N (2007) Logistical network design with robustness and complexity considerations. Int J Oper Prod Manag 37(3): 201–222
Ostrovsky M (2008) Stability in supply chain networks. Prod Oper Manag 98(3):897–923
Otto A (2003) Supply Chain Event Management: Three Perspectives. Int J Logist Manag, 14(2):1–13
Ozbayrak M, Papadopoulou TC, Samaras EA (2006) Flexible and adaptable planning and control system for an MTO supply chain system. Robot Comput Integrated Manuf 22(5-6):557–565
Peck H (2007) Supply chain vulnerability, risk, robustness, and resilience. In: Mangan J, Lalwani C, Butcher T (Ed) Global logistics and supply chain management. John Wiley, New York
Proth JM (2006) Scheduling: New trends in industrial environment. Information control problems in manufacturing. Annu Rev Control 31(1):157–166
Rice JB, Caniato F (2003) Building a secure and resilent supply network. Supply Chain Management Review, September/October:22–30
Ritchie B, Brindley C (2007) An emergent framework for supply chain risk management and performance measurement. J Oper Res Soc 58:1398–1411
Santoso T, Ahmed S, Goetschalckx M and Shapiro A (2005) A stochastic programming approach for supply network design under uncertainty. Eur J Oper Res 167(1):96–115
Sethi SP, Yan H, Zhang H (2005) Inventory and supply chain management with forecast updates. Springer
Sha DY and Che ZH (2006) Supply chain network design: partner selection and production/distribution planning using a systematic model. J Oper Res Soc 57:52–62

Sheffy J (2005) The resilient enterprise. MIT Press

Sheffy J and Rice Jr J B (2005) A supply chain view of the resilient enterprise. MIT Sloan Management Review 47(1):45–47

Shen ZM (2007) Integrated supply chain design models: a survey and feature research directions. J Ind Manag Optim 3(1):1–27

Simchi-Levi D, Wu SD and Zuo-Yun S (2004) Handbook of quantitative supply chain analysis. Springer, New York

Sokolov B, Yusupov R (2006) Risk integrated modelling for control in complex organizational-technical systems. Journal of Control and Informatics 1:1–22

Sterman J (1989) Modelling managerial behavior: misperceptions of feedback in a dynamic decision making experiment. Manag Sci 35(3):321–339

Swafford PM, Ghosh S, Murthy N (2008) Achieving supply chain agility through IT integration and flexibility. Int J Prod Econ 116(2):288–297

Tachizawa EM, Thomsen CG (2007) Drivers and sources of supply flexibility: an exploratory study. Int J Oper Prod Manag 27(10):1115–1136

Tayur S, Ganeshan R and Magazine M (1999) Quantitative models for supply chain management. Kluwer Academic Publishers, Boston

Van de Vonder S, Demeulemeester E and Herroelen W (2007) A classification of predictive-reactive project scheduling procedures. J Sched 10(3):195–207

Van Landeghem H, Vanmaele H (1996) Robust planning: a new paradigm for demand chain network design problems. Eur J Oper Res 94:362–376

Vickery S, Calantone S, Droge C (1999) Supply chain flexibility: an empirical study. J Supply Chain Manag 35(3):16–27

Wadhwa S, Saxena A, Chan FTS (2008) Framework for flexibility in dynamic supply chain management. Int J Prod Res 46 (6):1373-1404

Wildemann H (2007) Event Management in der Supply Chain. TCW, Munich

Williams Z, Lueg JE, LeMay SA (2008) Supply chain security: an overview and research agenda. Int J Logist Manag 19(2):254–281

Yan H, Yu Z and Cheng E (2003) A strategic model for supply chain design with logical constraints: formulation and solution. Comput Oper Res 30:2135–2155

Zschorn L (2006) An extended model of ATP to increase flexibility of delivery. Int J Comput Integrated Manuf 19(5):434–442

Chapter 7
STREAM: Stability-based Realization of Economic Performance and Management

A business that makes nothing but money
is a poor kind of business.
Henry Ford

7.1 Necessity for STREAM

The problem of uncertainty is one of the key problems in SCM. In the literature, a set of methods and models to decrease the negative consequences of the influence of uncertainty on SCs is described. At a conceptual level, a set of works on risk management exists, directed, mainly, to the strategic level of decision-making.

At the strategic level, increasing the SC stability becomes more and more important. At the tactical level of decision making, there is a large variety of developments on decreasing the influence of fluctuations of demand and mistakes in forecasts of demand (bullwhip effect), optimal inventory control, coordination of SCs, formation of a set of not final decisions (for example, the postponed differentiation) or methods of rolling planning.

At the operative level, various methods and models of scheduling and systems of SCMo and regulation in the case of the occurrence of deviations from the plan are presented. For the tactical and operational levels of decision making, in the literature, a number of techniques for the analysis of stability, flexibility and reliability of deliveries are presented.

Although there is a wealth of literature on planning economically optimal SCs, mitigating uncertainty in SCs, and event management in SCs, some crucial limitations with regard to the main focus of our research should be named. These are:

- the issues of SC economic performance and stability are not explicitly considered as a whole;
- the conceptual frameworks of SCM under uncertainty are mostly strategic and hardly supported by consistent engineering quantitative models and methods to ensure SC stability at the tactical and operations levels;
- lack of explicit integrated frameworks for SC analysis (monitoring and diagnosis) and decision making for SC performance recovery;
- the issues in SC planning and execution are mostly considered in isolation without explicitly linking scenario-driven and event-driven management; and
- the domination of qualitative methods in risk management while underestimating quantitative analysis techniques (sensitivity, stability, etc.) can be observed.

7.2 Economic Performance and Stability of Supply Chains: A Strategic Analysis

7.2.1 The "Waterline" Concept for Stability Analysis

Empirical analysis (see Fig. 7.1) of (1) the revenue decrease dynamics (both from material production or services and from stock exchange), (2) production/services volume decrease dynamics (due to demand fluctuations, supply breaks, and cash-to-cash cycle breaks) and (3) costs decrease dynamics shows that:

1. The impact of different perturbation factors is different at different economic cycle stages (growth, maturity, decrease, stagnation).
2. At all the economic cycle stages, the revenue and production decrease dynamics is different from cost reduction dynamics due to constant costs and middle- and long-term obligations.
3. Significant problems in the maintenance of SC stability occur when enterprises are not able to meet their obligations with regard to creditors, shareholders, deliveries and procurement, and personal costs.

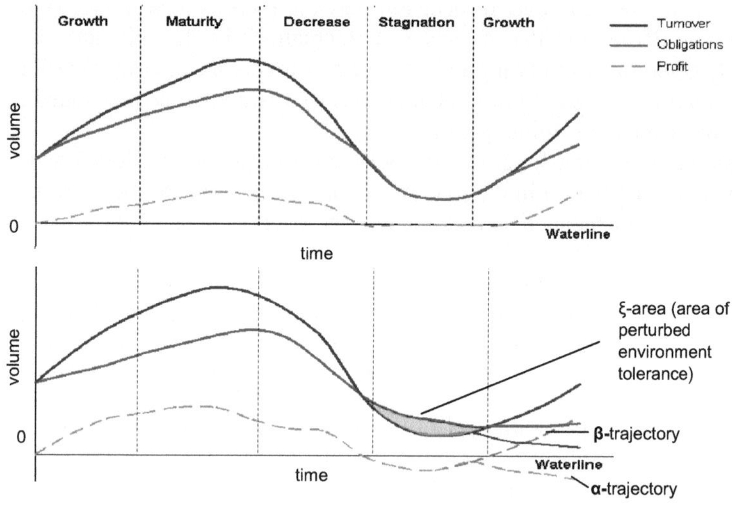

Fig. 7.1 The "waterline" stability concept and the area of perturbed environment tolerance

The results gained lead us to the proposition of a certain "waterline" that would characterize the ability of the SC to take its obligations

The *"waterline"* characterizes the so-called zero situation when SC enterprises are still able to meet their obligations but the revenue is equal to costs with zero profitability. In the case of market growth, it would be recommended to increase profitability and occupy a space over the "waterline".

At the stages of growth and maturity, perturbation factors such as demand fluctuations, technological breaks and theft/damage of cargo are the primary focus. In approaching the decrease/stagnation stages, the economic and political crisis should be taken into account. In any case, the profitability growth is to be followed by stability analysis to ensure a necessary control influence space in the case of disturbances.

In these settings, we propose to consider the ability of a SC to meet its obligations as the indicator of SC stability at the strategic level. At the tactical and operational levels, we will propose to consider the stability of SC plans in the further course of this book.

Another important aspect that is shown in Fig. 7.1 is the area of perturbed environment tolerance ξ. This area reflects the temporal interval and characteristics of SC functioning from the perturbation influence to the return to the planned trajectory or the selection of a new or a desired trajectory. In Fig. 7.1, the area of perturbed environment tolerance is shown for the situation when costs can not be reduced proportionally to the revenue and production decrease. Depending on the scale of this area and its placement with regard to the "waterline", the SC can lose stability (trajectory α) or return to the stable state (trajectory β).

Hence, in strategic SCP, the following factors should be taken into account:

1. Costs cannot be reduced proportionally to the revenue and production decrease. This leads to the occurrence of the area ξ, which we called the area of perturbed environment tolerance.
2. With regard to the scale of the area of perturbed environment tolerance and its placement to the "waterline", the SC can either lose stability or return to a stable state.
3. The area ξ does not practically occur for the SCs that are designed and planned taking uncertainty into account. In a theoretical formulation, this means that in these SCs there is a balance between the control area and the uncertainty area (see also Sect. 5.2).

7.2.2 *Factors of Stability Decrease and Maintenance*

The analysis of practical examples showed that the main factors of stability loss at the strategic level are as follows:

- specialization and orientation to one market (or one client);
- plans for unlimited profit growth;
- high credit obligations;
- lack of alternative suppliers;
- high dependency on stock exchange; and
- geographical concentration of SC facilities in only one region.

At the tactical and operational levels, the factors are as follows:

- weak coordination of plans and information about demand and supply;
- stockless processes;
- weak control of cargo security;
- technological breaks (machines, transport, information systems); and
- human errors and false information management.

To compensate for the disturbances, a large variety of control influences can be applied, i.e.:

- SC security management;
- capital reserves;
- strategic material inventories;
- market diversification and outsourcing;
- product lines' flexibility and modularity;
- safety stocks and time buffers;
- reserves of SC capacities; and
- SC coordination, monitoring, and event management.

All these measures are actually the adaptation reserves of SCs. They are characterized by different degrees of operativeness (i.e. using safety stocks or market diversification). This once more confirms the significance of time and dynamics in SC models. In the further sections of this chapter, we will consider diverse interrelations in this problem area. In Chap. 14, we will provide the generic mathematical model construction for stability analysis.

7.3 Terminological Basics of Global Stability Analysis at the Tactical–Operational Level

The terminological question is one of the most thankless in economics sciences. We would immediately like to make clear that the concepts and properties that we will consider were used and are used by other authors with other meanings. Moreover, the definitions of certain abstract categories may also appear to be abstract. This is quite natural because they usually acquire their concrete meaning in a concrete engineering environment. Nevertheless, we consider it necessary to introduce a logically interconnected system of definitions for correct research on the considered subject domain on the basis of SCM terminology and system-cybernetic theories. In conclusion, we will note that the properties of systems can be interpreted in both a narrow and in a broad sense. The main SC properties and their interrelations are presented in Fig. 7.2.

As mentioned above, these properties have been used in other literature, and sometimes with other meanings. The contribution of this framework is not how to name the property but to reflect all the related issues to ensure both SC economic performance and stability.

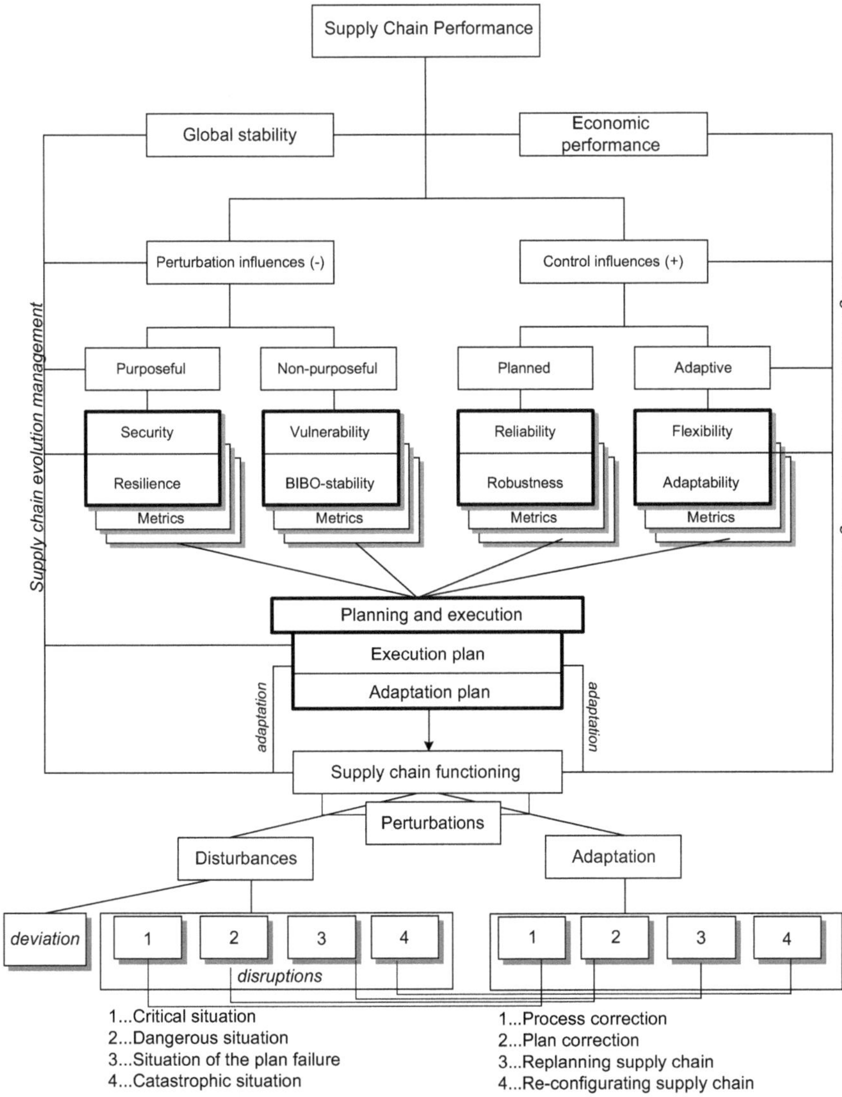

Fig. 7.2 Main properties with regard to the performance and stability of SC and their interrelations

The economic performance of a SC (in the wide interpretation) is a complex characteristic of the potential and real results of the SC functioning, taking into account the conformity of these results with the goals set by management.

The stability of a SC (in the wide interpretation, the SC *global stability*) is a complex property of a SC, characterizing the ability of a SC to maintain, realize and restore goal-oriented functioning in ever changing execution environment under influence of perturbation factors of unauthorized purposeful and non-purposeful characters.

SC *total performance* is defined by two components:

- the goal possibilities characterizing the potential ability of a SC to reach its goals in concrete environmental conditions; and
- the stability of a SC functioning.

The definition made above allows us to draw an important conclusion: the maintenance of a SC real performance according to the potential (planned) performance is based on the *maintenance of the* SC *stability.*

Control influence is an input positive influence on the object of the management, intended for the achievement of the management goals. The set of control influences can be divided into two categories: the construction of SC plans, taking into account uncertainty, and the regulation (adaptive) control influences at a stage of a SC realization.

The SC *functioning* (execution) is a process of control realization for achievement of SC goals at all three levels of planning.Unfortunately, any plan realization is related to the uncertainty along with the incompleteness of our knowledge of the future and the present (we cannot know all about our SCs).

Uncertainty is a property characterizing the incompleteness of our knowledge about the system's environment and its development.

Perturbation influence, non purposeful and purposeful (see Chap. 5).

Disturbance (perturbation influence) is an impossibility of the realization of the planned event (or a critical amount of events) according to the SC plan.

The efficient state of a SC is a state in which the SC is capable of carrying out the set functions (proceeding from the management goals), maintaining the values of key performance parameters within the demanded limits.

The disabled state of a SC is a state at which the SC does not carry out at least one of the functions (proceeding from the management goals) and the values of its key performance parameters move outside the demanded limits.

Deviation is a short-term transition of a SC from an efficient state to a disabled state, which does not lead to a loss of manageability. Deviation withdraws without external influences. Deviation is characterized by a supernumerary situation.

Supernumerary situation is a supernumerary mode of SC functioning in which several indicators of SC performance are outside the intervals of a regular mode in such limits where there is no disruption or catastrophe threat.

Disruption is the transition of a SC from a planned state to an unplanned state in which the achievement of the SCM goals without additional control influences is impossible.

Disruptions can be divided into four levels: a critical situation (disturbance of one process), a dangerous situation (disturbance of several processes), a situation of plan disruption (disturbance of many processes), and a catastrophic situation (disturbance of the overwhelming majority of processes).

Critical situation is a supernumerary mode of functioning in which the indicators of SC performance are outside the intervals of a regular mode in such limits where there is a real threat of disruption of the plan or a catastrophe.

Process correction is an end result of a critical situation; adaptation by means of operative changes in processes with the use of reserves (for example, safety stocks).

Dangerous situation is a supernumerary mode of functioning in which the indicators of SC performance are outside the intervals of a regular mode in such limits where the plan disruption or a catastrophe is almost inevitable. Liquidation of a dangerous situation is carried out on the basis of a plan correction.

Plan correction is an end result of a dangerous situation; adaptation by means of changes in control plans (for example, changes of the delivery time of production).

A situation of the plan disruption is a supernumerary mode of functioning during which the SC passes from efficient to such a disabled state where it is necessary to execute re-planning for the transition to an efficient state. Liquidation of a situation of the plan disruption is carried out on the basis of re-planning, i.e. a change of the tactical plan.

Replanning – an end result of a situation of disruption of the plan; adaptation by means of changes in tactical plans (for example, changes of manufacturing volumes).

Catastrophic situation is a supernumerary mode of functioning during which the SC passes from efficient to a disabled catastrophic state in which the transition to an efficient state is essentially excluded and/or is economically inefficient. Liquidation of a catastrophic situation is carried out on the basis of the changes in SCM goals and financial plans, i.e. changes in the strategic plan change. Actually, this situation leads the strategic management to the formation of new SCs and their management systems.

The security of SC*s* is the resistance to the external, not authorized actions developed to cause damage or to break a SC; a set of measures to protect the SC assets (product, facilities, equipment, information and personnel) from theft, damage or terrorism, and to prevent the introduction of unauthorized contraband, people or weapons of mass destruction into the SC.

The resilience of SC*s* is the property of SCs consisting of their ability to maintain a regular mode of functioning in predicted conditions of purposeful influence of de-stabilizing factors and to exclude the possibility of transition from a regular mode to a situation of the plan disruption or a catastrophe in not predicted conditions of influence of predicted destabilizing and/or unpredicted risk factors.

The *vulnerability of* SC*s* is the resistance to external perturbation influences (planned and not planned) of a casual character.

The stability of an SC (narrow interpretation; BIBO stability) is the property of the SC functioning; the state of a SC that is in a planned mode of functioning is stable if considering the fixed set of admissible control influences limited and small perturbation influences lead to limited and small changes of goal variables.

The reliability of SC*s* – see Chap. 6 and Sect. 7.4.1.

The robustness of SC*s* – a property of a SC consisting of its ability to continue its functioning at a certain level of perturbation influences.

The flexibility of SC*s* – see Chap. 6 and Sect. 7.4.1.

The adaptability of SC*s* – an ability of a SC to change its behaviour for prevention, improvements or acquisitions of new characteristics for the achievement of SC goals in the conditions of environment varying in time, the aprioristic information about which dynamics is incomplete.

7.4 Interrelations of Properties

In Table 7.1, the basic elements of the SC synthesis and analysis with regard to uncertainty are presented.

Table 7.1 Basic elements of SC synthesis and analysis with regard to uncertainty

Kinds of impacts	Control (adjustment) impacts		Perturbation impacts	
	Excessiveness		Purposeful	Occasional
	Reserves	Adaptation		
Strategic	Strategic material inventory Asset reserves	Market diversification Product modularity Outsourcing	Terrorism Piracy	Economical crises Political crises Natural disasters
Tactical and operational	Time buffers Safety stocks Additional warehouses Reservation of capacities IT-based coordination and monitoring	Unification of management functions "Rolling" or adaptive planning Not final decisions (e.g., postponement Virtual reserves (e.g., alternative suppliers) Dynamic pricing	Thefts and damage of cargo Financial mishandling	Bullwhip effect Technological disruptions Human errors
Business properties	Reliability	Flexibility	Security	Vulnerability
Attribute properties	Robustness	Adaptability	Resilience	BIBO stability
General property	SC global stability			

The conceptual model of properties' interrelations is based on conceptualizing the subject domain from uniform SCM and system-cybernetic positions by means of the interconnected considerations of:

- control and perturbation influences in SCs; and
- verbally describable properties of a SC as a business process and theoretically attributed properties of an SC as a complex system.

As shown in Table 7.1, two classes of control and perturbation influences are allocated, accordingly. To each of these classes, the basic concepts, business properties (verbally attributed properties) and formal mathematical properties (theoretically attributed) are correlated. The choice of these properties is mostly based on modern literature on SCM.

The SC stability can be defined by a set of verbally describable SC properties as a business process and theoretically attributed SC properties as a complex system (see Table 7.1).

On the basis of the literature analysis, we have suggested distinguishing four properties of an SC – business process security, vulnerability, reliability and flexibility and four theoretically attributed properties of an SC corresponding to them – resilience, BIBO-stability, robustness, and adaptability. The interrelation of the given properties is capable of giving a complete picture about various aspects of perturbation influences (purposeful and not purposeful) and control influences (in planning and in adaptation) in an SC.

7.4.1 Robustness, Adaptability and Adaptation

In technical systems, by robustness, a property of system reliability and "roughness" is understood. The analysis of *robustness* allows one to answer the following basic question: "What level of perturbation influences is the SC capable of sustaining?"

We suggest using the robustness as a formal attribute of an estimation of SCs reliability. The creation of SC reliability is based on introducing certain redundancy (reserves) in a SC, for example, time puffers, safety stocks, additional warehouses, reservation of capacities and IT-based coordination and monitoring. An SC may be said to be "robust" if it is capable of coping well with variations (sometimes unpredictable variations) in its operating environment with minimal damage, alteration or loss of functionality. Robustness in the concept STREAM characterizes the SC plan concerning its basic indicators of economic performance and reliability.

Another characteristic of SCs is *adaptability*. Adaptability is a formal attribute of the SC flexibility. It is possible to establish multi-variant and modular production, to carry out unification of the SCM functions, to use methods of "sliding" or adaptive planning, to form a set of not final decisions (e.g. postponement), and to create so-called virtual reserves of adaptation by means of an alternative suppliers pool (as known form the VE concepts).

In the previous Chapter, we defined reliability and flexibility and considered their interrelations. While considering reliability and flexibility into a SC, it is necessary to understand the SCM strategy. Examples of such a strategy can be:

- the maintenance of the SC reliability (i.e. the introduction of a maximum level of redundancy into an SC to reduce to a minimum the necessity for regulation in an SC; the so-called strategy of risk prevention);

- maintenance of the SC flexibility (i.e. the basic emphasis is made not on the investment in the SC reserves, but on the effectiveness of "management by exception"); and
- transfer of risk to the third parties (i.e. investments are made not in the SC strengthening but in the payment of penalties on the basis of contracts with the insurance companies; the so-called strategy of risk financing or acceptance).

The control influences providing both the reliability of the plan and its flexibility can be used for the adaptation of an SC.

Generally, *adaptation* is understood as the property of a system consisting of the continuous changes of its functioning and the abilities to function in unpredicted conditions by a goal-oriented adjustment of the process parameters and/or structures. Adaptation answers the following basic question: "What actions, when, and by whom should be undertaken for the liquidation of consequences of deviations and disturbances in an SC for the recovery of the planned or transition to a new control mode of SCs, providing the achievement of the management goals for the performance of SCs and the satisfaction of the customer requirements?"

SC adaptability is tightly interrelated with SC complexity. In practice, new advertisement companies in SCs start almost every week. If the SCs are designed in too complex a manner, they are unable to react quickly enough and to be managed effectively. Hence, manageability also becomes interconnected with SC complexity and adaptability. The developed concept of complex SC adaptation is composed of five levels. Each level represents a certain control loop corresponding to a certain class of disturbances in the SC functioning. Table 7.2 provides a systematic view of the levels of complex adaptation concept.

Table 7.2 Levels of complex adaptation concept

	Adaptation level	What is adapted?	How can it be adapted?	Management horizon
1	Parametric adaptation	SC parameters	Capacities reconfiguration, rush orders, etc.	Operative
2	Structural-functional adaptation	SC structures	Operations reallocation, supplier changing	Operative-tactical
3	Goal adaptation I	SC goals	Project goal adaptation, e.g. delivery delay	Tactical
4	Model adaptation	SC models	Introduction of new parameters, structures, restrictions and goals	Tactical-strategic
5	Goal adaptation II	SC strategies	Management goal adaptation	Strategic

The control loop (1) is based on parametrical SC adaptation (correction of processes) in a case when the elimination of the deviations is revealed as a result of SC functioning monitoring by updating some parameters of SC functioning (e.g. terms of operations execution, a stock rate, etc.). In the case of the impossibility of updating an SC by parametrical changes, it is necessary to carry out corresponding structural transformations (a loop 2 – structural SC adaptation). The given stage puts higher requirements onto information systems supporting decision making and demands the complex problem analysis in close interaction of the involved SC participants. If structural-functional adaptation does not bring about the desirable effect, regulation by updating the goal parameters (for example, the deliveries re-termination, increase in costs etc.) is necessary (control loop (3)).

It is necessary to note, however, that, owing to the influence of various objective and subjective factors, internal and environmental, there are changes not only in the SC plans, but also in the conditions of plan realization. This means that initial models of SC planning and an execution can cease to be representative and adequate. A feature of the proposed adaptation concept is control loops (4) and (5), intended for the adaptation of SCM models, and also the SCM goals. The model in loop (4) is a metamodel describing the adaptation of SCM models according to changing conditions of SC functioning and the acquisition of the new information on the system. Loop (5) represents the highest level of SC adaptation when the disturbances in SC execution are so serious that the achievement of the primary goals of top management is np longer possible.

7.4.2 Resilience and BIBO Stability

Let us turn to the "negative" side of the framework. Resilience and BIBO stability reflect the system's ability to return to its original (or desired) state after being disturbed. The concept of stability plays a fundamental role in systems and control theory. In control theory, stability is usually signified by BIBO (*bounded input bounded output*) stability (see, e.g. Stefani *et al.* 2002), i.e. if we give the system a bounded input (e.g., a simple impulse), the system produces a bounded output. In mathematics, stability theory deals with the stability of differential equations and dynamical systems solutions.

With regard to BIBO stability, Warburton *et al.* (2004) provided a stability boundary for the continuous time APIOBPCS (Automatic Pipeline, Inventory and Order Based Production Control System) SC ordering decision. The study proposed stability criteria of APIOBPCS via Pader approximation, the Routh Hurwitz array and the Nyquist criteria. The correct stability criterion compared with that of Riddals and Bennett (2002) has been calculated via Bellman and Cooke's theorem.

Disney *et al.* (2006) provided an extensive overview of control theory based approaches for security analysis. The paper conducted an analysis of an ordering and inventory control algorithm in both continuous and discrete time for a production and inventory control system employing a generalized order-up-to policy.

Dashkovskiy *et al.* (2005) proposed a small-gain type stability criterion for large-scale networks based on Lyapunov's stability.

Stability can be estimated by means of different approaches – Nyquist, Hurwitz, Routh, or with the cross-over model of McRuer (see, e.g. Disney and Towill 2002), who have analysed the stability of a discrete time system by mapping it into the plane using the Tustin transform and the Routh–Hurwitz array). The stability analysis is especially useful in situations that do not allow the construction of stochastic models. The stability analysis allows the proving of the execution plan feasibility, the selection of a plan with a sufficient degree of successful accomplishment probability from a set of alternative plans and the determination of SC bottlenecks and steps to strengthen them.

The primary question, which BIBO stability analysis answers, is the following: "Is a SC able to return to the initial (planned) state or to stay for a certain period of time within the admissible functioning area under the pressure of appearing disturbances?"

Resilience is a category that is usually related to SC security. According to the studies by Rice and Caniato (2003), Sheffy (2005), Glickman and White (2005) and Peck (2007), resilience refers to the ability to react to unexpected disruptions and to restore normal SC operations after being disturbed.

Within the STREAM concept, resilience characterizes the ability of the system to continue performance of its functions in the case of the occurrence of various deviations and disturbances caused by purposeful destroying influences, and BIBO stability – by non-purposeful perturbation influences.

Recent studies (Rice and Caniato 2003, Sheffy 2005, Peck 2007, Ponomarov and Holcomb 2009) indicate frameworks for SC resilience. Here, it is necessary to pay attention again to the aspect that, unlike technical systems, people cope with SCs, instead of automatic monitors. In this connection, important conclusions arise: the SC cannot be resilient and stable, continue functioning, return to the planned states or pass into a new condition, and thus remain thus effective without the realization by managers of corresponding managerial influences. The decisions taken by people, instead of automatic machines, provide basis that leads to the SC functioning under the influence of various perturbation factors. Resilience and stability in such a context are closely connected with the SC adaptation.

The property of stability appears to be connected to the volume of the area of possible control influences, the expansion of which leads to a stability increase. However, with cardinal changes in this area, the system develops, acquires new properties and parameters, and, hence, possesses other stability properties. A similar system change can be reflected in the form of spasmodic change of its trajectory in the state space. Such behaviour is investigated in the theory of dynamic systems with the use of the bifurcation point concept (Prigogine and Nicolis 1977). From here, it follows that the analysis of dynamic SC properties should be made in certain preliminary areas of change of structural parameters and output variables, since on different sites of a trajectory in the state space (between the bifurcation points), SCs, generally have a variety of dynamic properties.

With regard to the stability of non-linear systems, asymptotic stability is applied. For the stability of linear systems, exponential stability is used. In physics, the mechanisms of instability are also used. A related property is the system stabilizability. That means that, when all the uncontrollable states have stable dynamics, the system may be potentially stabilized.

Stability can be expressed numerically; one can also talk about "more or less" stable plans. A SC cannot be stable "in general". The stability analysis includes the following main aspects: a fixed variety of possible control adjustment actions, a certain period of time, a certain set of perturbations considered and algorithm stability. It is essential that system stability is defined according to certain classes of disturbances under consideration (see Fig. 7.3).

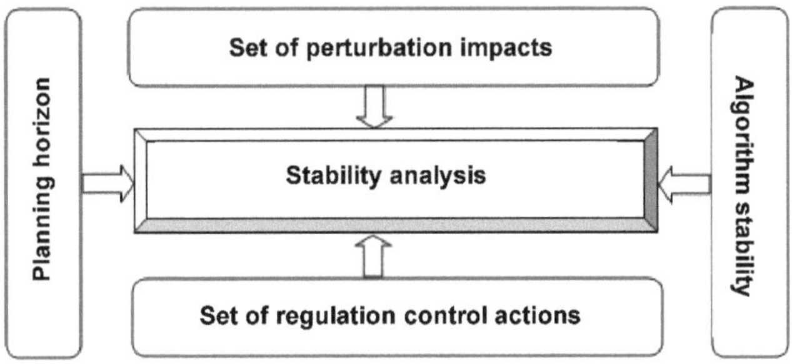

Fig. 7.3 Stability analysis framework

An aspect of stability analysis is SC oscillation analysis in terms of system dynamics. Sterman (2000) introduces three classes of oscillations: damped oscillations, expanded oscillations, and chaotic oscillations. SC stability analysis is carried out within a certain period of time, because of time delays between the occurrence of disturbance factors and their impact on the SC.

A special feature of SC stability analysis consists of adjustment actions elaborated by managers. A SC differs from a physical system. The latter is remarkable for its planning mechanisms, which have elements of subjectivism. That is why it becomes necessary to broaden the sense of "stability" when considering an SC.

7.4.3 Supply Chain Global Stability and Manageability

Literature analysis leads us to the conclusion that the concept of stability plays a fundamental role in systems and control theory. While synthesizing a system of any nature, ensuring system stability and, more precisely, the stability's continued existence is a first-order requirement. In its development, stability comes to be interpreted in different ways beginning with the classical BIBO stability up to non-quantified "conceptual" stability. The understanding of stability depends greatly on the system considered as well as on the methods and goals of system analysis.

In SCM, the issues of stability have attracted increased attention and interest in recent years (e.g. Daganzo 2004, Warburton *et al.* 2004, Ostrovsky 2008). Besides the above-mentioned research on BIBO-stability, different research streams on SC stability can be identified.

Cheng and Wu (2006) elaborated a multiproduct, multicriteria supply-demand network equilibrium model. For the case with multiple criteria, they derive the necessary and sufficient conditions for network equilibrium in terms of a vector variational inequality by Gerstewitz's function when the cost function is vector valued. Quyang (2007) analysed the bullwhip effect in multi-stage SCs operated with linear and time-invariant inventory management policies and shared SC information. Ostrovsky (2008) studied matching in vertical networks, generalizing the theory of matching in two-sided markets. Yang *et al.* (2008) investigated how the stability affects the alliance performance in a SC in the context of manufacturing firms. However, although there is a wealth of literature on stability in SCs, some crucial limitations in regard to the main focus of our research should be clarified.

First, the issues of SC economical efficiency and stability are not explicitly considered as a whole. Only a few studies have dealt with this domain. Van de Vonder *et al.* (2007) evaluated several predictive–reactive resource-constrained project-related scheduling procedures under the composite objective of maximizing both the schedule stability and the timely project completion probability. Son and Venkateswaran (2007) proposed a novel architecture that allows a multi-scale federation of interwoven simulations and decision models to support the integrated analysis of stability and performance in hierarchical production planning for supply networks.

Second, the known approaches do not consider the active (autonomous and goal-oriented) behaviour of the SC organizational units. Such an enhancement of the control theory-based models is mandatory, because SCs evolve through managerial actions and not through laws of physics and mechanics. *Third,* the issues of SC planning and execution are mostly considered in isolation without explicitly linking scenario-driven and event-driven management.

Without launching into a discussion on terminological issues, we will consider the SC property to approach the real SC performance with the planned one under the colliding SC processes in the real perturbed execution environment with regard to the variety of execution and goal criteria as the SC *global stability (see exact definition in Sect. 7.3).*

This understanding of SC global stability is tightly interconnected with such SC properties as controllability and observability, which play a crucial role in the design of control systems via the state space. These two conditions are of a dual nature and determine the existence of a complete solution to the control system synthesis problem (Ogata 1997, Lalwani *et al.* 2006). In general, the concept of *controllability* reflects the ability to move a system around in its entire configuration state space using only certain admissible manipulations. Depending on the type of models and systems considered, the definition of controllability varies (Cobb 1984, Koumboulis and Mertzios 1999).

However, limitations to the application of the controllability concept and models from the automatic control theory to the SCM domain exist. As we discussed in Chap. 5, SC systems move and evolve not through mechanical laws or automatic signals but through management decisions. In this setting, we see two problems that require an amplification of the controllability concept with regard to the SCM domain.

First, the SC models do not provide explicitly defined input–output structures. A possible way to consider the controllability in these settings is the behavioural frameworks of system dynamics (Polderman and Willems 1998). Kaneko *et al.* (2006) presented an approach to investigate the controllability and the achievability of discrete event systems within Willems's behavioural framework.

Second, the above-mentioned issue of achievability (or attainability) of a system's goals plays a significant role. These goals can be achieved only if there are enough control actions to guide the system to the achievement of its goals in a real execution environment with negative perturbation influences. Indeed, the achievement of the balance between the available control actions subject to a certain scale of perturbation influences makes the SC controllable and leads to its stabilization. Hence, both state and output controllability should be considered simultaneously.

Based on the two issues stated above, we propose to consider the aspects of SC controllability and achievability as one property, named manageability. We prefer to use the term manageability instead of controllability with regard to the SCM domain. The SC *manageability* may be defined as a general system property to generate, implement, analyse and adjust managerial actions to lead the system to the achievement of its goals. When considering the above-mentioned concept of SC global stability and the manageability concept, it becomes evident that these concepts are mutually interrelated. The manageability is tightly interconnected with complexity and optimality (see also Sect. 5.3). Even the investigations into this triangle may potentially provide new insights into SCM and engineering.

The formal models of SC global stability estimation will be considered in Chaps. 10 and 14. Finally, we note that the proposed term "SC global stability" should not be confused with stability as a whole and the hyperstability of Ashby (1956), Lyapunov (1966), and Popov (1973). We introduce the concept of SC global stability for a SC as a system that *evolves by management actions* and understanding under the global stability as follows: for the achievement of the SCM goals, *balancing positive control and negative perturbation influences* is preconditioned.

The idea of this approach to stability is to a certain extent similar to the issue of SC performance and risk management (Ritchie and Brindley 2007). In both cases, the problem consists of analysing potential SC goals and uncertainty that may negatively affect the achievement of these goals in a real perturbed execution environment. However, while risk management tries to estimate this constellation from the prospective point of view (from a certain planning instant of time up to the end of the planning horizon considered), the global stability approaches the issues from a *dynamic* point of view.

The dynamic interpretation of stability is the main distinguishing feature of the proposed approach. This dynamic interpretation consists of the following components:

- establishing concrete links to concrete processes, operation and parameters of SC execution dynamics through explicit interconnecting with the structure and operations dynamics models;
- considering not only the likelihood of a project's success but also the concrete adjustment steps for handling and adapting SCs in the case of disruptions based on the explicit interconnection of the stability and adaptability; and
- explicit interlinking of the planning, monitoring, and adjustment models.

This problem statement requires other mathematics, as in classical non-linear stability. However, the ideas of the global asymptotic stability to consider stability as a dynamic SC property that emerges through controlled adaptability on the basis of feedback loops (Casti 1979) fit into the SCM domain very well (see also Sect. 14.3). These interrelations will be considered later.

7.5 Example of Decision Making in the Case of Perturbation Influences

Let's consider the general conceptual scheme of decision making on SC planning under uncertainty (see Fig. 7.4).

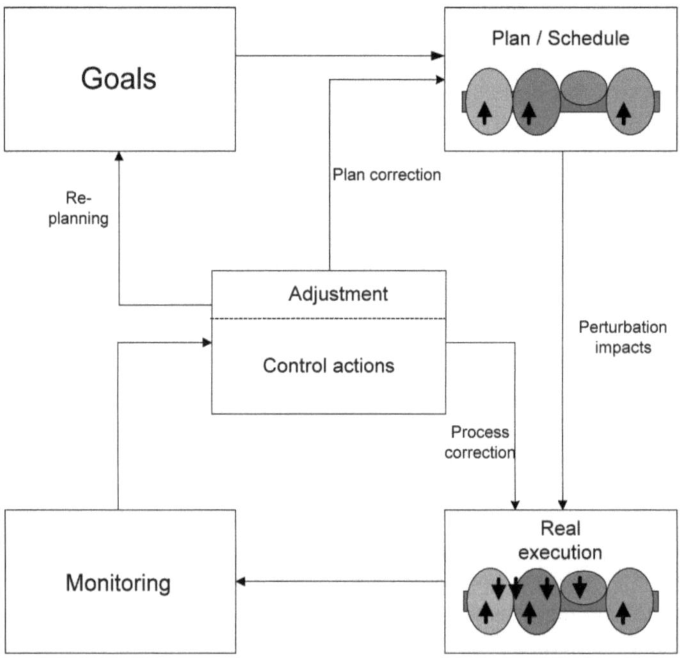

Fig. 7.4 The SC planning system with positive and negative feedbacks (from Ivanov 2009)

On the basis of the goals of the higher level of a SC, plans are formed. During SC functioning, the conformity of the actual course of events with the planned values is analysed. In the case of deviations (negative perturbation impacts), the necessary managerial influences (positive control influences) on the correction of processes and plans as for as SC configuration upgrade will be taken. The given scheme describes the basic functions of SCM under conditions of interaction with a perturbed environment. Such a scheme can be realized on each of three levels of decision-making (strategic, tactical and operative). The interrelation of these three management levels is very important, since it allows us to realize the construction of realistic plans balanced with each other, and to provide feedback for the adaptation of these plans taking into account the real performance of SC processes. Let us consider a concrete example of the analysis of the perturbation influences and the adaptation of an SC. In Fig. 7.5, the logic scheme of the analysis concerning the perturbation influences on SCs is presented.

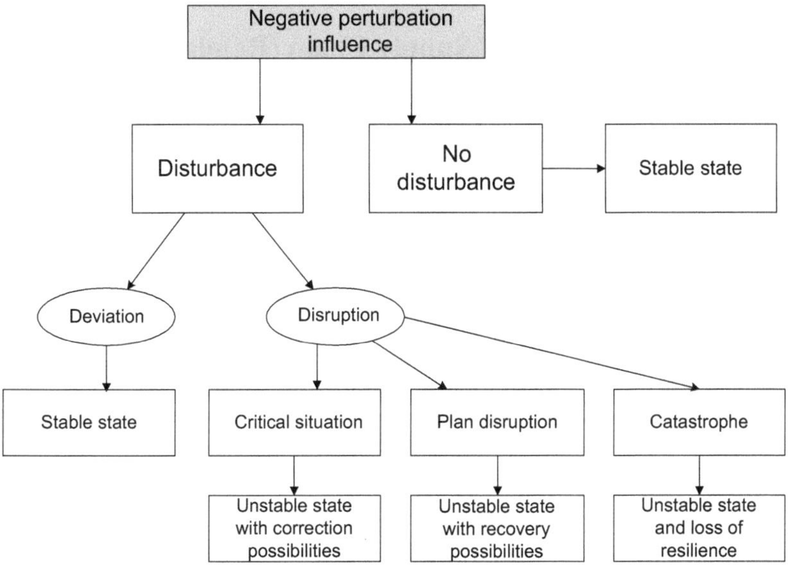

Fig. 7.5 The logic scheme of the analysis concerning the perturbation influences on SCs and their adaptation possibilities

Let's observe an example. As a result of the perturbation influence "Delay in delivery from the supplier", the SC can appear in an efficient or in a disabled state. In the case of a disabled condition, it is a question of disturbance in the performance of the SC function "Start manufacturing a product lot" owing to the absence of the materials (a deviation in the parameter "Start term of manufacturing"). Further, this disturbance can be classified as a disruption or deviation.

In the case that the broken function nevertheless can be executed (i.e. the parameter "End term of manufacturing a product lot" does not deviate from the planned value) without correcting managing influences, for example, on the basis of the safety stock use, it is a question of *deviation*. If the function can be executed only with the application of control influences, it is a question of *disruption*, which should be eliminated on the basis of managerial decisions on the recovery of an efficient SC state (for example, process corrections on the basis of the urgent material acquisition from other suppliers; see parametrical adaptation).

In the case of the impossibility of performance of all the planned orders according to the goals (delivery time, costs, quality), *plan correction*, for example, the redistribution of the resources between various orders, the attraction of additional suppliers, etc. is necessary. If there is a situation of plan disruption, goal adaptation is necessary, i.e. *replanning* (for example, changes in delivery times). An extreme case of failure is a catastrophic situation, when the recovery of the SC plans (at all three levels of management) is impossible and a new strategy and control system for the SC is necessary. In this case, it is possible to talk about the loss of a SC's resilience and stability.

7.6 General Algorithms of Supply Chain (Re)planning Under Uncertainty

7.6.1 General Algorithm of Supply Chain Planning

Let us consider the general logic scheme of decision making on the choice of a SC plan, taking into account SC reliability and flexibility. It consists of 10 steps, which will be considered below.

Step 1. The uncertainty analysis and risks identification
In the first step, a decision maker analyses the uncertainty and identifies the risks. Methods of risk management can be widely used in this step.

Step 2. The risks analysis in an SC
At this stage, there is a linkage of the identified risks to concrete parts and events in a SC. The influence of risk on a SC, in particular on key operations in an SC, will be defined along with revealing critical stages and operations in an SC. The given analysis can be realized on the basis of expert methods and with the use of the sensitivity theory.

Step 3. The development of managerial actions scenarios in the case of disturbances in a SC
At the given stage, the revealed "bottlenecks" in an SC and the potential perturbation influences are brought into correspondence with certain control influences. Scenarios of the actions of managers (for example, as EMP (Event Management Plan)) are developed here.

Step 4. Introduction of redundancies for the SC strengthening
The given stage is intended for the creation of certain reliability and flexibility reserves (safety stocks, reserve channels, a pool of alternative suppliers, a system of

information coordination, the formation of a set of not final decisions) for the SC strengthening, especially its bottlenecks and key operations. A range of SC variants with different levels of redundancy are under construction empirically, each of which is estimated in the following stage.

Step 5. SC BIBO stability analysis
At the given stage, there is an estimation of different SC configuration and plans to different areas of reliability and adaptability under the influence of different areas of perturbation influences.

The stability analysis allows the definition of the admissible borders of deviations in SC execution parameters and of the possibility of returning the SC to a planned (or wished for) state after disturbance. As a result, some zones of stability are defined, to each of which there corresponds a certain level of necessary control influences. The given idea is presented in Fig. 7.6.

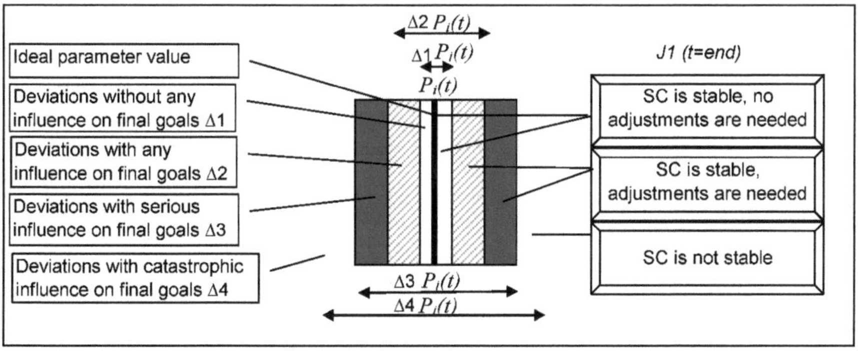

Fig. 7.6 SC parametrical stability analysis

Figure 7.6 shows an extract of one execution performance parameter $\Delta\ p_i$ and one final goal J_1. Various deviations $\Delta_{1....}\Delta_4$ from the ideal parameter values $p_i(t)$ correspond to three classes of SC states: the SC state is stable and no adjustments are needed (deviation Δ_1), the SC state is stable, but adjustments are needed (deviations Δ_2 and Δ_3), and the SC has lost its stability (deviation Δ_4, the final goal cannot be achieved any more under the current conditions, i.e. considering a fixed variety of possible adjustment actions, a certain period of time, and a certain set of deviations as well as algorithm stability). Such an analysis must be performed for each execution parameter at all the relevant instants of time and regarding each final goal. In simple issues it can be realized via expert methods and analytical algorithms. Practical complex problems require other techniques to take into account dynamics and multiple criteria. This approach will be presented in Chap. 14.

Stability analysis can be applied to the unpredictable events problem. In this case, it is not important what has caused a disturbance, but it is important that we can estimate the influence of the disturbance on the SC operation parameters. Further, these disturbances can be correlated with certain earlier admissible values of parameter deviations calculated for other "planned" events and to take corresponding measures of adaptation.

Step 6. An estimation of costs for the elimination of redundancies and disturbances

At the given stage, there is a cost estimation of various measures for the strengthening and adaptation of SCs. The block of costs for the maintenance of SC reliability and flexibility along with the "productive" costs (e.g., the total costs of ownership) form the basis for an estimation of cumulative costs in a SC as a result. It is possible to present the results of this step in the following form (see Fig. 7.7).

On the basis of expert methods, there is an elimination of part of the alternative plans considered in stage 5 (for example, owing to unrealistic costs for the reliability and flexibility maintenance or plans with an unacceptably level of stability).

<table>
<tr><th></th><th colspan="2">SC Reliability</th><th colspan="2">SC Flexibility</th><th colspan="2">SC Adaptation</th><th>SC Processes</th><th>Total SC Costs</th></tr>
<tr><td rowspan="2">Supply Chain No.1</td><td>Reserve 1</td><td rowspan="2">Costs</td><td>Adaptation 1</td><td rowspan="2">Costs</td><td>Adjustment 1</td><td rowspan="2">Costs</td><td rowspan="2">TCO</td><td rowspan="2">Total SC Costs</td></tr>
<tr><td>Reserve 2</td><td>Adaptation 2</td><td>Adjustment 2</td></tr>
<tr><td rowspan="2">Supply Chain No.2</td><td>Reserve 1</td><td rowspan="2">Costs</td><td>Adaptation 1</td><td rowspan="2">Costs</td><td>Adjustment 1</td><td rowspan="2">Costs</td><td rowspan="2">TCO</td><td rowspan="2">Total SC Costs</td></tr>
<tr><td>Reserve 2</td><td>Adaptation 2</td><td>Adjustment 2</td></tr>
<tr><td>...</td><td></td><td></td><td></td><td></td><td></td><td></td><td></td><td></td></tr>
<tr><td rowspan="2">Supply Chain No.M</td><td>Reserve 1</td><td rowspan="2">Costs</td><td>Adaptation 1</td><td rowspan="2">Costs</td><td>Adjustment 1</td><td rowspan="2">Costs</td><td rowspan="2">TCO</td><td rowspan="2">Total SC Costs</td></tr>
<tr><td>Reserve 2</td><td>Adaptation 2</td><td>Adjustment 2</td></tr>
</table>

Fig. 7.7 SC cost assessment with regards to the reliability and flexibility

Step 7. The formation of a set of alternative SCs

At the given stage, the set of alternative SCs received after the performance of step 6 is formed.

Step 8. The final analysis of the SC stability

The same as in step 5, but on the narrowed set of alternatives and taking into account costs for SC reliability, flexibility, and adaptation.

Step 9. The results' calculation and analysis according to the SC economic performance and stability

The given stage consists of the analysis of the alternative SCs generated in step 8 concerning the level of SC economic performance and stability. The results' calculation and analysis according to the SC economic performance and stability can be represented in the form of Table 7.3.

Table 7.3 The simple analysis of alternative SC chains

	Turnover	Costs	Profit	Stability	Profit in the worst case
SC 1	J_{11}	J_{21}	J_{31}	J_{41}	J_{51}
SC 2	J_{12}	J_{22}	J_{32}	J_{42}	J_{52}
SC 3	J_{13}	J_{23}	J_{33}	J_{43}	J_{53}

Each of the estimated alternative SCs as a result is characterized by certain indicators of profitability and costs. Table 7.3 reflects the results with regard to one level of control impacts and one level of perturbation impacts. In Table 7.4, an overview of the total results is presented.

Table 7.4 The complex analysis of alternative SC chains

	Perturbation impact 1			Perturbation impact 2			Perturbation impact 3		
	Control Influence 1	Control Influence 2	Control Influence 3	CI 1	CI 2	CI 3	CI 1	CI 2	CI 3
Supply Chain 1	Turnover	Turnover	Turnover	Turnover	Turnover	Turnover	Turnover	Turnover	Turnover
	Costs	Costs	Costs	Costs	Costs	Costs	Costs	Costs	Costs
	Stability	Stability	Stability	Stability	Stability	Stability	Stability	Stability	Stability
Supply Chain 2	Turnover	Turnover	Turnover	Turnover	Turnover	Turnover	Turnover	Turnover	Turnover
	Costs	Costs	Costs	Costs	Costs	Costs	Costs	Costs	Costs
	Stability	Stability	Stability	Stability	Stability	Stability	Stability	Stability	Stability
Supply Chain 3	Turnover	Turnover	Turnover	Turnover	Turnover	Turnover	Turnover	Turnover	Turnover
	Costs	Costs	Costs	Costs	Costs	Costs	Costs	Costs	Costs
	Stability	Stability	Stability	Stability	Stability	Stability	Stability	Stability	Stability

Table 7.4 reflects the results with regard to all the considered levels of control impacts and all the considered levels of perturbation impacts. The decision-makers can analyse different interrelations of SC performance and stability, and select the most preferable one from a number of alternatives in accordance with the individual risk perception. The final choice of a SC configuration or plan occurs on the basis of managerial individual preferences and the risk perception.

Step 10. The final choice of a SC configuration or plan

The final choice of a SC configuration or plan occurs on the basis of managerial individual preferences and the risk perception.

7.6.2 General Algorithm of Supply Chain Replanning

At the stage of SC execution, SC monitoring and regulation are carried out. There is a gathering of the primary information on the movement and security of deliveries on the basis of various technologies (for example, RFID or bar codes). These actual data are transferred to the level of analytical information systems. There, there is an initial processing of the information, its analysis concerning conformity to plans, and the notification of participants about possible deviations on the basis of the SC monitoring system. These data are transferred to the process level where, on the basis of the event management method, control influences for the elimination of the arising deviations will be elaborated (see Fig. 7.8).

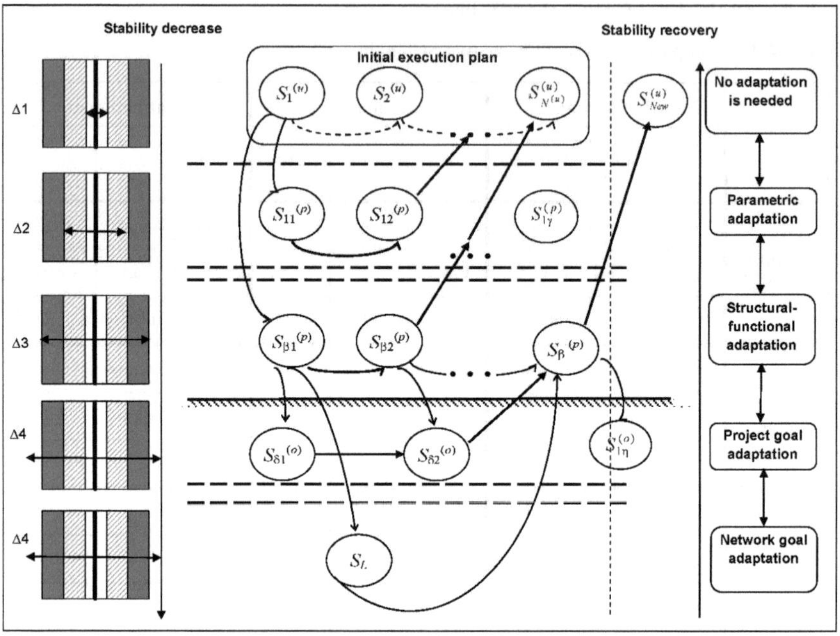

Fig. 7.8 SC operability analysis (from Ivanov *et al.* 2010)

Let's consider the general logic scheme of decision making on the elimination of disturbances in SCs.

Step 1. The analysis of conformity of actual and planned goals
At the given stage, a comparison of the actual values of parameters and the goals of SC execution with the planned values is carried out. If the arising deviations in aggregate do not lead to a loss of stability and the SC maintains a stable state, necessities for correcting control influences are not present. Otherwise, a transition to step 2 is necessary.

Step 2. Alerting managers about the necessity for taking regulating decisions
In the case where perturbation influences lead to a loss of stability and the SC loses its stable state, regulating control has an influence. On the basis of the actual stability analysis data and the planned scenarios of recovering the SC operability, a certain set of actions on the restoration of a planned (or wished for) course of events is proposed to managers.

Step 3. Decision-making on the SC adaptation
Taking operative decisions is based on a system comparison of various kinds of control influences with various levels of parameter deviations of the SC gained on

the basis of the stability analysis. For this analysis, the technique of integration of the SC adaptation concept with the stability analysis is proposed.

Figure 7.8 depicts different variants of system behaviour changes in the case of any perturbation impacts on the system state $S_1^{(u)}$ of the initial execution plan. The perturbation impacts may cause various execution parameter deviations $\Delta\, p_i$ and an operability decrease regarding the final goals $J\ (t{=}end) = \{J_1, \ldots, J_c\}$. To match the system stability analysis and recover the SC operability and global stability, let us apply the above-mentioned complex SC adaptation concept (see Fig. 7.9).

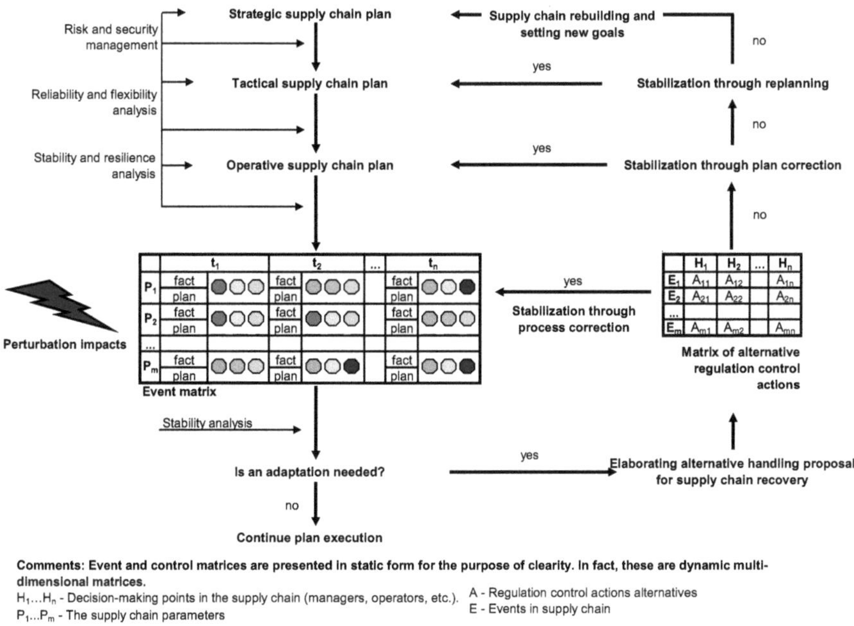

Fig. 7.9 Integrated stability and adaptation analysis (from Ivanov 2009)

As shown further in Fig. 7.9, the SC functions are explicitly divided into the planning (process management level) and execution (event management level) stage. This is an essential point, because the process will be presented as events. This will be considered later. At the planning stage, different reserves to mitigate uncertainty and to ensure SC security are built. This results in a number of alternative SC plans with different values of economic performance and stability indices. The planning ends with the simulation of different execution scenarios for different SC configurations and plans with a subsequent evaluation of these alternatives by managers according to their individual risk perceptions.

After planning, the stage of SC operations execution follows. At the physical security level, cargo movement control takes place. The data from primary control devices (e.g. RFID) are transmitted, accumulated and evaluated within the information systems level. At the interface between the information systems level and the event management level, based on SCEM tools, SC monitoring and reinstating (adaptation) take place. This results in decisions on SC processes, plans or goal correcting, amending or replacing on the basis of the disturbances that occurred and the control actions that existed.

Each element of the event matrix is a characteristic of events at a certain moment of time. In the left part of the elements in the matrix, the actual values and planned goals, and in the right, the result of a comparison of these values from the point of view of the SC stability maintenance, calculated on the basis of stability analysis models and algorithms, are presented accordingly. The given matrix represents the SCMo results, providing a complex representation about the reaction of various SC parts to perturbation influences. On the basis of the matrix analysis, the SC stability is analysed. If the arising deviations in aggregate do not lead to a loss of stability and the SC maintains a stable state, necessities for correcting control influences are not present. Otherwise, adaptation is necessary.

Each adaptation level characterizes a certain control loop in accordance with oscillations and deviations and corresponds to certain management actions. We distinguish parametric adaptation (i.e. rush orders), structural-functional adaptation (i.e. supplier structure changing), project goals' adaptation (i.e. delivery delay), and SC goals' adaptation (i.e. network profit changing), as well as SC strategy and models' adaptation. This makes it possible to match the results of the stability analysis and the actual execution analysis. It also provides a decision maker with a tool to make decisions about the SC adjustment in a real-time mode. The concept presented amplifies the application area of stability analysis to the SC performance adjustment and makes it possible to increase the quality of decision-making of the SC reconfiguration and adjustment.

References

Ashby WR (1956) An Introduction to Cybernetics. Chapman & Hall, London

Casti JL (1979) Connectivity, complexity and catastrophe in large-scale systems. Wiley-Interscience, New York and London

Cheng TCE, Wu YN (2006) A multiproduct, multicriterion supply-demand network equilibrium model. Oper Res 54(3):544–554

Cobb JD (1984) Controllability, observability and duality in singular systems. IEEE Trans Automat Contr 29:1076–1082

Daganzo CF (2004) On the stability of supply chains. Oper Res 52(6):909–921

Dashkovskiy S, Rüffer BS, Wirth FR (2005) A small-gain type stability criterion for large scale networks of ISS systems. Decision and Control and European Control Conference CDC-ECC, Seville:5633–5638

Disney SM, Towill DR (2002) A discrete linear control theory model to determine the dynamic stability of vendor managed inventory supply chains. Int J Prod Res 40(1):179–204

Disney SM, Towill DR, Warburton RDH (2006) On the equivalence of control theoretic, differential, and difference equation approaches to modeling supply chains. Int J Prod Econ 101:194–208

Glickman TS, White SC (2006) Security, visibility, and resilience: The keys to mitigating supply chain vulnerabilities. Int J Logist Syst Manag 2(2):107–119

Ivanov D (2009) Adaptive aligning of planning decisions on supply chain strategy, design, tactics, and operations. Int J Prod Res, in press

Ivanov D, Sokolov B, Kaeschel J (2010) A multi-structural framework for adaptive supply chain planning and operations control with structure dynamics considerations. Eur J Oper Res 200:409–420.

Kaneko O, Misaki T, Fujii T (2006) On controllability of discrete event systems in a behavioral framework. J Comput Syst Sci 45(4):513–517

Kleindorfer PR, Saad GH (2005) Managing disruption risks in supply chains. Prod Oper Manag 14(1):53–68

Koumboulis FN, Mertzios BG (1999) On Kalman's controllability and observability criteria for singular systems. Circ Syst Signal Process 18(3):269–290

Lalwani CS, Disney S, Towill DR (2006) Controllable, observable and stable state space representations of a generalized order-up-to policy. Int J Prod Econ 101:172–184

Lyapunov AM (1966) Stability of motion. Academic Press, New-York and London

Ogata K (1997) Modern control engineering. Prentice Hall International, London

Ostrovsky M (2008) Stability in supply chain networks. Prod Oper Manag 98(3):897–923

Polderman JW, Willems JC (1998) Introduction to mathematical systems theory: a behavioral approach. Springer, New York

Ponomarov SY, Holcomb MC (2009) Understanding the concept of supply chain resilience. Int J Logist Manag 20(1):124–143

Popov VM (1973) Hyperstability of control systems. Springer, Berlin and New York

Prigogine I, Nicolis G (1977) Self-Organization in Non-Equilibrium Systems. John Wiley and Sons, New York

Quyang Y (2007) The effect of information sharing on supply chain stability and the bullwhip effect. Eur J Oper Res 182:1107–1121

Rice JB, Caniato F (2003) Building a secure and resilient supply network. Supply Chain Management Review, September/October:22–30

Riddalls CE, Bennett S (2002) The stability of supply chains. Int J Prod Res 40(2):459–475

Ritchie B, Brindley C (2007) An emergent framework for supply chain risk management and performance measurement. J Oper Res Soc 58:1398–1411

Sheffi Y (2005) The resilient enterprise. MIT Press, Massachusetts

Son Y-J, Venkateswaran J (2007) Hierarchical supply chain planning architecture for integrated analysis of stability and performance. Int J Simulat Process Model 3(3):153–169

Stefani RT, Shahian B, Savant CJ, Hostetter GH (2002) Design of feedback control systems. Oxford University Press, Oxford

Sterman JD (2000) Business dynamics: systems thinking and modeling for complex world. McGraw-Hill/Irwin, New Jersey

Van de Vonder S, Demeulemeester E and Herroelen W (2007) A classification of predictive-reactive project scheduling procedures. J Sched 10(3):195–207

Warburton RDH, Disney SM, Towill DR, Hodgson JPE (2004) Further insights into "The stability of supply chains". Int J Prod Res 42(3):639–648

Yang J, Wang J, Wong CWY, Lai KH (2008) Relational stability and alliance performance in supply chain. Omega 36:600–608

Chapter 8
Quantitative Modelling of Supply Chains

We should not pretend to understand the world only by the intellect.
The judgment of the intellect is only part of the truth.
Carl Gustav Jung

It is impossible to describe the unlimited variety of real world
with limited mathematical means.
Boris Sokolov

8.1 Operations Research

OR on the SCM can be divided into three primary approaches to conducting SC modelling. These are optimization, simulation, and heuristics.

8.1.1 Optimization

Optimization is an analysis method that determines the best possible method of designing a particular SC. Optimization methods for SCM have been a very visible and influential topic in the field of OR. Tayur *et al.* (1999), de Kok and Graves (2004) and Simchi-Levi *et al.* (2004) provide a systematic summary of OR on quantitative models of the SCM, especially for inventory management, tactical planning decisions, and supply contracts.

The formulation of strategic production–distribution models for SCM has been widely investigated. Most of these formulations are introduced in the form of MILP models. Beginning with the seminal work of Geoffrion and Graves (1974) on multi-commodity distribution system design, a large number of optimization-based approaches have been proposed for the design of supply networks (Vidal and Goetschalckx 1997). Arntzen *et al.* (1995) develop a MIP model, called GSCM (Global SC Model), that can accommodate multiple products, facilities, stages (echelons), time periods and transportation modes.

Beamon (1998), Tayur *et al.* (1999), Goetachalckx *et al.* (2002), de Kok and Graves (2004), Simchi-Levi *et al.* (2004), Harrison (2005), Chopra and Meindl (2007), Shen (2007) and Chandra and Grabis (2007) provided a systematic summary of operations research on quantitative models for SCD. A good overview of mathematical programming approaches is presented in Chandra and Grabis (2007).

Graves and Willems (2005) develop a dynamic program with two state variables to solve the SC configuration problem for SCs that are modelled as spanning trees and applied it to optimizing the SC configuration for new products. Meepetchdee and Shah (2007) develop a framework of logistical network design with robustness and complexity considerations and used an MILP model for concept implementation. Yan *et al.* (2003) propose a strategic production–distribution model for SC design with consideration of BOM formulated as logical constraints in an MIP model. Safaei *et al.* (2009) report on an approach to integrated multi-site production-distribution planning in SC by hybrid modelling.

The drawback of using optimization is difficulty in developing a model that is sufficient detailed and accurate in representing complexity and uncertainty of SCM, while keeping the model simple enough to be solved (Harrison 2005). Furthermore, most of the models in this category are deterministic and static. Additionally, those that consider stochastic elements are very restrictive in nature.

An interesting approach to computational issues in mathematical programming is the Lagrange relaxation (Amiri 2006, Chandra and Grabis 2007). This is an attempt to simplify the problem while usually constraints are introduced into the objective function with a penalty function (the so-called dual control)). As a result, a relaxed problem of the original problem is obtained. The relaxed problem is solved to get an upper bound (for maximization problems) of the original problem. Any feasible solution of the original problem provides a lower bound. Iterative heuristic algorithms are used in searching for the optimal solution of the original problem in this narrowed range. The upper and lower bounds are continuously updated.

The main advantage of the optimization approach is the idea of *optimality* and striving for the best solution. However, the optimal approach should be taken very carefully. The problems of applying optimization-based decision-making are tightly interrelated with complexity, uncertainty and multiple objectives. The optimal approaches are very "fragile" and presume certain problem dimensionality, fullness and certainty of the model. Besides, the optimal solutions are usually very sensitive to deviations. Moreover, the decision making is actually tightly interconnected with dynamics and should be considered as an adaptive tuning process. A particular feature of complexity in SCs is multi-criteria decision making by managers with their own preferences that, in their turn, are always changing. Hence, it becomes impossible to build any general selection function for multi-criteria decision making.

Indeed, with regard to SCs as complex systems, optimization can be realistically considered as the direction for best solutions and the ideology of decision-making. Finding optimal solutions is possible but it can be very time-consuming. However, these optimal solutions can be used for benchmarking to estimate the quality of solutions obtained by heuristics or simulation models. Unless mitigating circumstances exist, optimization is the preferred approach for SCM. However, in reality, only a few partial SCM problems (mostly of a strategic nature) may be correctly addressed by optimization.

8.1.2 Simulation

Simulation is imitating the behaviour of one system with another. By making changes to the simulated SC, one expects to gain understanding of the dynamics of the physical SC. Simulation is an ideal tool for further analysing the performance of a proposed design derived from an optimization model. One promising area is the study of combining simulation methods with optimization methods in an iterative way. An overview of the existing hybrid techniques can be found in Chandra and Grabis (2007).

SC simulation utilizes three classes of *software tools* (Schenk *et al.* 2009):

- all-purpose discrete event simulation tools;
- specific SC simulation software; and
- SCM software with simulation functionality.

A similar classification and lists of specific software tools are contained in Banks *et al.* (2002). The concept of discrete-event simulation and tools have been around for quite a while and have been described in Law and Kelton (2000) and Schriber and Brunner (2007). The second and third tool classes may include both commercial off-the-shelf software (COTS) and customized developments. Stäblein *et al.* (2007) describe specific SC simulation software, and Nissen (2000) the integration of simulation functionalities in SCM software.

Two classes of *dynamic models*, namely continuous and discrete models, are widely used to depict process sequences in flow systems. Continuous models are based on differential equations and are most frequently applied as system dynamics models to reproduce manufacturing and logistics processes (Sterman 2000, Scholz-Reiter *et al.* 2006). System dynamics models are developed relatively rarely since planners often find the models in this class too abstract or are unable to generate them. Since these models normally relate to real problems relatively roughly and very abstractly, they are hereafter referred to as macroscopic models (Tolujew and Reggelin 2008). Operative planning routinely applies microscopic discrete-event models. The principles and tools of discrete-event simulation (Law and Kelton 2000) are utilized to implement discrete models. Since workstations, technical resources, carriers and units of goods are represented as individual objects in most cases, event-oriented models may also be referred to as microscopic models (Tolujew and Reggelin 2008).

Both approaches have several deficits. *Macroscopic models* are very abstract and therefore do not represent very accurately the numerous different logistics objects (products, resources, etc.) and the control strategies that need to be considered when solving practical problems (Schenk *et al.* 2008).

Microscopic simulation models represent real world objects with a high level of detail. Consequently, models in this class are often very complicated and slow and their creation and implementation are time-consuming. Both microscopic and macroscopic models are frequently unsuited to handling analysis and planning tasks under time constraints when production and logistics systems are disrupted and therefore to ensuring logistics processes are dependable and stable (Schenk *et al.* 2008).

A large variety of *hybrid techniques* have been developed to combine simulation and optimization. These techniques can be divided into sequential and simultaneous hybrid models with regard to interactions between an optimizer and a simulator (Koechel and Nielaender 2005, Chandra and Grabis 2007, Safaei *et al.* 2009). A good example of a combination of discrete-event simulation, agent simulation and system dynamics is the tool AnyLogic.

Another interesting hybrid approach within the simulation itself is mesoscopic simulation. The term *mesoscopic simulation* was first introduced in logistics in traffic simulation (Kates *et al.* 1988) and appeared in the SCM domain in (Marthaler *et al.* 2003) for modelling based on differential equations but not for a new modelling paradigm. Nevertheless, a clear definition of the term mesoscopic has not yet been developed. In traffic simulation, the term mesoscopic is often applied to refer to a combination of macroscopic and microscopic simulations. Tolujew and Alcala (2004) and Schenk *et al.* (2009) reported a mesoscopic simulation approach that integrates two classes of dynamic models that are widely used to reproduce process sequences in flow systems, namely macroscopic and microscopic models.

The distinct advantage of the mesoscopic approach presented in Schenk *et al.* (2009) is its easier, faster and less laborious creation of models that are easier to reconfigure than microscopic models but nevertheless allow the modelling of the dynamic characteristics of the analysed processes on a level equivalent to discrete-event simulation. In Fig. 8.1, mesoscopic and microscopic simulation views are presented.

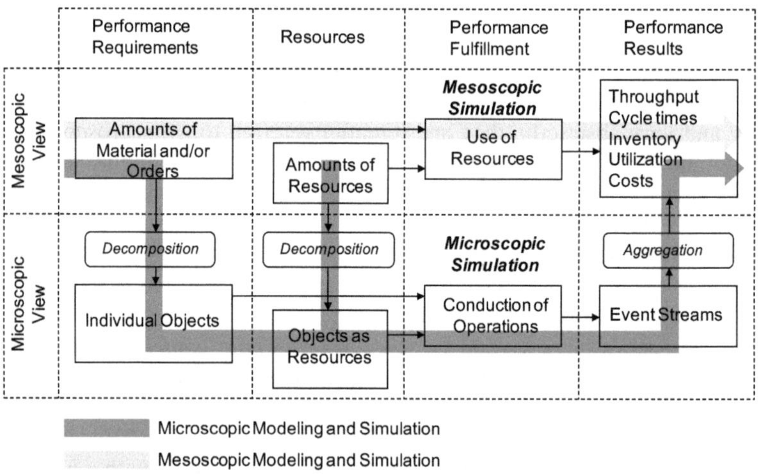

Fig. 8.1 Mesoscopic and microscopic simulation views (from Schenk *et al.* 2008)

Instead of individual flow objects, the mesoscopic approach monitors quantities of objects that belong to a logical group (e.g. a batch, a delivery etc.). The results are not obtained by counting individual objects but by using mathematical formulas to calculate the results as continuous quantities in every modelling time step Δt. The basic components of the mesoscopic modelling approach are multichannel funnels, multichannel delay elements, sources and sinks. Furthermore, this approach differentiates different product types.

As depicted in Fig. 8.1, microscopic simulation is often employed to arrive at pure mesoscopic results from problems presented in the pure mesoscopic view. This "detour" is quite complicated and costly because it involves the decomposition and aggregation of data. Data loss and deformation seem unavoidable. Direct and dynamic transformation of mesoscopic input data (performance requirements and/or resources) into mesoscopic performance results without the aid of event-driven process modelling is the advantage of mesoscopic simulation.

The philosophy behind the mesoscopic approach can be described by the phrase "discrete time/continuous quantity". The representation of individual flow objects (typical of discrete-event simulation) that reproduce goods, persons, means of transportation or job orders is dispensed with. Instead, only numbers are employed, which are used in the model to represent respective quantities of objects or materials and can be modified with mathematical formulas in every step of the discrete simulation time. This type of mesoscopic modelling and simulation is a method to complete planning tasks quickly in production and logistics systems, it being possible in principle to reproduce the dynamic properties of the processes being analysed on a level that corresponds to classical simulation.

However, simulation modelling has some limitations with regard to the complicated designing of new models, problems in results' interpretation and the unknown quality of the found solution. Besides, the simulation models are very subjective. The resulting output depends on the subjectively constructed input. Finally, the simulation models are developed for concrete application cases and cannot (or only very conditionally) provide any general methodical regularity.

8.1.3 Heuristics

Heuristics are intelligent rules that often lead to good, but not necessarily the best, solutions. Heuristic approaches typically are easier to implement and require fewer data. However, the quality of the solution is usually unknown. Unless there is a reason not to use the optimization, heuristics is an inferior approach. In SC settings, nature-based heuristics such as genetic algorithms (Huang *et al.* 2005) and ACO (Ant Colony Optimization) (Teich 2003) are usually applied.

In SCs, the concurrent open shop problems are encountered most frequently. It is well known that most scheduling problems of this class are NP hard due to high dimensionality (Roemer 2006). That is why heuristics (e.g. genetic algorithms) are usually applied instead of optimization. They do not guarantee the optimal solution but allow a permissible result to be found within an acceptable period of time. The quality of this solution with regard to the potential optimum, however, remains unknown. Second, the multiple objective problems are still a "bottleneck" of the heuristics.

8.2 Control Theory

In OR, improvements in SC planning and scheduling are algorithmic (Kreipl and Pinedo 2004). However, in recent years, the works on SCM have been broadened to cover the whole SC dynamics. In these settings, control theory is becoming of even greater interest to researchers and practitioners (Disney and Towill 2002, Braun *et al.* 2003, Anderson *et al.* 2006, Sethi and Thompson 2006, Disney *et al.* 2006, Lalwani *et al.* 2006, Ivanov *et al.* 2007, 2009, 2010, van Houtum *et al.* 2007, Sethi and Thompson 2006, Wang *et al.* 2007). In these studies, a number of valuable control frameworks, models and algorithms have been developed to the SCM domain.

8.2.1 Control Theory Basics

Control theory is a multi-disciplinary scientific discipline that contains powerful conceptual and constructive tools to conduct research on the dynamic problems of the flexible (re)distribution of a variable set of jobs to a variable set of resources (Kalinin and Reznikov 1987, Stefani *et al.* 2002).

The control theory is tightly interconnected with cybernetics. Wiener (1948) proposed summing up the whole area of regulation and communication theory under the term "cybernetics": "We have decided to call the entire field of control and communication theory, whether in the machine or in the animal, by the name Cybernetics".

The renovation of cybernetics has two sources. The first source lies in the attempts to revise the methodological backgrounds of cybernetics. Maruyama (1963) had paid attention to the systems in which the mutual causal effects are deviation-amplifying. Economic, social, and biological examples were considered. In contrast to Weiner's cybernetics with deviation-counteracting systems, the studies of deviation-amplifying mutual causal relationships were called "the second cybernetics".

Von Foester (1974) defined "the second-order cybernetics" with awareness that an observer is an element of the system. The studies considered processes resulting in an increase in biological and social complexity. Stafford Bear, in his work since 1974, has emphasized that investigation into complexity problems should evolve Ashby's law of requisite variety (Ashby 1956).

Unfortunately, the logically relevant chain of fundamental notions of cybernetics – control – informational processes – universal transformer of information (computer, cybernetic machine) was split. An expansion of computer technologies caused the illusion of their ability to solve any problem. The imperfection of these technologies has already caused catastrophes that allowed scientists to proclaim the establishment of a "risk society" rather than an "informational" one. This inspires a renewed interest in the theoretical background of control problems.

Within the control theory, different control approaches exist, i.e. optimal control, adaptive control, and intelligent control. With regard to the SCM domain, we will limit the literature review to optimal control and adaptive control. One of the most popular techniques of optimal control is the *model predictive control* (MPC) in connection with a PID controller. MPC has been a preferred algorithm for robust, multi-variable control that has been widely used in the process industries. The popularity of MPC stems from the relative ease with which it can be understood, and its ability to handle input and output constraints (Prett *et al.* 1989).

MPC is a control strategy based on the explicit use of a process model to predict the process output (performance) over a long period of time (Camacho and Bordons 2004). The model attempts to predict the control variables for a set of time periods. Predicted control variables depend on disturbance forecasts (i.e. demand, prices and interest rates) and also on a set of given parameters that are known in the control literature as control inputs.

In MPC, a system model and current and historical measurements of the process are used to predict the system behaviour at future time instants. A control-relevant objective function is then optimized to calculate a sequence of future control moves that must satisfy the system constraints. A PID controller is intended to correct errors between a measured process variable and a desired set point by producing a corrective action that can adjust the processes. As a control-oriented framework, an MPC-based planning scheme has the advantage that it can be tuned to provide acceptable performance in the presence of significant uncertainty, forecast error, and constraints on inventory levels, production, and shipping capacity. However, the "optimal control" structure in MPC is only a means to achieve such a result, as it does not optimize a true performance index of the closed-loop control system.

Applications of MPC to multi-echelon production–inventory problems and SCs have been examined previously in the literature (Tzafestas *et al.* 1997, Bose and Pekny 2000, Braun *et al.* 2003). These approaches are conceptually different and require less detailed knowledge in comparison with cost-optimal stochastic programming solutions, which require many "what-if" cases to be run and examined by highly skilled professionals. MPC offers the same flexibility in terms of the information sharing, network topology and constraints that can be handled. The MPC approach is not simply to run this planning with safety-stock hybrids more frequently, but rather to develop decision policies based on control-theoretic concepts and apply these to SCs (Wang *et al.* 2007).

Perea-Lopez *et al.* (2003) used MPC to manage a multi-product, multi-echelon production and distribution network. The multi-product batch plants represent a unique feature and are modelled with an MILP representation. A two-layered optimization-based control approach to multi-product SC networks is presented by Seferlis and Giannelos (2004). The control strategy applies multivariable MPC principles to the entire network while maintaining the safety inventory levels through the use of dedicated feedback controllers for each product and storage node. The optimization-based controller aims at maximizing customer satisfaction with the lowest operating costs. The inventory controllers are embedded within the optimization framework as additional equality constraints.

Another well-known approach that is presented as a robust manner of making decisions under uncertainty is to solve the planning problem using stochastic optimization (Tsiakis *et al.* 2001, Bonfill *et al.* 2005). A solution with the maximum expected performance is obtained by including estimated scenarios in the formulation; these estimated scenarios are generated by representing uncertain parameters as random variables. Braun *et al.* (2003) developed a de-centralized MPC implementation for a six-node, two-product, three-echelon demand network problem developed by Intel Corporation that consists of interconnected assembly/test, ware-house, and retailer entities.

More recently, Anderson *et al.* (2006) presented a model for stochastic optimal control for staffing and backlog policies in a two-stage customized service SC. Mestan *et al.* (2006) addressed the optimal operation of multi-product SC systems using MPC. The SC considered is a hybrid system managed by continuous/discrete dynamics and logic rules. For optimization, the SC is modelled within the framework of mixed logical dynamical systems and the overall profit is optimized. Puigjaner *et al.* (2008) proposed an MPC approach that comprises stochastic and optimization models. A scenario-based multi-stage stochastic MILP model has been employed to address the SC dynamics problem. Wang *et al.* (2007) addressed challenges of SCM in semi-conductor manufacturing resulting from high stochastic and non-linearity in throughput times, yields, and customer demands. They presented the advantages of the control-oriented receding horizon formulation behind MPC for three benchmark problems. The effects of tuning, model parameters, and capacity have been shown by comparing system robustness and multiple performance metrics in each case study.

Adaptive control (AC) is a control strategy with some form of recursive system identification. Usually, a parametric adaptive control is considered (Sastry and Bodson 1994). Research in AC has a long and vigorous history. Kalman (1960) developed the concept of a general self-tuning controller with explicit identification of the parameters of a linear, single-input, single-output plant and proposed to use these parameters' estimation to update an optimal linear quadratic controller. In the 1960–70s, due to the establishment of Lyapunov's stability and proving convergence in AC systems, stochastic control made giant strides with the understanding of dynamic programming due to Bellmann and others (Bellmann 1972). In the 1980s, adaptive schemes for different applied domains appeared (Goodwin *et al.* 1980). However, the AC approach has not found a wide application in the SCM domain so far, except in some work (e.g. Scholz-Reiter *et al.* 2004, Dashkovsky *et al.* 2005). The main cause of this is that the AC techniques are intended for technical systems with an automatic controller and automatic reactions of milliseconds. This is not the case in SCs, but the main principles of the AC (not the formal techniques) can enrich the control framework for the SCM that will be shown later in this study.

8.2.2 Advantages of Control Theory Application in the SCM Domain

Let us turn to the analysis of advantages and disadvantages of either of the considered control approaches for the SCM domain. Both the MPC and AC can be considered as a robust, flexible decision framework for dynamically managing processes and meeting customer requirements in SCs. However, the application of the MPC and AC approaches in the SCM domain is challenged by the following issues.

Let us consider a general control loop of a dynamic system as known from the control theory (see Fig. 8.2).

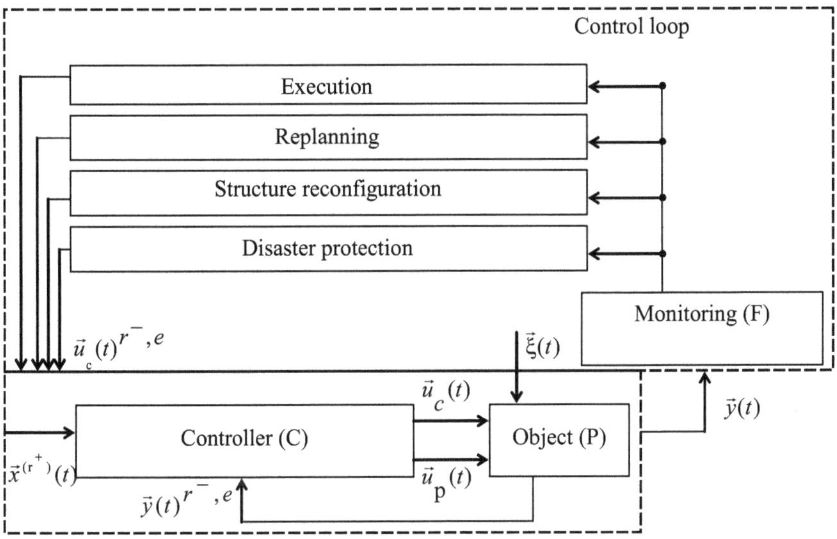

Fig. 8.2 General control feedback multi-loop of a dynamic system

Y(t) is the system output measured by a monitoring system *F* with regard to compliance with the input variables $\vec{u}_p(t)$ that in turn are subject to SC goals $\vec{x}\ (t)$ of a superordinated level (e.g. service level and costs). Based on current information from the *F* negative perturbation influences ξ (e.g. new demand forecasts, disruptions in current processes), an adapted input $\vec{u}_c(t)$ will be given to the controller *C* that is responsible for adjustment control actions *u* to the system under control *P (the* SC*)*.

Actually, widespread incremental SC planning and scheduling deal only with the lower part of Fig. 8.2 (without the *F*-driven feedback system). Such an approach can be justified for such problems as those where a single schedule computation should be fulfilled. These problems may be of either a very strategic nature or a very operative nature. In the most tactical–operational problems that refer to the SC dynamics to be under control, the negative feedback is mandatory.

One of the efficient approaches to implement control of the systems in this class is adaptive planning (Okhtilev *et al.* 2006, Andreev *et al.* 2007, Ivanov and Ivanova 2008). Adaptive planning uses not only simple open time slots (in contrast to incremental planning) but employs conflict-driven plan changes during the system execution. Adaptive planning implies problem resolution and redefinition through the learning process, rather than problem solving. This is the *first* strong contribution of the control theory to the SCM domain – this lets us interpret planning and scheduling not as discrete operations, but as a continuous adaptive process.

The feedback systems are preferably characterized by disturbance elimination, process execution under uncertainty even if the model structure does not completely correspond to real processes and the stabilization of unstable processes. The possibility of covering the whole SC dynamics and the permanent changes in the SC processes and environment without the strong necessity to accomplish total "remodelling" is ultimately the *second* strong contribution of control theory to the SC planning and scheduling domain.

Based on the feedback loops, the integration of the planning and scheduling stages is possible. The system output $Y(t)$ under control can be expressed either as goals of order realization in the SC (i.e. lead time) or as goals at a superordinated level (i.e. service level and SC costs). Indeed, a SC manager is usually more interested in achieving the desired service level and SC costs than minimizing the lead time of a number of production orders. Of course, theoretically, a set of interlinking operational (the lead time) and tactical (the service level) goals can be constructed so that the operational goals may be subject to optimization. However, these constellations will be very specific to different concrete SCs. Hence, we prefer to consider primarily the tactical goals as being subject to optimization and control.

This means that not only a problem solution in a fixed environment (the system under control) but also a simultaneous consideration of system formation and management problems' solution in this system are possible. This aspect is of significant practical importance. In practice, the challenge is not to calculate optimal schedules to optimize local order fulfilment parameters but to schedule SCs subject to the achievement of SC goals with regard to profitability and stability (see also Chap. 4).

The modern control theory provides the possibility to model and solve more realistic planning problems (incorporating dynamism and uncertainty) as systems can be modelled in terms of dynamic multi-structural macro-states, based on the simultaneous consideration of the management as a function of both states and structures (Okhtilev *et al.* 2006). This approach has been called as structure dynamics control theory. Applications of this theory to the SCM domain have been presented in Ivanov *et al.* (2007, 2009) and Ivanov (2009). Besides, a wide range of different SC properties such as stability, controllability, and observability can be reflected within the control theory (Disney *et al.* 2006, Lalwani *et al.* 2006, Ivanov *et al.* 2009).

8.2.3 Shortcomings and Requirements for Extension

With regard to Fig. 8.2, two critical points of any control strategy can be distinguished:

- the controller;
- the output values' adjustment; and
- the input parameters' correction.

First, the controller *C* is in the classical theory of automatic control an automat. In SCs, the controllers are human beings, managers and operators. There are many of them, from different enterprises with different interests and goals. Stadtler (2009) emphasized that SC members have their own objectives; thus, the system becomes competitive while each member wins by cooperating. The information systems can propose some handling alternatives for managers, but it will depend on people only for adjustment measures, where, and how they will be implemented. The SC tuning by means of the controller, unlike in automatic systems, occurs not within milliseconds but within minutes, hours, days, weeks, months or years, with a delay between the deviation occurrence and decision making depending on the disruption character. This is why the formal models of automatic AC and the automatic PID controller from the MPC approach cannot be applied to the SCM domain. Besides, the known allocation of the MPC to the SCM domain comes mostly from the process industry with continuous control models that are difficult to apply to many other branches with discrete operations.

Moreover, people do not strive for a 100% guarantee of the result; they consciously tend to take risks. Some literature (e.g., Sokolov 2006, Peck 2007) points out the problem of contradiction between objective risk (determined by experts, applying quantitative scientific means) and perceived risk (perception of managers). Actually, the objective risk treatment is rooted in technical science where 100% reliability is mandatory. In socio-economic systems, like SCs, a value of 95% as an orientation for SCs is empirically suggested (e.g., Sheffi 2005). Different managers perceive risk to different extents, and these perceptions can change in the same manager due to changes in his environment. That is why the models for SCs should not strive for a unique optimal solution but allow the formation of a number of alternative solutions with different degrees of efficiency and risk.

Second, the issue of process adjustment should be addressed. The MPC approach is very strong with regard to the updated future forecasting while the AC approach has a more developed methodological basis for the SC tuning based on the information about the past. Both for the MPC and for the AC, the complexity of non-linear differential models and algorithms should be taken into account. The problem of differentiation presentation is that SC decision- making is of a discrete nature. Hence, the combination of MPC control and AC control as well as a combination of control theory with OR techniques and MAS can be considered as ways to develop control frameworks that take into account particular features of the SCs. Such an extended framework will be presented in the further course of this book.

8.3 Complex Adaptive Systems and Multi-agent Systems

The paradigms of CAS)and MAS are popular simulation techniques in the SCM domain. The term *CAS* was coined by Holland, Gell-Mann and others (Gell-Mann 1995, Holland 1995). A CAS is presented as a dynamic network of many agents and decentralized control. The coherent behaviour in the system arises from competition and cooperation among the agents. Hence, the overall behaviour of the system is the result of a huge number of decisions made every moment by many individual agents.

CAS has been extensively applied by researchers in different management and organizational disciplines (Anderson 1999, Anderson and Tushman 2001, Lissack and Letiche 2002, Zhang 2002, Richardson 2004, 2005, 2007, Amaral and Uzzi 2007). Aldunate *et al.* (2005) considered specifically the interrelations of complexity and disaster recovery.

With regard to CAS and SCM, Choi *et al.* (2001) claimed that emergent patterns in a supply network can be managed much better through positive feedback than through negative feedback from control loops. However, the authors conclude that managers must appropriately balance the control and the emergence areas. Hence, the application of CAS should be considered as being combined with the control theory.

Surana *et al.* (2005) investigated how various concepts, tools and techniques used in the study of CAS can be exploited in the SCM domain. In the study by Pathak *et al.* (2007), a theory-based framework is developed that combines aspects of CAS theory, industrial growth theory, network theory, market structure and game theory. This framework specifies categories of rules that may evoke different behaviours in the two fundamental components of any adaptive supply networks, i.e. the environment and the firms in that environment. The framework is implemented as a multi-paradigm simulation utilizing software agents and it joins discrete-time with discrete-event simulation formalisms.

MAS reflect the ideas of CAS at the software entities level. A number of researchers have attempted to apply agent technology to manufacturing enterprise integration, SCM, manufacturing planning, scheduling and control, material handling, etc. The past research on utilization of the MAS for the SCM problems (Swaminathan *et al.* 1998, Fox *et al.* 2000, Shen *et al.* 2001) has mostly dealt with agent-based frameworks and software architectures, where agents are autonomous, goal-oriented software processes. Another agent-based model has been elaborated by Kaihara and Fujii (2008) to reflect SCs' abilities to adapt. Ahn *et al.* (2003) suggested a flexible agent system for SCs that can adapt to the changes in transactions introduced by new products or new trading partners. In many studies, MAS and CAS are interrelated implicitly and explicitly. Tilebein (2006) considered CAS as the general adaptation framework and MAS as the solver technique.

However, neither of the approaches is free of limitations. The tools and techniques of CAS are based on the fields of non-linear dynamics, statistical physics, and information theory. The formal features of these models are very specific and require a very solid mathematical background; hence, only a very limited circle of researchers into SCM can apply them. Second, the models of natural complex systems that form the basis of CAS differ from discrete and human-decision-driven SCs. Hence, the ideas of CAS can be applied to the SCM domain; however, the formal aspects of models are subject to a special analysis.

Though techniques for solving large-scale problems in a decentralized way were widely developed in MAS, most of them are based on weakly-grounded heuristic principles. The theoretical background of the MAS is still under development and does not allow the consideration of the MAS as a general SC modelling framework. Current applications of agent-based paradigms are limited to software. It is mostly underestimated that these paradigms offer a valuable theoretical perspective on decentralized network management.

8.4 Critical Analysis

In Table 8.1, the analysis of the advantages and shortcomings of the above-mentioned modelling approaches with regard to the SCM domain is presented.

Table 8.1 Analysis of the SC modelling approaches

Approach	Application areas	Advantages	Disadvantages
Systems Science and Control Theory	General methodological basics of complex systems synthesis and analysis	Well-elaborated basis of complex dynamic systems synthesis and analysis	Elaborated for technical and biological systems; for SC modelling an extension is required
Analytical methods	Simple problems of SC design and vendor selection	Clarity Optimal solutions	Not flexible enough Modelling uncertainty and dynamics is not sufficient enough
Heuristics	Complex problems, i.e. dynamical SC structuring	Complex problems solution, also with incomplete information	Do not guarantee optimal solutions; the quality of a found solution with regard to potential optimum usually remains unknown
Simulation	System investigations in dynamics	Modelling complex dynamic systems Graphics	Complicated elaboration Complicated results interpretation Do not guarantee optimal solutions

Over the last decade, a wealth of valuable OR approaches for SC strategic, tactical and operational planning have been extensively developed (Simchi-Levi *et al.* 2004, de Kok and Graves 2004, Chopra and Meindl 2007). With regard to OR, the following shortcomings of SCP can be revealed from the dynamics point of view. *First*, problems of high dimensionality and NP-hard problems exist whether reduced to a simple dimensionality or heuristics are applied. *Second*, complex dynamics of real SC execution cannot be reflected in single-class models. *Third*, models of planning and control are not explicitly interconnected in terms of uncertainty. The works on SCM need to be broadened to cover the SC dynamics. In these settings, control theory and systems analysis are becoming of even greater interest to researchers and practitioners

The *first* strong contribution of the control theory to the SCM domain is the interpretation of planning and scheduling not as discrete operations but as a continuous adaptive process. *Second*, the possibility of covering the whole SC dynamics and the permanent changes in SC processes and environment without the strong necessity to accomplish the total "remodelling" is also ultimately a strong contribution of the control theory in the SC planning and scheduling domain. *Finally*, the control theory in the SC planning and scheduling domain allows the consideration of goal-oriented formation of SC structures and the solution of problems in this system as a whole.

However, the control theory application in the SCM domain also has its challenges and limitations. *First*, SCs evolve through management actions. The controller is, in the classical theory of automatic control, an automat that reflects mechanical laws or identification signals. In SCs, the controllers are human beings, managers and operators. This results in subjective decision making with time delays between receiving the signal and taking a control action. *Second*, the complexity of non-linear differential models and algorithms should be taken into account. Although the conceptual aspects of control theory may be applied to the SCM domain, the formal mathematical aspects are too complex and specific to become widespread in the SCM domain. Another problem of differentiation presentation is that SC decision-making is of a discrete nature. This is necessary to develop a constructive way to overcome these inhibitions by means of (1) the formulation of a problem input and output in a discrete form while running the model in a continuous form and (2) the transformation of non-linearity to the model constraints while presenting the dynamic model itself as a linear system. *Third*, in SCs, it is practically impossible to develop a model structure with the defined input–output interrelations. Hence, the control theory laws may not work in these settings. One of the possible approaches is to apply *behavioural frameworks* (Polderman and Willems 1998).

References

Ahn H J, Lee H, Park SJ (2003) A flexible agent system for change adaptation in supply chains. Expert Syst Appl 25:603–618

Aldunate RG, Pena-Mora F, Robinson GE (2005) Collaborative distibuted decision making for large scale disaster relief operations: drawing analogies from robust natural systems. Complexity 11(2):28–38

Amaral LAN, Uzzi B (2007) Complex system – a new paradigm for the integrative study of management, physical, and technological systems. Manag Sci 53:1033–1035

Amiri A (2006) Designing a distribution network in a supply chain system: formulation and efficient solution procedure. Eur J Oper Res 171:567–576

Anderson P (1999) Complexity theory and organization science. Organ Sci 10:216–232

Anderson P, Tushman ML (2001) Organizational environments and industry exit: the effects of uncertainty, munificence and complexity. Ind Corp Change 10:675–711

Anderson Jr EG, Morrice DJ, Lundeen G (2006) Stochastic optimal control for staffing and backlog policies in a two-stage customized service supply chain. Prod Oper Manag 15(2):262–278

Andreev M, Rzevski G, Skobelev G, Shveykin P, Tsarev A and Tugashev A (2007) Adaptive planning for supply chain networks. In: Marik V, Vyatkin V, Colombo AW (Ed) Holonic and multi-agent systems for manufacturing. Springer, Berlin, Heidelberg

Arntzen BC, Brown GG, Harrison TP, Trafton L (1995) Global supply chain management at digital equipment corporation. Interfaces 25(1):69–93

Ashby WR (1956) An introduction to cybernetics. Chapman & Hall, London

Banks J, Buckley S, Jain S, Lendermann P and Manivannan M (2002) Opportunities for simulation in supply chain management. Proceedings of the 2000 Winter Simulation Conference: 1245–1251

Beamon BM (1998) Supply chain design and analysis: models and methods. Int J Prod Econ 55(3):281–294

Bellmann R (1972) Adaptive control processes: a guided tour. Princeton Univ. Press, Princeton, New Jersey

Bonfill A, Espu˜na A, Puigjaner L (2005) Addressing robustness in scheduling batch processes with uncertain operation times. Ind Eng Chem Res 44:1524–1534

Bose S, Pekny J (2000) A model predictive control framework for planning and scheduling problems: a case study of consumer goods supply chain. Comput Chem Eng 24:329–335

Braun MW, Rivera DE, Flores ME, Carlyle WM, Kempf KG (2003) A model predictive control framework for robust management of multi-product, multi-echelon demand networks. Annu Rev Contr 27:229–245

Camacho EF, Bordons C (2004) Model predictive control. Springer, London

Chandra C, Grabis J (2007) Supply chain configuration. Springer, New York

Choi TY, Dooley KJ, Rungtusanatham M (2001) Supply networks and complex adaptive systems: control versus emergence. J Oper Manag 19(3):351–366

Chopra S and Meindl P (2007) Supply chain management. Strategy, planning and operation. Pearson Prentice Hall, New Jersey

Dashkovskiy S, Rüffer BS, Wirth FR (2005) A small-gain type stability criterion for large scale networks of ISS systems. Proceedings of the 44th IEEE Conference on Decision and Control and European Control Conference CDC-ECC, Seville:5633–5638

Disney SM, Towill DR (2002) A discrete linear control theory model to determine the dynamic stability of vendor managed inventory supply chains. Int J Prod Res 40(1):179–204

Disney SM, Towill DR, Warburton RDH (2006) On the equivalence of control theoretic, differential, and difference equation approaches to modeling supply chains. Int J Prod Econ 101:194–208

Foerster von H (1974) Cybernetics of Cybernetics. Paper delivered at 1970 annual meeting of the American Society for Cybernetics. University of Illions, Urbana

Fox MS, Barbuceanu M, Teigen R (2000) Agent-oriented supply chain management system. Int J Flex Manuf Syst 12:165-188

Gell-Mann M (1995) Complex adaptive systems. In: Morovitz H, Singer JL (Ed) The mind, the brain, and complex adaptive systems. Reading: Westview Press, New York

Geoffrion A, Graves G (1974) Multicommodity distribution system design by Benders decomposition. Manag Sci 29(5):822–844

Goetschalckx M, Vidal C J and Dogan K (2002) Modeling and design of global logistics systems: a review of integrated strategic and tactical model and design algorithms. Eur J of Oper Res 143(1):1–18

Goodwin GC, Graebe SF, Salgado ME (1980) Control system design. Prentice-Hall, New Jersey
Graves SC and Willems SP (2005) Optimizing the supply chain configuration for new products. Manag Sci 51(8):1165–1180
Harrison TP (2005) Principles for the strategic design of supply chains. In: Harrison TP, Lee H L, Neale JJ (Ed) The practice of supply chain management. Kluwer academic publishers, Boston
Holland JH (1995) Hidden order: how adaptation builds complexity. Helix Books, Cambridge
Huang G, Zhang Y and Liang L (2005) Towards integrated optimal configuration of platform products, manufacturing processes, and supply chains. J Oper Manag 23:267–290
Ivanov D (2009) DIMA – a research methodology for comprehensive multi-disciplinary modelling of production and logistics networks. Int J Prod Res 47(5):1133–1155
Ivanov D, Ivanova M (2008) A framework of adaptive control for production and logistics networks. In: Scholz-Reiter B (Ed) Dynamics in logistics. Springer, Berlin
Ivanov D, Arkhipov A, Sokolov B (2007) Intelligent planning and control of manufacturing supply chains in virtual enterprises. Int J Manuf Tech Manag 11(2):209–227
Ivanov D, Kaeschel J, Sokolov B (2009) Structure dynamics control-based framework for adaptive reconfiguration of collaborative enterprise networks. Int J Manuf Techn and Manag 17(1-2):3–41
Ivanov D, Sokolov B, Kaeschel J (2010) A multi-structural framework for adaptive supply chain planning and operations control with structure dynamics considerations. Eur J Oper Res 200:409–420.
Kaihara T, Fujii S (2008) Supply chain management for virtual enterprises with adaptive multi-agent mechanism. Int J Manuf Tech Manag 14(3-4):299–310
Kalinin V, Reznikov B (1987) Systems and control theory: structural-mathematical approach. VIKI, Leningrad (in Russian)
Kalman RE (1960) A new approach to linear filtering and prediction problems. J Basic Eng 82(1):35–45
Kates R, Bogenberger K, Hoops M (1988) Mesoscopic simulation with ANIMAL: optimal utilization of downstream traffic detector data and the propagation of information. In: Schreckenberg W (Ed) Traffic and granular flow. Springer, Berlin Heidelberg New York
Koechel P, Nielaender U (2005) Simulation-based optimisation of multi-echelon inventory systems. Int J Prod Econ 93-94:505v513
Kok de AG and Graves SC (2004) Supply chain management: design, coordination and operation. Elsevier, Amsterdam
Kreipl S and Pinedo M (2004) Planning and scheduling in supply chains: an overview of issues in practice. Prod Oper Manag 13(1):77–92
Lalwani CS, Disney S, Towill DR (2006) Controllable, observable and stable state space representations of a generalized order-up-to policy. Int J Prod Econ 101:172–184
Law AM and WD Kelton (2000) Simulation Modeling and Analysis, 3rd Ed. McGraw-Hill, New York
Lissack MR, Letiche H (2002) Complexity, emergence, resilence, and coherence: gaining perspective on organizations and their study. Emergence 4(3):72v94
Marthaler D, Armbruster D, Ringhofer C (2003) A mesoscopic approach to the simulation of semiconductor supply chains. Simulation 79(3):157–162
Maruyama M (1963) The second cybernetics. Deviation amplifying mutual causal process. Am Sci 51:164–179
Meepetchdee Y and Shah N (2007) Logistical network design with robustness and complexity considerations. Int J Oper Prod Manag 37(3):201–222
Mestan E, Turkay M, Arkun Y (2006) Optimization of operations in supply chain systems using hybrid systems approach and model predictive control. Ind Eng Chem Res 45:6493–6503
Nissen V (2000) Simulationsmöglichkeiten in SAPs Supply Chain Management Werkzeug Advanced Planner & Optimizer. Inform Manag Consult 15:79–85

Okhtilev M, Sokolov B, Yusupov R (2006) Intelligent technologies of complex systems monitoring and structure dynamics control. Moskau, Nauka (in Russian)

Pathak SD, Day JM, Nair A, Sawaya WJ, Kristal MM (2007) Complexity and adaptivity in supply networks: building supply network theory using a complex adaptive systems perspective. Decis Sci 38 (4):547–580

Peck H (2007) Supply chain vulnerability, risk, robustness, and resilience. In: Mangan J, Lalwani C, Butcher T (Ed) Global logistics and supply chain management. John Wiley, New York

Perea-López E, Ydstie BE, Grossmann IE (2003) A Model Predictive Control strategy for supply chain optimization. Comput Chem Eng 27:1201–1218

Polderman JW, Willems JC (1998) Introduction to mathematical systems theory: a behavioral approach. Springer, New York

Prett CE, Garc´ıa DM, Morari M (1989) Model predictive control: theory and practice-a survey. Automatica 25(3):335–348

Puigjaner L, Lainez JM (2008) Capturing dynamics in integrated supply chain management. Comput Chem Eng 32:2582–2605

Richardson KA (2004) Systems theory and complexity: Part 1. E:CO 6(3):75–79

Richardson KA (2005) Systems theory and complexity: Part 3. E:CO 7(2):102–114

Richardson KA (2007) Systems theory and complexity: Part 4. The evolution of systems thinking. E:CO 9(1):166

Roemer T (2006) A note on the complexity of the concurrent open shop problem. Journal of Scheduling 9(4):389–396

Safaei AS, Moattar Husseini SM, Farahani RZ, Jolai F, Ghodsypour SH (2009) Integrated multi-site production-distribution planning in supply chain by hybrid modelling. Int J Prod Res, DOI: 10.1080/00207540902791777, in press.

Sastry S, Bodson M (1994) Adaptive control: stability, convergence, and robustness. Prentice-Hall, New Jersey

Schenk M, Tolujew J, Reggelin T (2008) Mesoskopische Simulation von Flusssystemen - algorithmisch steuern und analytisch berechnen. In: Nyhuis P (Ed) Beiträge zu einer Theorie der Logistik. Springer, Berlin

Schenk M, Tolujew J, Reggelin T (2009) Comparison of three methods of implementation of mesoscopic flow models. In: Ivanov D, Meinberg U (Ed) Logistics and supply chain management. Trends in Germany and Russia. Cuvillier Verlag, Goettingen

Scholz-Reiter B, Hoehns H, Hamann T (2004) Adaptive control of supply chains: building blocks and tools of an agent-based simulation framework. CIRP Annals Manufacturing Technology 53(1):353–356

Scholz-Reiter B, Delhoum S, Zschintzsch M, Jagalski T, Freitag M (2006) Inventory control in shop floors, production networks and supply chains using system dynamics. In: Wenzel S (Ed) 12. ASIM-Fachtagung. SCS Publishing House, Erlangen

Schriber J, Brunner T (2007) Inside discrete-event simulation software: how it works and why it matters. Proceedings of the 2007 Winter Simulation Conference 113–123

Seferlis P, Giannelos NF (2004) A two-layered optimisation-based control strategy for multi-echelon supply chain networks. Comput Chem Eng 28:799–809

Sethi SP, Thompson GL (2006) Optimal Control Theory: Applications to Management Science and Economics, Second Edition. Springer, Berlin

Sheffy J (2005) The resilient enterprise. MIT Press, Massachusetts

Shen W, Norrie DH, Barthes JP (2001) Multi-agent systems for concurrent intelligent design and manufacturing. London, Taylor & Francis Group

Shen ZM (2007) Integrated supply chain design models: a survey and feature research directions. J Ind Manag Optim 3(1):1–27

Simchi-Levi D, Wu SD and Zuo-Yun S (2004) Handbook of quantitative supply chain analysis. Springer, New York

Sokolov B, Yusupov R (2006) Risk integrated modelling for control in complex organizational-technical systems. Journal of Control and Informatics 1:1–22

Stäblein T, Baumgärtel H, Wilke J (2007) The supply net simulator SNS: an artificial intelligence approach for highly efficient supply network simulation. In: HO Günther, DC Mattfeld, L Suhl (Ed) Management logistischer Netzwerke. Physica Verlag, Heidelberg
Stadtler H (2009) A framework for collaborative planning and state-of-the-art. OR Spectrum 31: 5–30
Stefani RT, Shahian B, Savant CJ, Hostetter GH (2002) Design of Feedback Control Systems. Oxford, University Press Oxford
Surana A, Kumara S, Greaves M, Raghavan UN (2005) Supply-chain networks: a complex adaptive systems perspective. Int J Prod Res 43(20):4235–4265
Swaminathan JM, Smith SF and Sadeh NM (1998) Modeling supply chain dynamics: a multiagent approach. Decis Sci 29(3):607–632
Tayur S, Ganeshan R and Magazine M (1999) Quantitative models for supply chain management. Kluwer Academic Publishers, Boston
Teich T (2003) Extended Value Chain Management (EVCM). GUC-Verlag, Chemnitz
Tilebein M (2006) Decentralized supply chain management: a view from complexity theory. In: Blecker T, Kersten W (Ed) Complexity management in supply chains: concepts, tools and methods. Erich Scmidt Verlag, Berlin
Tolujew J, Alcalá F (2004) A mesoscopic approach to modeling and simulation of pedestrian traffic flows. In: Horton G (Ed) 18th European simulation multiconference. SCS International, Ghent
Tolujew J, Reggelin T (2008) Mesoskopische Simulation: Zwischen der kontinuierlichen und ereignisdiskreten Simulation. In: Rabe M (Ed) Advances in simulation for production and logistics applications. Tagungsband zur 13. Fachtagung. Fraunhofer IRB-Verlag, Stuttgart
Tsiakis P, Shah N and Pantelides CC (2001) Design of multi-echelon supply chain networks under demand uncertainty. Ind Eng Chem Res 40:3585–3604
Tzafestas S, Kapsiotis G, Kyriannakis E (1997) Model-based predictive control for generalized production planning problems. Comput Ind 34:201–210
Van Houtum G Y, Scheller-Wolf A, Yi J (2007) Optimal control of serial inventory systems with fixed replenishment intervals. Oper Res 55(4):674–687
Vidal C, Goetschalckx M (1997) Strategic production-distribution models: a critical review with emphasis on global supply chain models. Eur J Oper Res 98:1–18
Wang W, Rivera D, Kempf KG (2007) Model predictive control strategies for supply chain management in semiconductor industry. Int J Prod Econ 107:56–77
Wiener N (1948) Cybernetics or control and communication in the animal and the machine. John Wiley & Sons Inc., New York
Yan H, Yu Z and Cheng E (2003) A strategic model for supply chain design with logical constraints: formulation and solution. Comput Oper Res 30:2135–2155
Zhang W-B (2002) Theory of complex systems and economic dynamics. Nonlinear Dynam Psychol Life Sci 6(2):83–101

Chapter 9
DIMA – Decentralized Integrated Modelling Approach

Let us enrich ourselves with our mutual differences.
Paul Valery

9.1 General Basics of DIMA

A SC is characterized by uncertain interactions of the elements and distributed goals. SCs can be described by various models (static and dynamic, stochastic and deterministic, analytical and simulation, etc.), which are interconnected. SCs are also characterized by a set of interrelated structures. Furthermore, the SC elements are active. Their activities are based on their own interests and goals. *Active* elements cause the necessity for balancing SC partner interests, a large number of uncontrolled factors, and formalizing difficulties. Besides, the SC execution is accomplished by permanent changes in the internal network properties and the environment. It requires SC adaptation to the current execution environment. So, reflections of SC configuration, planning and execution models are needed. The other issue of SC modelling is the interlinking of conceptual and mathematical models in order to achieve adequate, scaleable and representative models and providing application independence. Such a complex nature of SCs requires a combined application of different disciplines and modelling approaches (see Fig. 9.1).

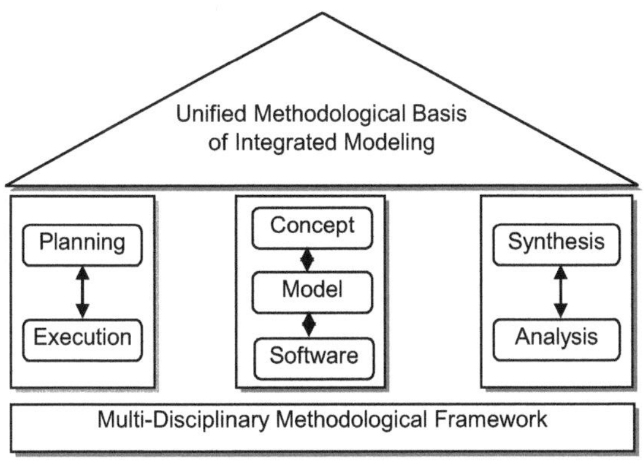

Fig. 9.1 The DIMA vision of SC integrated modelling (from Ivanov 2009a)

The basics of the DIMA methodology were developed in (Ivanov 2006, 2009a) to contribute to comprehensive SC modelling and to establish foundations for network theory as called for by an increasing number of researchers (Beamon 1998, Barbasi 2005, Kuehnle 2008). This methodology is also closely interconnected with the methodology of IDSS (see Fig. 9.2).

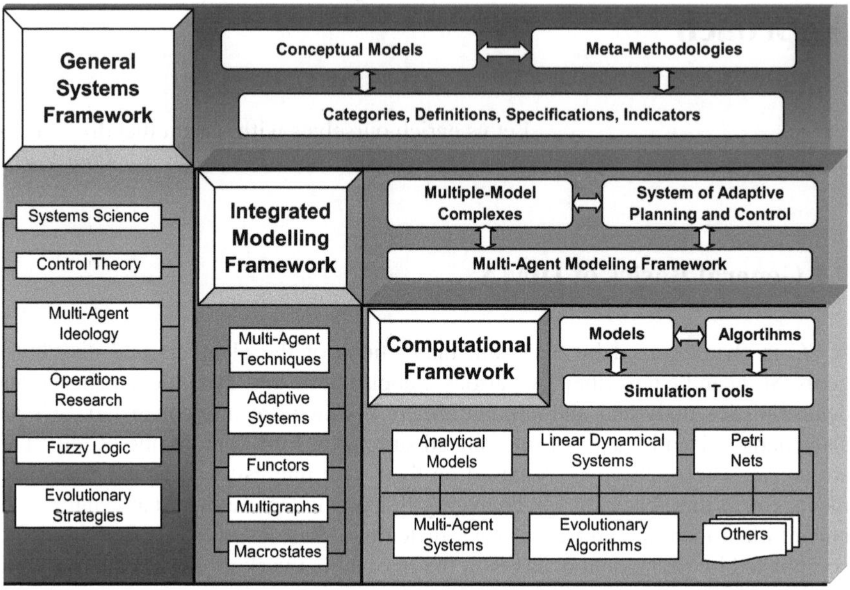

Fig. 9.2 Integrated modelling "concept-model-computation" (from Ivanov 2009a)

The main parts of the DIMA methodology are: the general systems framework (GSF), the integrated modelling framework (IMF), and the computing framework. The GSF defines conceptualized business models, meta-methodologies and a set of categories, definitions, specifications and performance indicators, which are developed during the integration of various theoretical frameworks (Ivanov *et al.* 2005, 2006, Okhtilev *et al.* 2006). The IMF defines the rules of the integrated multi-disciplinary mathematical model building. It proposes constructive methods and techniques of (1) how to combine various model classes and (2) how to model interconnectedly the partial SC problems. The computing framework integrates the building of mathematical models and algorithms, and their implementation as software. The DIMA methodology represents a multi-disciplinary modelling framework, which meets the particular features of SC modelling. The approach creates a unified methodological basis of SC integrated modelling, from the conceptual level, mathematical modelling, up to algorithms and simulation tools.

The main principles of the DIMA are as follows:

- SC elements' activity;
- multiple modeling;
- multiple objective optimization;
- integration; and
- decentralization.

9.2 Active Modelling Objects

SCs are characterized by decentralization resulting in different goals and uncertain interactions of their elements. The SC elements are *active* (they can compete and have conflicting aims, interests and strategies). Their activities are based on their own interests and goals.

The preliminary investigations confirm that the most convenient concept for the formalization of SC control processes is the concept of an *active modelling object* (AMO). In a general case, it is an artificial object, moving in space and time and interacting (by means of information, financial or material flows) with other AMO and the environment. This idea is based on the concept of active moving objects (Kalinin and Sokolov 1985).

An AMO is based on the moving, interaction with the environment and other AMO, functioning of the main (goal-oriented) and auxiliary facilities, and resource consumption (replenishment). The joint execution of these functions, the interaction being the main one, provide for new AMO characteristics.

The proposed structure of an AMO can be interpreted widely. For example, the multi-agent ideology can be considered as a basis for modelling the active elements. In Ivanov *et al.* (2007), we introduced the specific description of these so-called agent-based AMO. The agents' description and interactions are shown in detail in Ivanov and Kaeschel (2003), Ivanov (2006) and Ivanov *et al.* (2007a,b).

We consider agents as part of the generic model constructions (Ivanov *et al.* 2007). Thus, the elements (enterprises) of models are active and goal-oriented, act autonomously, are reactive and adaptive, and communicate with other agents. They are part of a multidisciplinary complex of models used not only at the simulation stage, but also at the levels of conceptual modelling, formalization and mathematical modelling (see Fig. 9.3).

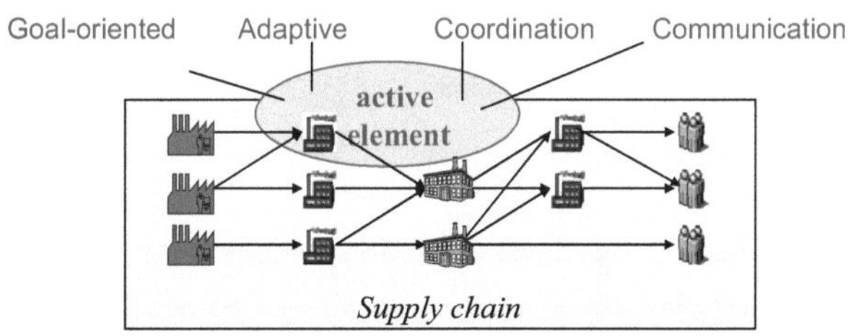

Fig. 9.3 Agents as active objects in SC models

We introduce the specific description of these active objects in terms of multi-agent theory. For formal representations of agents, three main functions are usually used (Kaihara 2004): the production function, profit function, and bidding function. The agents try to fill up the capacities of each competence so as to maximize the discrepancy between price and costs. In order to take into account so-called soft factors (e.g. reputation, trust, etc.), we also considered a reputation function of agents. The formalization of agents and the functional and interaction models have been considered in Ivanov (2006) and Ivanov *et al.* (2007a).

9.3 Integration Views

Integration is considered from four perspectives of the system modelling: the integration of various modelling approaches and frameworks, the integration of planning and execution models, the integration of decision-making levels, and the implementation of integration throughout: "conceptual model → mathematical model → computation" (see Figs. 9.4 and 9.5).

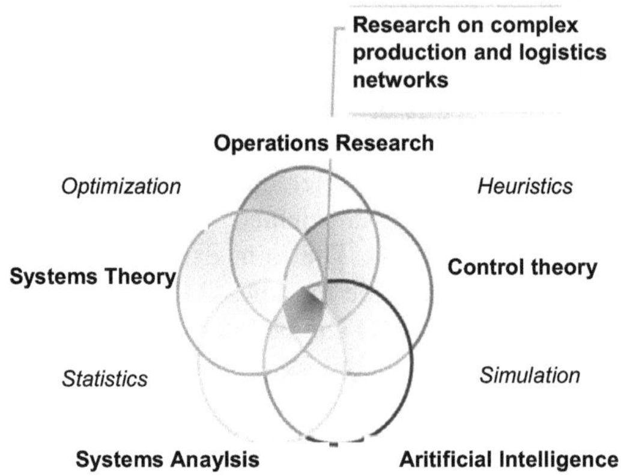

Fig. 9.4 Mathematical integration view in DIMA

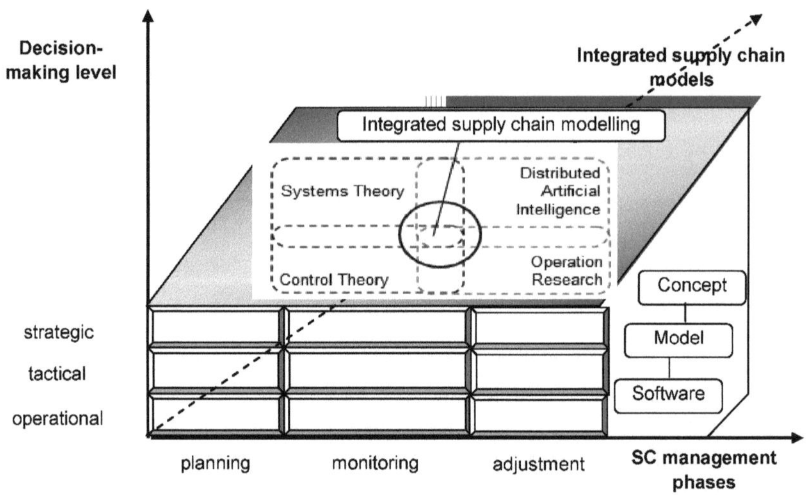

Fig. 9.5 Decision-making integration view of DIMA (from Ivanov *et al.* 2010)

SC strategy, design, planning and operations are interlinked while constructing ideal SC states as well as while reconfiguring SCs in relation to a current execution environment. Tables 9.1–9.3 depict the interlinking of problems and data at SC strategic, tactical, and operational levels (Ivanov 2009b).

Table 9.1 Classification of SC design decisions and decision-supporting methods

Strategic decisions	Input data	Output data	Decision-supporting methods
Background information: corporate strategy; SC strategy; financial strategy; marketing competitive strategy			
SC goals	Profit Assets Reliability Flexibility	How the optimal compromise of the SC goals can be achieved? How to deal with multiple criteria?	Analytical hierarchy process Pareto optimality Heuristics
Production programme design	Product variety Stock keeping units Bill-of-material Demand Response time Time-to-market	Which products, in what quantity, variety, batches, etc. to produce? Product availability Technological plan building	Linear programming System dynamics Ontology analysis Graph theory
Cooperation and coordination design	Levels of coordination: orders forecasting point-of-sale	How the enterprises will collaborate? What information systems must be used?	MAS Fuzzy logic Game theory
Distribution and production design	Location data (costs, geo-referenced data, taxes etc.) Demand Inventories Process data (capacities, costs, etc.) Movement data (transportation costs, time, mode, and capacity)	How many facilities of what capacities and of what location are needed? Suppliers' selection and their allocation to plants How the transportation should be organized? How to deal with demand uncertainty? How to secure the SC with regard to purposeful disruptions, e.g. theft or terrorism?	MIP Information modelling Dynamic programming Decision analysis Assignment methods System dynamics Discrete-events simulation Stochastic optimization MAS Heuristics

Table 9.2 Classification of SC planning decisions and decision-supporting methods

Tactical decisions	Input data	Output data	Decision-supporting methods
Background information: collaboration strategy; product structure; suppliers' structure; facilities' structure; distribution and production structure			
Distribution and production plans	Demand Costs Capacities Inventory Production volume	Demand forecasts Cycle and safety inventories Capacity utilization Product availability	Statistics and probability theory MIP Dynamic programming Queues theory MAS Evolutionary heuristics
Replenishment plans	Demand Costs Capacities Inventory Volume	Economic order quantity (EOQ)	Linear programming Statistics and probability theory
Shipment plans	Geographic data Transport data (costs, time, mode, capacity)	Routing	Combinatorial methods MAS Evolutionary heuristics

Table 9.3 Classification of SC operational and execution decisions and decision-supporting methods

Operative decisions	Input data	Output data	Decision-supporting methods
Background information: distribution, production, replenishment and shipment plans			
Scheduling	Delivery date, place and price Batch size	Production scheduling Routing	Simulation Evolutionary heuristics
Available-to-Promise / Capable-to-Promise (ATP/CTP)	Customer orders Inventories, capacities	ATP/CTP response	ATP/CTP algorithms
Monitoring	Demand / Supply Capacities Inventory Production volume	Comparison plan–fact	Operative analysis methods
Adjustment	Deviations and disruptions	Adaptation steps	Adaptive planning and control
Deliveries to customers	Delivered products Payments	Performance management	SC reference model (SCOR)

Let us consider some examples. Transportation and inventory are primary components of the order fulfilment process in terms of cost and service levels. Therefore, companies must consider the important interrelationships among transportation, inventory and customer service in determining their policies. The suppliers' selection is linked not only to their capacities, costs, etc. but also to their collaboration abilities with each other and with the focal enterprise. Therefore, coordination between the various players in the chain is the key to its effective management. Pricing and inventory decisions (Muriel and Simchi-Levi 2004) as well as product, distribution and production decisions are also matched together.

9.4 Multi-model Complexes and Qualimetry of Models

While describing a complex system mathematically, it is almost impossible to consider only one system model (Casti 1979). There exist a number of models and each of them reflects one particular domain of the system behaviour. Besides, each of the models has its own mathematical structure. Cross-linked SC planning and execution problems require the combined application of various modelling techniques (optimization, statistics, heuristics and simulation).

At different stages of the SC life cycle, a particular problem can be solved by means of *different modelling techniques* due to the changeability of the data nature, structure and values, as well as requirements for output representation. The selection of a solution method depends on the data fullness, problem scale, one or multiple objectives, requirements on output representation and the interconnection of a problem with other problems. Different approaches from the OR, control theory, and agent-based modelling have a certain application area and a certain solution procedure. The isolated application of only one solution method leads to a narrowing in problem formulation, overdue constraints and sometimes unrealistic or impracticable goals.

In the DIMA methodology, it is understood under multiple modelling that various modelling approaches like control theory, OR, systems analysis, agent-based modelling and the psychology of decision-making are not isolated, but are considered as a united modelling framework.

As discussed above, modelling adequacy cannot be ensured within a single model thus multi-model complexes should be used. Each class of models can be applied to the objects for analysis of their particular aspects (see Fig. 9.6). The integration and combined application of various models is implemented by means of multiple model complexes (Ivanov *et al.* 2007b, Ivanov 2009a), which are based on the application of functors (Mesarovich and Takahara 1975, Sokolov and Yusupov 2004). The coordinated application of different models improves modelling quality, as disadvantages of one model class are compensated for by advantages of other classes, which enable the examination of the determined classes of modelling tasks.

The *multiple-model complexes* allow problem examination and solution in different classes of models, and result representation in the desired class of models (the concept of "virtual" modelling). This becomes possible under the terms of collective application of structural–mathematical and categorical–functorial conceptions of the models' architecture. A study by Sokolov and Yusupov (2004) demonstrated the capabilities of the categorical–functorial approach to qualimetry of models by an example of a polymodel description. In Fig. 9.6, we provide an example of a multiple model complex for the SCM domain.

The *qualimetry* is rooted in the quality science and reflects the concept of comprehensive quality, which, according to the ISO 8402-2000 international standard, means a totality of characteristics of an object that determine its capability to satisfy the established or supposed requirements.

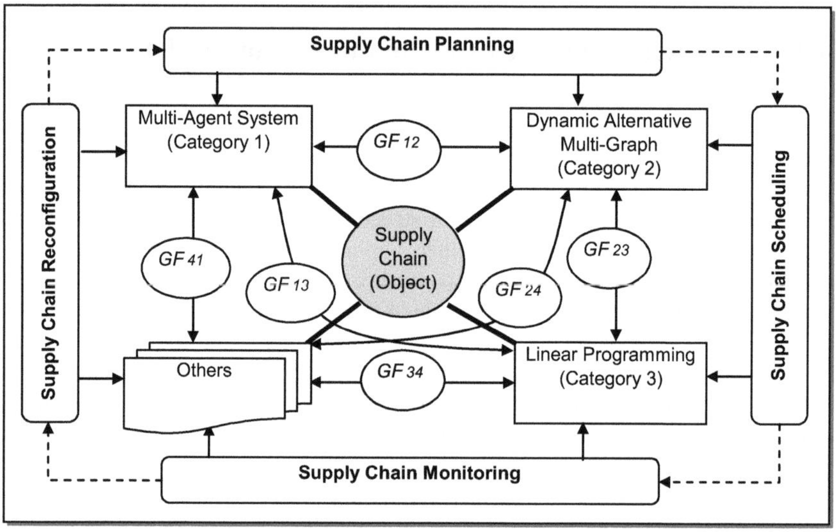

Fig. 9.6 Example of a multi-model complex (from Ivanov *et al.* 2007a)

A category is a mathematical construction that consists of a class of objects and a class of morphisms (arrows). Categories provide one of the most convenient methods for describing objects within the framework of the developed qualimetry of models since, *first*, the conceptual level of the representation of objects in the given theory has to be invariant relative to the method of description of their internal structure and, *second*, an object in a category is integral, since its internal structure is not considered. Finally, objects acquire properties only in a certain category and these properties reveal themselves only in comparison with other objects. Functors establish relationships between different categories. A one-place covariant functor can be characterized as a mapping from one category to another that preserves their category structure (Mesarovich and Takahara 1975).

Let us discuss an example of how to apply the multiple model complexes (see Fig. 9.7). The interconnection between different models is ensured by means of *functors* (GF). The problem of the SC analysis and synthesis is mostly formalized using either graph (network) models or models of *linear and integral program-*

ming. As a rule, the problem of analysing and synthesizing programmes for SC execution is formalized with the help of *dynamic models*. However, the problems of coordination and consistency of the results remain open. To obtain a constructive solution to these problems, we propose to use a functorial transition from the category of digraphs (*Cat*Φ) that specifies the models of execution of operations to the category of dynamic models (*Cat*D), which describes the processes of SC execution.

Cat Φ (category of static models $\Delta_{\Theta}^{(s)}, \Theta = 1,\ldots,\theta$)

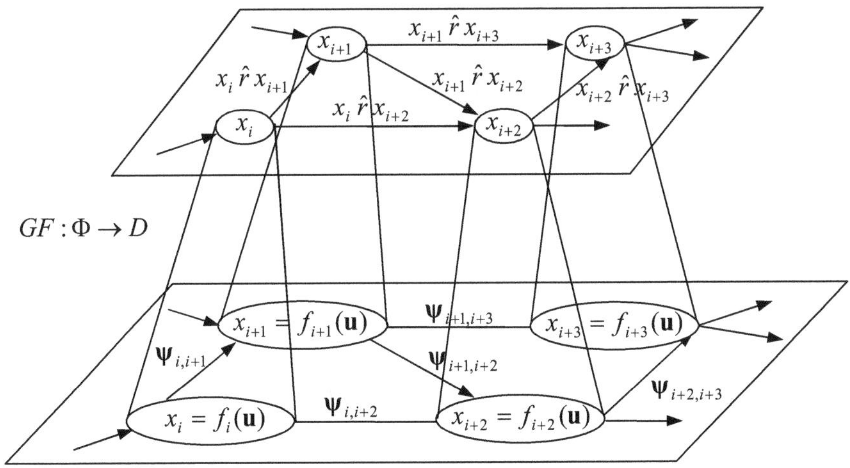

Cat D (category of dynamic models $\Delta_{\Theta}^{(d)}, \Theta = 1,\ldots,\theta$)

Fig. 9.7 Functorial transition from the category of static models in the category of dynamic models (from Okhtilev *et al.* 2006, Ivanov *et al.* 2007b)

The simplified mathematical model of the above-mentioned transition can be presented as follows:

$$\Delta^{(d)} = \Bigg\{ \mathbf{u} \,|\, \frac{dx_i}{dt} = \sum_{j=1}^{n} u_{ij}; \sum_{i=1}^{\overline{n}} u_{ij}(t) \le 1; \sum_{j=1}^{n} u_{ij} \le 1; u_{ij}(t) \in \{0,1\};$$

$$\sum_{j=1}^{n} u_{ij} \left[\sum_{\varepsilon \in \Gamma_{1i}^{-}} (a_\varepsilon - x_\varepsilon(t)) + \prod_{\varphi \in \Gamma_{i2}^{-}} (a_\varphi - x_\varphi(t)) \right] = 0; i = 1,\ldots,\overline{n};\ j = 1,\ldots,n, \qquad (9.1)$$

$$t \in (T_0, T_f] = T;\ x_i(T_0) = 0;\ x_i(T_f) = a_i \Bigg\}$$

where x_i is a variable characterizing the state of the process $\overline{B}^{(i)}$, u_{ij} is a control action ($u_{ij}(t) = 1$, if the resource $B^{(j)}$ is used for the process $\overline{B}^{(i)}$), a_i, a_ε, a_φ are given quantities (end conditions), values of which should have the corresponding variables $x_i(t)$, $x_\varepsilon(t)$, $x_\varphi(t)$ in the end of the planning interval at the instant of time $t = T_f$, t is the running instant of time, T_0 is the start instant of time of the planning horizon, T_f is the end instant of time of the planning horizon, T is a planning horizon, $\sum_{j=1}^{n} u_{ij} \sum_{\varepsilon \in \Gamma_{1i}^-} (a_\varepsilon - x_\varepsilon(t)) = 0$ are constraints "and" which mean the condition of the total processing of all the predecessor operations, $\sum_{j=1}^{n} u_{ij} \prod_{\varphi \in \Gamma_{i2}^-} (a_\varphi - x_\varphi(t)) = 0$ are constraints "or" which mean the condition of the processing of at least one of the predecessor operations and Γ_{1i}^-, Γ_{i2}^- are the sets of processes which immediate precede the process $B^{(i)}$.

Finally, let us propose the following generalized inter-model coordination via a choice structure with a multi-preference relation (Okhtilev *et al.* 2006). This structure allows the combination of analytical, simulation, and knowledge-based approaches to modelling monitoring and control in different applied areas (Kalinin and Sokolov 1985, 1987, 1996, Ivanov *et al.* 2004, Okhtilev *et al.* 2006):

$$\begin{aligned} \Big\{ & Q^{(\Theta)}(\bar{s},(\Omega,F,\bar{\lambda}_{\bar{\mu}}),\{\Delta_{\hat{\rho}}^{(\Theta)}\}_{\hat{\rho}\in\hat{I}_1},\{\Delta_{\hat{\eta}}^{0(\Theta)}\}_{\hat{\eta}\in\hat{I}_2},\{\bar{r}_{i_1}^{\alpha(\Theta)}(\omega)\}_{i_1\in\hat{\Gamma}},\{\bar{r}_{i_2}^{\beta(\Theta)}(\omega)\}_{i_2\in\hat{\Gamma}_1}, \\ & \{\overline{W}_{\hat{e}}\}_{\hat{e}\in\overline{\Phi}_1},\{\overline{W}_{\hat{k}}\}_{\hat{k}\in\overline{\Phi}_2},\{F^{\tilde{k}(\Theta)}(\omega)\}_{\tilde{k}\in\hat{\Gamma}_2}\Big\}_{\Theta\in\hat{I}}, \end{aligned} \tag{9.2}$$

where each mathematical structure $Q^{(\Theta)}(\bar{s},(\Omega,F,\bar{\lambda}_{\bar{\mu}})$ defines a class of selection models (mathematical, logic, and algebraic, deterministic or uncertain, etc.), Ω is a space of events (the set of uncertainly), F is a sigma-algebra over the space Ω, $\bar{\lambda}_{\bar{\mu}}$ is a measure over (Ω, F), $\{\Delta_{\hat{\eta}}^{0(\Theta)}\}_{\eta\in\Xi_2}$ is a collection of the main basic sets of alternatives (each basic set corresponds to a mathematical structure $Q^{(\Theta)}(\bar{s},(\Omega,F,\bar{\lambda}_{\bar{\mu}})$: $\{\Delta_{\hat{\rho}}^{(\Theta)}\}_{\hat{\rho}\in\hat{I}_1}$ is a set of auxiliary alternatives to be used mostly in coordination choice tasks, $\{\bar{r}_{i_1}^{\alpha(\Theta)}(\omega)\}_{i_1\in\hat{\Gamma}}$ is a set of preference relations to be used for the selection of the best alternatives via the structures $\{Q^{(\Theta)}\}_{\Theta\in\hat{I}}$, $\{\bar{r}_{i_2}^{\beta(\Theta)}(\omega)\}_{i_2\in\hat{\Gamma}_1}$ is a set of the relations defining constraints to be satisfied when an alternative is selected, $\{\overline{W}_{\hat{e}}\}_{\hat{e}\in\overline{\Phi}_1},\{\overline{W}_{\hat{k}}\}_{\hat{k}\in\overline{\Phi}_2}$ are constructions formed of basic sets via Cartesian products and the generation of subsets (the first construction corresponds to the input scale of choice and the second one corresponds to the output scale) and $\{F^{\tilde{k}(\Theta)}(\omega)\}_{\tilde{k}\in\hat{\Gamma}_2}$ is a set of rules for constructing the resulting choice functions and preferences relations.

The model construction (Eq. 9.2) allows us to approach from unified positions the issues of analysis and rational selection of solution methods with regard to the vector optimization problems as well as game, multi-stage, and group selection problems. In general, the proposed problem statement in terms of multiple objectives and uncertainty is composed of (1) building a number of alternative solutions and (2) selecting the alternative that returns an extreme value to the selection function. Based on the principles of *external amendments* and *non-final solutions*, it becomes possible to eliminate the criteria-driven and model-driven uncertainty and to transform the problem of uncertainty to its deterministic equivalent. This idea will be depicted in dynamic scheduling models in Chap.12.

The investigations have shown that, in the framework of the considered polymodel description, not only are the functionality conditions held, but also the conditions of the general position of the adjacency mapping (Sokolov and Yusupov 2004). The proposed general scheme of inter-model coordination needs detailed elaboration of the main classes of dynamic models, such as system dynamics, logical dynamics models, Petri nets and dynamic models of SC operations.

9.5 Decentralization

Decentralization in the DIMA methodology considers the main principle of management and decision-making in SCs. This means that the models contain elements of decentralized decision-making and SC elements' activity. Decisions about SCM are not established and optimized "from above" but are a product of iterative coordinating activities of the enterprises (agents) in a SC and a SC coordinator (e.g. an OEM or a 4PL provider) with possible *asymmetries and dominations*. Let us consider as an example the planning procedure for one customer order with a possibility of flexible SC structuring. The SC planning cycle is

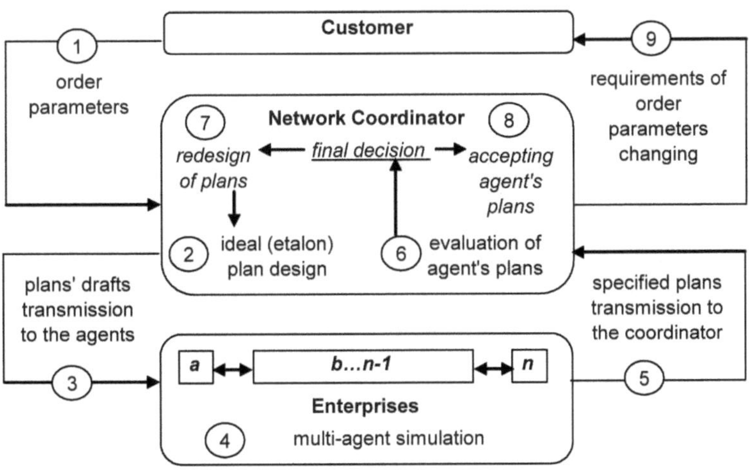

Fig. 9.8 SC planning with flexible partner selection (from Ivanov *et al.* 2010)

In this case, the SC planning problem consists of determining a pool of currently available partners in accordance with the technological details of the product (1), the synthesis and evaluation of alternative SC structures from a pool of selected executors according to the various project stages, as well as the selection and scheduling of the configured SC (2, 3). The SC planning is performed as an iterative process of matching the interests of the network coordinator and enterprises (4–9). In some cases, such balancing is impossible and it is necessary to change the customer requirements (10).

For generating optimal plans, methods of control theory, in particular dynamical system optimization, are used. In real-world settings, the selection of suppliers and scheduling must be performed under time pressure. That is why heuristics can be used to obtain timely, satisfactory results. After that, the communication stage starts. The agents evaluate the proposals of the coordinator and generate their own propositions, which can be either accepted by the coordinator (in comparison with the ideal and heuristics plans) or rescheduled for further balancing. The integration of optimization and heuristics techniques allows estimating the quality of agents' heuristic decisions regarding the optimal solution.

9.6 Integrated Modelling for Supply Chain Adaptive Planning and Execution

The large variety of SCM issues can be classified into the subclasses of SC analysis and synthesis. The models of SC analysis can be divided into SC design analysis as well as SC operative monitoring (see Fig. 9.9).

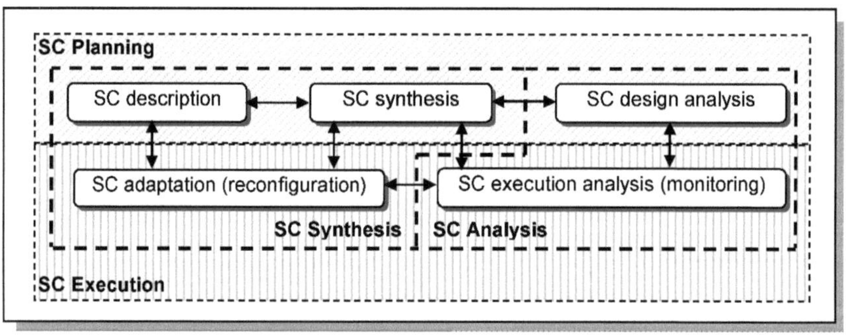

Fig. 9.9 Complex of conceptualized models for the problem of the SC dynamical structural-functional synthesis and reconfiguration (adopted with changes from Ivanov 2009a)

The models of SC synthesis are composed of SC configuration and reconfiguration models. Most of the SC problems are cross-linked (i.e. models of SCMo and reconfiguration, static and dynamic models of the SC (re)configuration, SC synthesis and analysis models). Conceptual, mathematical and information models are also interconnected with each other.

The SCs must be configured according to the project goals and reconfigured in dynamics according to the current execution environment. More typically, questions are centred on rationalizing the SCs in response to permanent changes in the SC itself and its environment (Harrison 2005). The general modelling structure of the SC dynamical structural–functional synthesis and reconfiguration is shown in Fig. 9.10 (Ivanov 2006).

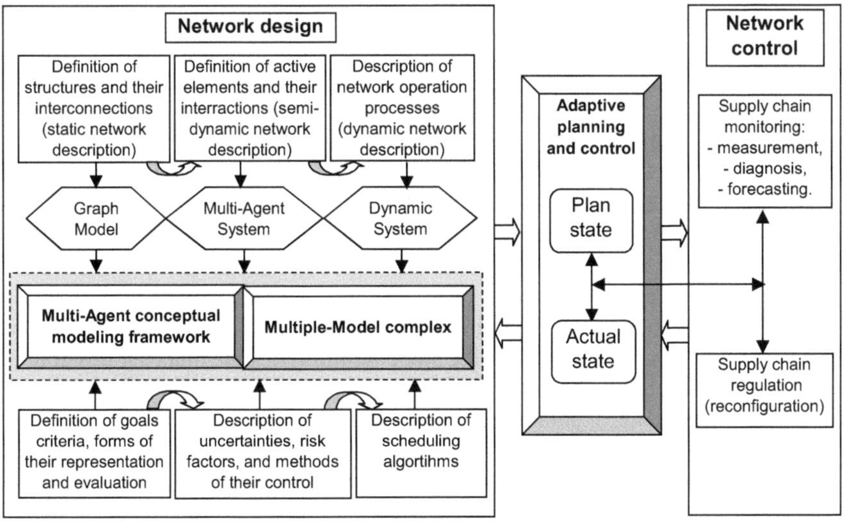

Fig. 9.10 The general modelling schema of the SC dynamical structural-functional synthesis and reconfiguration (from Ivanov *et al.* 2007b)

Let us consider the main steps of this schema referring also to the main phases of the SCM in Fig. 9.10. The modelling starts with the static graph-theoretical network description. The SC can be described graph-theoretically as a directed graph — a digraph. Then the elements of the organizational graph (enterprises) are described as active agents in terms of multi-agent theory. So the model of enterprise interactions can be constructed. In this stage, we combine graph-theoretical modelling with agents to describe active elements of the graph as well as to implement modelling dynamics of the SC objects' collaboration.

The goal of the coordinator consists of project description and structuring according to the proper level of decomposition. Any project can be presented as several consecutive and/or parallel operations. In complex projects, different operations have different degrees of importance to the final result. That is why the importance level of each operation must be taken into account. When a coordinator is in charge of several projects, there may be situations when the same competency is needed for different projects at the same time. These projects should not be considered independently, because they compete for the same resource. In order to analyse such "linked" projects together, we propose to present them as a joined structure scheme that we call a technological network. The model of the technological network is an oriented graph. Its heads are considered competencies that are necessary for project realization, and its edges serve to show the logical sequence of operations. The weights of each head reflect the volume of the competency that is needed for the project's performance (e.g. the total working time to execute a concrete technological operation on a concrete machine). Finally, at the end of this step, the coordinator obtains properly structured projects that are considered as a technological network. The next step is dynamic network description. In this stage, a set of interlinked dynamic models (see Chaps. 10 and 12) is formed to link the "ideal" planning results and the SC execution programmes under uncertainty.

The models of SC execution are comprised of SCMo and SC regulation (reconfiguration) models. SCMo is based on the monitoring of the SC macro-structural macro-states (MSMS). The monitoring plan execution consists of determining diagnostics moments (critical control points), when the analysis of the planned and factual parameters of the SC execution (demand, inventories, jobs' starting and ending, and stability) is carried out. The particular feature of the SC monitoring in terms of macro-states is that, at each monitoring stage, the control parameters are extracted from the parameter vector of the DAMG. The mathematical description of the DAMG has been considered in Okhtilev *et al.* (2006). The extracting rules depend on the management goals at the stage monitored. This makes it possible to consider all of the SC execution parameters described in the DAMG, and to extract the necessary control parameters in the current execution situation.

SC reconfiguration (real-time replanning) is comprised of deviations analysis, the elaboration of compensating control actions and the construction of a new plan and production of appropriate correcting actions for the transition from the actual SC state trajectory to the planned one at a given time interval or by the final time. The model of SC reconfiguration is interconnected to the planning model. It is also based on the dynamical alternative multi-graph. While selecting a new SC, it is also essential to take into account a number of specific requirements (preferences of the SC focal enterprise or 4PL provider and suppliers at different SC levels). The essence of decision making about a SC reconfiguration, that is to say about a SC plan transition from a current state to a desired state, is to ensure that the agents' interests and the interests of a SC coordinator comply with each other.

References

Barbasi AL (2005) Network theory - the emergence of the creative enterprise. Science 308:639–641

Beamon BM (1998) Supply chain design and analysis: models and methods. Int J of Prod Econ 55(3):281–294

Casti JL (1979) Connectivity, complexity and catastrophe in large-scale systems. Wiley-Interscience, New York and London

Harrison TP (2005) Principles for the strategic design of supply chains. In: Harrison TP, Lee H L, Neale JJ (Ed) The practice of supply chain management. Kluwer academic publishers, Boston

Ivanov D (2006) DIMA – Decentralized Integrated Modeling Approach. Ein Ansatz zur interdisziplinären Modellierung von Produktions- und Logistiknetzwerken. Verlag der GUC, Chemnitz

Ivanov D (2009a) DIMA – a research methodology for comprehensive multi-disciplinary modelling of production and logistics networks. Int J Prod Res 47(5):1133–1155

Ivanov D (2009b) An adaptive framework for aligning (re)planning decisions on supply chain strategy, design, tactics, and operations. Int J Prod Res, DOI: 10.1080/00207540902893417, in press

Ivanov D, Kaeschel J (2003) Vernetzte Planung und kontinuierliche Lieferkettenoptimierung. PPS-Management 4:29–32

Ivanov D, Arkhipov A, Sokolov B (2004) Intelligent Supply Chain Planning in Virtual Enterprises. In: Camarihna-Matos LM (Ed) Virtual Enterprises and Collaborative Networks. Kluwer academic publishers, Boston

Ivanov D, Kaeschel J, Arkhipov A, Sokolov B and Zschorn L (2005) Quantitative models of collaborative networks. In: Camarihna-Matos LM, Afsarmanesh H, Ortiz A (Ed) Collaborative networks and their breeding environments. Springer, New York

Ivanov D, Kaeschel J, Arkhipov A, Sokolov B (2006) A conceptual framework for modeling complex adaptation of collaborative networks. In: Proceedings of the 7th international conference on virtual enterprises. Helsinki, 25-27 September.

Ivanov D, Arkhipov A, Sokolov B (2007a) Intelligent planning and control of manufacturing supply chains in virtual enterprises. Int J Manuf Tech Manag 11(2):209–227

Ivanov D, Kaeschel J, Sokolov B (2007b) Integrated modeling of agile enterprise networks. International Journal of Agile Systems and Management 2(1):23–49

Ivanov D, Sokolov B, Kaeschel J (2010) A multi-structural framework for adaptive supply chain planning and operations control with structure dynamics considerations. Eur J Oper Res. 200: 409–420

Kaihara T (2004) Enterprise negotiation algorithm with walrasian virtual market. In: Camarihna-Matos L (Ed) Virtual enterprises and collaborative networks. Kluwer Academic Publishers, Boston

Kalinin VN (1982) The problem of optimal control of active moving objects. Differ Equat 17(2):1350–1355

Kalinin VN, Sokolov BV (1985) Optimal planning of the process of interaction of moving operating objects. Differ Equat 21(5):502–506

Kalinin VN, Sokolov BV (1987) A dynamic model and an optimal scheduling algorithm for activities with bans of interrupts. Autom Rem Contr 48(1-2):88–94

Kalinin VN, Sokolov BV (1996) Multiple-model description of control processes for airspace crafts. J Comput Syst Sci Int 6:192–199

Kuehnle H (2008) A system of models contribution to production network (PN) theory. J Intell Manuf 18(5):543–551

Mesarovic MD and Takahara Y (1975) General systems theory: mathematical foundations. Academic Press, New York, Can Francisco, London.

Muriel A and Simchi-Levi D (2004) Supply chain design and planning – application of optimization techniques for strategic and tactical models. In: Kok de AG, Graves SC (Ed) Supply chain management: design, coordination and operation. Amsterdam, Elsevier

Okhtilev M, Sokolov B, Yusupov R (2006) Intelligent technologies of complex systems monitoring and structure dynamics control. Nauka, Moscow (in Russian)

Sokolov BV, Yusupov RM (2004) Conceptual foundations of quality estimation and analysis for models and multiple-model systems. J Comput Syst Sci Int 6:5–16

Chapter 10
Structure Dynamics Control and Multi-model Analysis

It is not the strongest of the species that survive,
nor the most intelligent, but the one most responsive to change.
Charles Darwin

The trouble with our times is that
the future is not what it used to be.
Paul Valery

10.1 On the Control Approach: Conceptual and Mathematical Issues

In recent years, the work on SCM has been broadened to cover SC dynamics. In these settings, control theory is becoming of even greater interest to researchers and practitioners (Disney and Towill 2002, Braun *et al.* 2003, Anderson *et al.* 2006, Disney *et al.* 2006, Sethi and Thompson 2006, Lalwani *et al.* 2006, Ivanov *et al.* 2007, 2009, van Houtum *et al.* 2007, Wang et al., 2007).

This study joins this research stream. It considers SC planning and scheduling from the perspectives of adaptable, stable, and analysable plans and schedules that enable the achievement of management goals instead of "ideal" optimal schedules that fail in a real perturbed execution environment. In the approach presented, the SC optimization will be considered from the perspectives of the entire value chain and as a function of the achievement of management goals. The approach aims to provide advanced insights into SC dynamics and constructive ways to transit from simple open time slots incremental planning to dynamic, feedback-based adaptive and integrated SC planning and scheduling to implement adaptability, stability, and crisis-resistance throughout the value chain.

In Chap. 8, we have considered control theory and revealed its general advantages and disadvantages with regard to the SCM domain. Here we would like to continue this discussion on a more detailed level. We will consider the issues in the application of the control approach to the SCM domain from two perspectives: conceptual and mathematical.

Conventionally, the conceptual basics of control theory lead to the fields of automatic control, signal identification, and automatic regulation of technical systems. Unquestionably, the optimal control and systems analysis provide a number of advanced insights into the dynamics, stability, adaptability, non-linearity, and high-dimensionality of complex systems. However, these techniques have not been widely applied to complex business systems. There are some reasons for this.

First, the decision-making in business systems is of a discrete nature. In technical control systems, it is assumed that the control u can be selected continuously in time. The problem becomes even more complex as, though the decision-making in business systems is discrete, the processes and flows nevertheless remain continuous. Hence, the mathematical models of classical optimal control need domain-specific modifications. As the mathematics of the optimal control is very complex, these modifications can only be made by specialists who can deal with both differential equations and business processes.

Second, the mathematics of optimal control also has its limitations. In the 1960–1970s, significant advances were made in optimal control techniques (Pontryagin 1961, Lee and Markus 1967, Moiseev 1974, Bryson and Ho 1975). Beginning from the formulation of explicit links between resources and operations, the logic of operations execution has also been introduced into the models. The non-linearity has been considered in the right parts of the differential equations. The main problem was caused by the step functions (Moiseev 1974, Bryson and Ho 1975, Aida-Zade 2005) and the arising sectionally continuous functions. The fact that the derived function from the step function is infinity has negatively influenced the further development of optimal control techniques for planning and scheduling in complex systems.

Third, the narrow understanding of control as a regulation function has often negatively influenced researchers in the application of control theory to management domains. In fact, control theory and related disciplines, such as systems analysis, provide enough concepts and tools to consider control not only in this narrow interpretation, but also in a wide interpretation as a function that integrates the planning and execution of complex systems. In this wide interpretation, the control becomes closely related to the management while automatic decisions are integrated with human decision-making.

Summarizing, our investigations have shown that the tasks of structural–functional synthesis and the analysis of SCs cause the following problems:

- the problem of high dimensionality and non-linearity of models that describe SC structures and operations;
- the problem of describing uncertainty factors; and
- the problem of multi-objective decision-making.

In this chapter as well as in the following chapters, we will take up the above-mentioned challenges and develop novel frameworks, models, and solvers. *First*, we will consider planning and scheduling as an integrated function within an adaptive framework. *Second*, we will formulate the planning and scheduling models as optimal control problems, taking into account the discreteness of decision-making and the continuity of flows. By special techniques for process dynamics models and constraint formulation, e.g. by transferring the non-linearity into the left part of the differential equations, we will show how to transform the non-linear operations dynamic model into a linear one. In doing so, the dimensionality of problems can be reduced and discrete optimization methods of linear programming can be applied for solution within the general dynamic non-linear model. For solving the problem, Pontryagin's maximum principle will be applied and the Lagrange multipliers will be presented in dynamic form. The mathematical aspects will be considered in more detail in related models in the following chapters.

In concluding this section, we would like to emphasize that the main motivation for this research approach is to combine the possibilities of different decision-making techniques, such as OR, control theory, systems analysis, and agent-based modelling to achieve new quality in the decision-making support, e.g. in applying the proved fundamentals of control theory to the SCM domain, the conventional OR-based modelling techniques for SCM can be enriched by new viewpoints on dynamics, stability, adaptability, consistency, non-linearity, and high-dimensionality of the complex system.

The mathematics of the optimal control can help in revealing new conformities to natural laws that remain unrevealed within the OR field. Hence, the conventional SCM problems may be considered from a different viewpoint and new problems may be revealed and formulated.

On the other hand, the optimal control models may serve as an orientation for assessing the solution qualities of agents and heuristics as well as for managing SCs in dynamics by following optimization guidelines (this does not ultimately presume the finding of a strong optimum) rather than by relying on methodically weak-grounded approaches.

10.2 Basics of Structure Dynamics Control

In this chapter, the general technology of SC tactical–operational planning and the technology of SDC will be considered, which include the following stages:

- structural–functional synthesis of the SC structures, plans, and schedules;
- structural and parametric adaptation of planning models and algorithms to the past and present states of the SC and to the past and present states of the environment; and
- SC scheduling, control programme construction for SC structure dynamics, simulation of possible scenarios of SC functioning according to the schedule, and structural and parametric adaptation of the schedule, models, and algorithms to future SC and environment states (predicted via simulation models)

The unified description of various control processes lets one synthesize simultaneously different SC structures. The proposed approach lets one establish dependence relation between control technology applied to SCs and the SCM goals. The mathematical models presented in this paragraph are an extended application for the SCM domain of the models for complex technical systems (Kalinin and Sokolov 1985, 1987, 1996, Skurikhin *et al.* 1989, Sokolov and Yusupov 2004; Okhtilev *et al.* 2006).

One of the main SC features is the multiple structure design and changeability of structural parameters because of objective and subjective factors at different stages of the SC life cycle. In other words, SC structure dynamics are constantly encountered in practice (see Fig. 10.1).

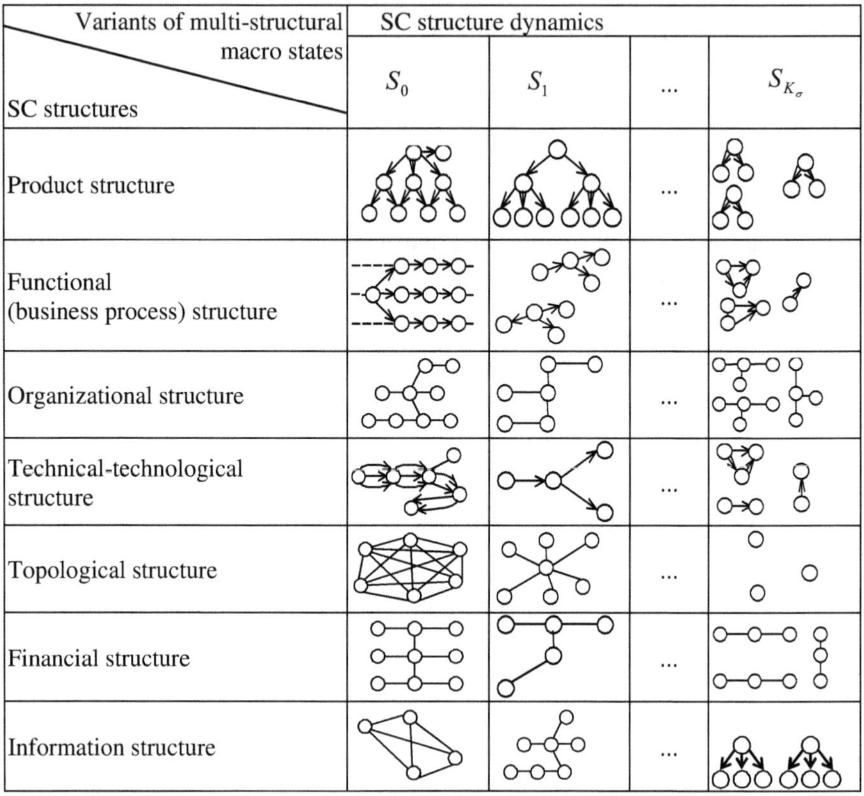

Fig. 10.1 SC multi-structural composition and structure dynamics (from Okhtilev *et al.* 2006)

In Fig. 10.1, S is an SC *multi-structural macro-state* and δ is a current number of the SC multi-structural macro-states in dynamics $\delta = 1,...,K_\sigma$. The multi-structural macro-state of an SC is composed of different structures and their interrelations.

At different stages of the SC evolution, the elements, parameters, and structural interrelations change. In these settings, an SC can be considered a *multi-structural process* (Ivanov 2009, Ivanov *et al.* 2010).

The main SC structures are the following:

- product structure (bill of materials);
- functional (structure of management functions and business processes);
- organizational (structure of facilities, enterprises, managers, and workers);

- technical–technological (structure of technological operations for production and structure of machines, technical devises, etc.);
- topological (geographical) structure;
- information (structure of information flows according to the coordination strategy); and
- financial (structure of SC costs and profit centres).

In the theory of SDC (Okhtilev et al., 2006), the control is composed of both state and structure control. The proposed approach to the problem of SC control in the terms of general context of SC SDC enables:

- common goals of SC functioning to be directly linked with those implemented (realized) in SC control process;
- a reasonable decision and selection of adequate consequence of problems solved and operations fulfilled related to structural dynamics to be made (in other words to synthesize and develop the SC control method); and
- a compromise distribution (trade-off) of a restricted resources appropriated for SDC to be found voluntary.

The approach to SDC problems is based on the functional–structural approach for describing objects of any different nature. At the same time the problems of SDC are the generalization of structural–functional synthesis problems, which are traditionally formulated in automation of complex technical-organizational systems.

Let us introduce the main definitions.

SC *macro-state* is a general SC state in which one or a number of SC objects can function.

Structural state is an SC macro-state that reflects the current states of objects in an SC structure as well as interrelations between these objects.

Multi-structural macro-state is an SC macro-state that reflects the current states of objects and structures in SCs as well as interrelations between them.

Structure dynamics is a process of SC structure transition from one planned macro-state to another.

Structure dynamics control is a process of producing control inputs and implementing the SC transition from the current macro-state to a planned one.

According to the specifics of the SDC problems, they belong to the class of the SC structural–functional synthesis problems and the problems of programme construction for SC development. The main disadvantage of the problems belonging to the above class is that optimal control programmes for SC main elements and subsystems can be implemented only when the lists of functions and algorithms for control and information processing in these subsystems and elements are known. The distribution of the functions and algorithms among the SC elements and subsystems, in its turn, depends upon the actual control laws for these elements and subsystems. The described contradictory situation is complicated by the changes of SC parameters and structures caused by different factors during the SC life cycle. Currently, the class of problems being reviewed is not examined thoroughly enough.

The synthesis (selection) of technical–organizational structure (in our case, an SC structure (structures)) is usually reduced to the following general optimization problem (Zvirkun and Akinfeev 1993):

$$\overline{S}\left\{\left[\bar{f} \subset \overline{F}(\overline{\pi})\right]\overline{R}\left[\overline{m} \subset \overline{M}\right]\right\} \rightarrow extr\,, \tag{10.1}$$

$$\overline{\pi} \subset \overline{P}\,, \tag{10.2}$$

$$\bar{f} \subset \overline{F}(\overline{\pi})\,, \tag{10.3}$$

$$\overline{m} \subset \overline{M}\,, \tag{10.4}$$

where $\overline{P}$ is a set of feasible control algorithms and $\overline{F}$ is a set of interrelated functions (tasks, operations) that can be performed by the system. For each subset $\overline{\pi} \subset \overline{P}$, there exists the set $\overline{F}(\overline{\pi})$ from which the realizations sufficient for the given principles (algorithms) should be chosen, i.e. it is necessary to choose $\bar{f} \subset \overline{F}(\overline{\pi})$, $\overline{M}$ is a set of SC possible elements, and the map $\overline{R}$ takes $\overline{F}$ to $\overline{M}$.

It is stated that the optimal map $\overline{F}$ returns an extremum to some objective function (functions) $\overline{S}$ under given constraints. The modifications of the considered problem concern the aspects of uncertainty and multi-criteria decision-making. The complexity of the synthesis problem (Eqs. 10.3 and 10.4) is mainly caused by its high dimensionality that is by the number of variables and constraints in the detailed problem statement. That is why the methods of decomposition, aggregation and sub-problem coordination are widely used. Another peculiarity complicating the problem is the integer-valued variables.

In this chapter and in Chap. 12, we will present a complex of SC control models. These models give unified technology for an analysis and optimization of various processes concerning SC planning and execution. The main advantage of the constructed models is that the structural and functional constraints for SC control are defined explicitly. The common conceptual basis facilitated the construction of a complex of unified dynamic models for SC control. The models describe the functioning SC along with the collaboration processes within them. The unified description of various control processes allows us to synthesize simultaneously different SC structures. Moreover, it lets us establish the dependence relationship between the control technology applied to SCs and the SCM goals. This statement is exemplified by an analysis of SC functional abilities and goal abilities. It is important that the presented approach extends new scientific and practical results obtained in the modern control theory into the SCM domain. By now, the prototype programme realizations of models have been developed and used for the evaluation of SC goal abilities and for the planning of SC operations (see Chap. 15).

10.3 A Multi-Model Description of Supply Chain Control Processes

Let us define data structures and interrelations between the data of different SC structures. For this purpose, we used the CASE tools, special languages, such as UML, and systems dynamics and ontology analysis. Because these are large-scale data models, we introduce only an extract here.

Class 1. Organizational structure: structure of enterprises, management departments, and workers (structure #1).

Subclass 1.1. Structure of enterprises: competencies, location, etc.

Subclass 1.1.1. Competencies: capacities, costs, reliability, quality.

Subclass 1.1.2. Collaboration of enterprises.

Class 2. Business process structure: coordinating parameters (demand, inventories, or orders), operations (distribution, production, replenishment; matched with subclasses), functions (in relations with the management departments) (structure #2).

Class 3. Product structure: product variety, demand, bill of material, etc. (structure #3).

Class 4. Technological structure: operations, machines (in relation to the technical devises of the subclass 1.1), quality data, etc. (structure #4).

Class 5. Topological structure (locations, movements, etc.) (structure #5).

Class 6. Financial structure (costs in correspondence to the classes 1–5) (structure #6).

Let us introduce the following basic sets and structures:

$B=\{B^{(j)},\ j\in N=\{1,\ldots,n\}\}$ is a set of internal objects, e.g. enterprises that are embodied in an SC and are necessary for its functioning;

$\overline{B}=\{\overline{B}^{(i)}, i\in\overline{N}=\{1,\ldots,\overline{n}\}\}$ is a set of external objects (customers, shareholders, creditors, logistics service providers) interacting with the SC (the interaction may be informational, financial or material);

$\widetilde{B}=B\cup\overline{B}$ is a set of objects in the SDC;

$\widetilde{C}=C\cup\overline{C}=\{C_1,C_2,\ldots,C_n\}\cup\{\overline{C}_1,\overline{C}_2,\ldots,\overline{C}_{\overline{n}}\}$ is a set of channels that are used for informational interaction;

$D=\left\{D_{\mu}^{(i)},\mu\in K_i^{(o)}=\{1,\ldots,s_i\},\ i=1,\ldots,\overline{n}\right\}$ is a set of interaction operations with the object $B^{(i)}$;

$\Phi R=\left\{\{\Phi S_{\pi}^{(j)}\}\cup\{\Phi N_{\pi'}^{(j)}\}, j\in N,\ \pi\in K_j^{(p,1)}=\{1,\ldots,k_j^{(p,1)}\},\pi'\in K_j^{(p,2)}=\{1,\ldots,k_j^{(p,2)}\}\right\}$ is a set of SC resources;

$\Phi S^{(j)}=\left\{\Phi S_{\pi}^{(j)},\pi\in K_j^{(p,1)}\right\}$ is a set of non-storable resources (time) of the object $B^{(j)}$;

$\Phi N^{(j)}=\left\{\Phi N^{(j)},\pi'\in K_j^{(p,2)}\right\}$ is a set of storable resources (materials) of the object $B^{(j)}$;

$P=\left\{\{P^{(j)}_{<\bar{\mu},\bar{\rho}>}\}\cup\{P^{(i,j)}_{<\bar{\mu},\bar{\rho}>}\}\right\}$ is a set of SC flows that are under consumption for different resources (financial flows, material flows and information flows);
$P^{(j)}=\left\{\{P^{(j)}_{<\bar{\mu},\bar{\rho}>}\}, j\in N, \bar{\mu}\in\widetilde{K}^{(o)}_i=\{1,...,\bar{s}_i\}, \bar{\rho}\in\widetilde{K}^{(f)}_i=\{1,...,\bar{p}_i\}\right\}$ is a set of flows produced by or necessary for the object $B^{(j)}$;
$P^{(i,j)}=\left\{\{P^{(i,j)}_{<\mu,\rho>}\}, i\in\overline{N}, j\in N, \mu\in K^{(o)}_i=\{1,...,s_i\}, \rho\in K^{(f)}_i=\{1,...,p_i\}\right\}$ is a set of flows (informational, financial or material) produced when the objects $B^{(i)}$ and $B^{(j)}$ interact.

Let $G=\left\{G_\chi, \chi\in NS\right\}$ be the set of structures that are being formed within the SC. To interconnect the structures let us consider the following DAMG:

$$G^t_\chi=\left\langle X^t_\chi, F^t_\chi, Z^t_\chi\right\rangle, \tag{10.5}$$

where the subscript χ characterizes the SCD structure type, $\chi\in NS=\{1,2,3,4,5,6\}$, the time point t belongs to a given set T; $X^t_\chi=\{x^t_{<\chi,l>}, l\in L_\chi\}$ is a set of elements of the structure G^t_χ (the set of DAMG vertices) at the time point t; $F^t_\chi=\{f^t_{<\chi,l,l'>}, l,l'\in L_\chi\}$ is a set of arcs of the DAMG G^t_χ and represent relations between the DAMG elements at time t; $Z^t_\chi=\{z^t_{<\chi,l,l'>}, l,l'\in L_\chi\}$ is a set of parameters that characterize relations numerically.

The graphs of different types are interdependent thus, for each operation the following maps should be constructed:

$$MM^t_{<\chi,\chi'>}: F^t_\chi\to F^t_{\chi'}. \tag{10.6}$$

Composition of the maps can be also used at time t:

$$MM^t_{<\chi,\chi'>}=MM^t_{<\chi,\chi_1>}\circ MM^t_{<\chi_1,\chi_2>}\circ\ldots\circ MM^t_{<\chi'',\chi'>}. \tag{10.7}$$

A multi-structural state can be defined as the following inclusion:

$$S_\delta\subseteq X^t_1\times X^t_2\times X^t_3\times X^t_4\times X^t_5\times X^t_6,\ \delta=1,...,K_\sigma. \tag{10.8}$$

Now we obtain the set of the SC multi-structural macro-states in dynamics:

$$S=\{S_\delta\}=\{S_1,...,S_{K_\sigma}\}. \tag{10.9}$$

Allowable transitions from one multi-structural state to another one can be expressed by means of the maps below:

$$\Pi^{t}_{<\delta,\delta'>} : S_{\delta} \rightarrow S_{\delta'} . \quad (10.10)$$

Here we assume that each multi-structural state at time $t \in T$ is defined by a composition (Eq. 10.6).

Now, the problem of SC with structure dynamics considerations can be regarded as a selection of multi-structural macro-states $S^{*}_{\delta} \in \{S_1, S_2, ..., S_{K_\sigma}\}$ and transition sequence (composition) $\Pi^{t_1}_{<\delta_1,\delta_2>} \circ \Pi^{t_2}_{<\delta_2,\delta_3>} \circ ... \circ \Pi^{T_f}_{<\delta',\delta>}$ $(t_1 < t_2 < ... < T_f)$, under some criteria of effectiveness, e.g. service level and costs.

Dynamics of the SC execution is presented as a DAMG to relate the above sets and structures. The DAMG is characterized by MSMS. The DAMG and the MSMS have been developed to meet the requirements on multi-structural design and to link planning and execution models, taking into account the structure dynamics.

In integrating the operation dynamics model and the structure dynamics model, the following general model construction can be presented. The goal is to find such $< U^{t}_{*}, S^{*T_f}_{\delta} >$ on the following constraints:

$$J_{\zeta}\left(X^{t}_{\chi}, F^{t}_{\chi}, Z^{t}_{\chi}, MM^{t}_{<\chi,\chi'>}, \Pi^{t}_{<\tilde{\delta},\tilde{\tilde{\delta}}>}, t \in (T_0, T_f]\right) \rightarrow \underset{<U^{t}, S^{*T_f}_{\delta}> \in \Delta^{(d)} \bigcup \Delta^{(s)}}{extr}, \quad (10.11)$$

$$\begin{aligned}\Delta^{(d)} \bigcup \Delta^{(s)} = \Big\{ &< U^{t}, S^{T_f}_{\delta} > \Big| R_{\tilde{r}}\left(X^{t}_{\chi}, F^{t}_{\chi}, Z^{t}_{\chi}, MM^{t}_{<\chi,\chi'>}, \Pi^{t}_{<\tilde{\delta},\tilde{\tilde{\delta}}>}\right) \le \tilde{\tilde{R}}_{\tilde{r}}; \\ &U^{t} = \Pi^{t_1}_{<\delta_1,\delta_2>} \circ \Pi^{t_2}_{<\delta_2,\delta_3>} \circ ... \circ \Pi^{t_2}_{<\tilde{\delta},\delta>} \Big\}\end{aligned}, \quad (10.12)$$

where U^{t} are control actions for synthesis, J_{ζ} are SC performance metrics (costs, service level, etc.), $\zeta \in \{1, ..., \Im\}$ is a set of the performance metric numbers, $\Delta^{(d)} \bigcup \Delta^{(s)}$ is a set of dynamic and static alternatives of SCs, $\tilde{r} \in \{1, ..., \tilde{R}\}$ is a set of business and information processes constraints numbers, $R_{\tilde{r}}$ is a set business and information processes constraints; $\tilde{\tilde{R}}_{\tilde{r}}$ are constants, which are known and $T = (T_0, T_f]$ is interval of time for SC synthesis. Other symbols have been explained above.

10.4 General Formal Statement of the Supply Chain Structure Dynamics Control Problem

Let us consider the generalized formal statement of the SC SDC problem. This statement is introduced to work out in detail the description (Eqs. 10.11 and 10.12). First we introduce basic sets and vectors:

$$
\begin{aligned}
&\mathbf{x}(t) = \left\| \mathbf{x}^{(g)\mathrm{T}}(t), \mathbf{x}^{(k)\mathrm{T}}(t), \mathbf{x}^{(o)\mathrm{T}}(t), \mathbf{x}^{(p)\mathrm{T}}(t), \right.\\
&\left. \mathbf{x}^{(f)\mathrm{T}}(t), \mathbf{x}^{(e)\mathrm{T}}(t), \mathbf{x}^{(c)\mathrm{T}}(t), \mathbf{x}^{(\nu)\mathrm{T}}(t) \right\|^{\mathrm{T}},\\
&\mathbf{y}(t) = \left\| \mathbf{y}^{(g)\mathrm{T}}(t), \mathbf{y}^{(k)\mathrm{T}}(t), \mathbf{y}^{(o)\mathrm{T}}(t), \mathbf{y}^{(p)\mathrm{T}}(t), \right.\\
&\left. \mathbf{y}^{(f)\mathrm{T}}(t), \mathbf{y}^{(e)\mathrm{T}}(t), \mathbf{y}^{(c)\mathrm{T}}(t), \mathbf{y}^{(\nu)\mathrm{T}}(t) \right\|^{\mathrm{T}},\\
&\mathbf{u}(t) = \left\| \mathbf{u}^{(g)\mathrm{T}}(t), \mathbf{u}^{(k)\mathrm{T}}(t), \mathbf{u}^{(o)\mathrm{T}}(t), \mathbf{u}^{(p)\mathrm{T}}(t), \right.\\
&\left. \mathbf{u}^{(f)\mathrm{T}}(t), \mathbf{u}^{(e)\mathrm{T}}(t), \mathbf{u}^{(c)\mathrm{T}}(t), \mathbf{u}^{(\nu)\mathrm{T}}(t) \right\|^{\mathrm{T}},\\
&\xi(t) = \left\| \xi^{(g)\mathrm{T}}(t), \xi^{(k)\mathrm{T}}(t), \xi^{(o)\mathrm{T}}(t), \xi^{(p)\mathrm{T}}(t), \right.\\
&\left. \xi^{(f)\mathrm{T}}(t), \xi^{(e)\mathrm{T}}(t), \xi^{(c)\mathrm{T}}(t), \xi^{(\nu)\mathrm{T}}(t) \right\|^{\mathrm{T}},\\
&\boldsymbol{\beta} = \left\| \boldsymbol{\beta}^{(g)\mathrm{T}}, \boldsymbol{\beta}^{(k)\mathrm{T}}, \boldsymbol{\beta}^{(o)\mathrm{T}}, \boldsymbol{\beta}^{(p)\mathrm{T}}, \boldsymbol{\beta}^{(f)\mathrm{T}}, \boldsymbol{\beta}^{(e)\mathrm{T}}, \boldsymbol{\beta}^{(c)\mathrm{T}}, \boldsymbol{\beta}^{(\nu)\mathrm{T}} \right\|^{\mathrm{T}}.
\end{aligned}
\tag{10.13}
$$

The formulas define a dynamic system describing SC SDC processes. Here, $\mathbf{x}(t)$ is a general state vector of the SC and $\mathbf{y}(t)$ is a general vector of output characteristics. Then $\mathbf{u}(t)$ and $\mathbf{v}(\mathbf{x}(t),t)$ are control vectors. Here, $\mathbf{u}(t)$ represents the SC control programmes (plans of SC functioning) and $\mathbf{v}(\mathbf{x}(t),t)$ is a vector of control inputs compensating for perturbation impacts $\xi(t)$. The vector $\boldsymbol{\beta}$ is a general vector of SC parameters. The subscripts define the types of models M.

We consider the following models:

- M_g – dynamic model of SC motion control
- M_k – dynamic model of SC channel control
- M_o – dynamic model of SC operations control
- M_f – dynamic model of SC flow control
- M_p – dynamic model of SC resource control
- M_e – dynamic model of SC operation parameters control
- M_c – dynamic model of SC structure dynamic control
- M_ν – dynamic model of SC auxiliary operation control

All the described vectors should meet space–time, technical, and technological limitations; in other words, the vectors should belong to given sets:

$$\mathbf{u}(t) = \| \mathbf{u}_{pl}^{\mathrm{T}}(t), \mathbf{v}^{\mathrm{T}}(\mathbf{x}(t),t) \|^{\mathrm{T}}; \mathbf{u}_{pl}(t) \in \mathbf{Q}(\mathbf{x}(t),t); \ \mathbf{v}(t)(\mathbf{x}(t),t) \in \mathbf{V}(\mathbf{x}(t),t), \quad (10.14)$$

$$\boldsymbol{\xi}(t) \in \boldsymbol{\Xi}(\mathbf{x}(t),t); \ \boldsymbol{\beta} \in \mathrm{B}, \quad (10.15)$$

$$\mathbf{x}(t) \in \mathbf{X}(\boldsymbol{\xi}(t),t), \quad (10.16)$$

where $\mathbf{Q}(\mathbf{x}(t),t)$, $\mathbf{V}(\mathbf{x}(t),t)$ and $\boldsymbol{\Xi}(\mathbf{x}(t),t)$ are correspondingly allowable areas for programme control, real-time regulation control inputs and perturbation inputs, B is an area of allowable values of parameters and $\mathbf{X}(\boldsymbol{\xi}(t),t)$ is an area of allowable states of SC structure dynamics. The dynamics of state and output vectors can be described by means of a transition function and an output one:

$$\mathbf{x}(t) = \tilde{\boldsymbol{\varphi}}(\mathbf{x}(t), \mathbf{u}(t), \boldsymbol{\xi}(t), \boldsymbol{\beta}, t), \quad (10.17)$$

$$\mathbf{y}(t) = \tilde{\boldsymbol{\psi}}(\mathbf{x}(t), \mathbf{u}(t), \boldsymbol{\xi}(t), \boldsymbol{\beta}, t). \quad (10.18)$$

The transition function $\tilde{\boldsymbol{\varphi}}(\mathbf{x}(t), \mathbf{u}(t), \boldsymbol{\xi}(t), \boldsymbol{\beta}, t)$ and the output function $\tilde{\boldsymbol{\psi}}(\mathbf{x}(t), \mathbf{u}(t), \boldsymbol{\xi}(t), \boldsymbol{\beta}, t)$ can be defined in an analytical or algorithmic form.

There are additional constraints for the initial state and the final state:

$$\mathbf{x}(T_0) \in \mathbf{X}_0(\boldsymbol{\beta}), \ \mathbf{x}(T_f) \in \mathbf{X}_f(\boldsymbol{\beta}). \quad (10.19)$$

Equation (10.19) determines the end conditions for the SC state vector $\mathbf{x}(t)$ at time $t = T_0$ and $t = T_f$ (T_0 is the initial time of the time interval in which the SC is being investigated, and T_f is the final time of the interval).

Let us introduce the following vector of the *multi-model quality functional* (total performance metric) to evaluate the SC performance in the operation period:

$$\mathbf{J}_{\Theta}(\mathbf{x}(t), \mathbf{u}(t), \boldsymbol{\xi}(t), \boldsymbol{\beta}, t) = \left\| \mathbf{J}^{(g)\mathrm{T}}, \mathbf{J}^{(k)\mathrm{T}}, \mathbf{J}^{(o)\mathrm{T}}, \mathbf{J}^{(p)\mathrm{T}}, \mathbf{J}^{(f)\mathrm{T}}, \mathbf{J}^{(e)\mathrm{T}}, \mathbf{J}^{(c)\mathrm{T}}, \mathbf{J}^{(v)\mathrm{T}} \right\|^{\mathrm{T}}. \quad (10.20)$$

The problem of SC SDC includes tasks of three main classes:

Class A problems (problems of structured analysis, problems of SC structure dynamics analysis with or without perturbation impacts); *for constraints* (Eqs. 10.14–(0.19), $t \in (T_0, T_f]$ *it is necessary to obtain* $\mathbf{x}(t)$, $\mathbf{y}(t)$, $J_G(t)$, where

$J_G(t)$ is the generalized performance metric which is constructed by multi-criteria procedures.

Class B problems (estimation (observation) of problems, monitoring problems, problems of SC structural state identification); *for constraints* (Eqs. 10.14–10.19), $\mathbf{y}(\tilde{t})$, $\tilde{t} \in (T_0, T_f]$ *it is necessary to obtain* structural state estimation $\hat{\mathbf{x}}(t')$ and structure parameters estimations $\hat{\boldsymbol{\beta}}$; here $t', t \in (T_0, T_f)$.

Class C problems (problems of control input selection and problems of SC parameter selection, i.e. multi-criteria control problems for SC structures, modes and parameters, and multi-criteria problems of SC structural–functional synthesis); *for constraints* (Eqs. 10.14–10.19), $t \in (T_0, T_f]$ and performance metrics in Eq. 10.23 *it is necessary to obtain* $\mathbf{u}_{pl}(t)$, $\mathbf{v}(\mathbf{x}(t), t)$, $\boldsymbol{\beta}$ such that the generalized functional $J_G = J_G(\mathbf{J}(\mathbf{x}(t), \mathbf{u}(t), \mathbf{v}(x(t), t), \xi(t)))$ possesses its extreme values.

10.5 Generalized Dynamic Model of Supply Chain Control Processes (*M* Model)

In this section, a possible interconnection scheme of SC control models is presented. This generalized model provides a unified technology for an analysis and optimization of various processes concerning SC planning and execution.

The detailed mathematical formulation of these models has been presented by Kalinin and Sokolov (1985, 1986) and Okhtilev *et al.* (2006). Here let us provide the generalized dynamic model of SC control processes (M model):

$$\begin{aligned} M = \{&\mathbf{u}(t) \mid \dot{\mathbf{x}} = \mathbf{f}(\mathbf{x}, \mathbf{u}, t); \mathbf{h}_0(\mathbf{x}(T_0)) \le \mathbf{O}, \\ &\mathbf{h}_1(\mathbf{x}(T_f)) \le \mathbf{O}, \mathbf{q}^{(1)}(\mathbf{x}, \mathbf{u}) = \mathbf{O}, \mathbf{q}^{(2)}(\mathbf{x}, \mathbf{u}) < \mathbf{O}\}, \end{aligned} \tag{10.21}$$

$$\begin{gathered} J_\vartheta = J_\vartheta\big(\mathbf{x}(t), \mathbf{u}(t), t\big) = \varphi_\vartheta\big(\mathbf{x}(t_f)\big) + \int_{T_0}^{T_f} f_\vartheta\big(\mathbf{x}(\tau), \mathbf{u}(\tau), \tau\big) d\tau, \\ \vartheta \in \{g, k, o, f, p, e, c, v\}, \end{gathered} \tag{10.22}$$

where $\mathbf{x} = \left\| \mathbf{x}^{(g)\mathrm{T}}, \mathbf{x}^{(k)\mathrm{T}}, \mathbf{x}^{(o)\mathrm{T}}, \mathbf{x}^{(p)\mathrm{T}}, \mathbf{x}^{(f)\mathrm{T}}, \mathbf{x}^{(e)\mathrm{T}}, \mathbf{x}^{(c)\mathrm{T}}, \mathbf{x}^{(v)\mathrm{T}} \right\|^{\mathrm{T}}$ is a vector of the SC generalized state, $\mathbf{u} = \left\| \mathbf{u}^{(g)\mathrm{T}}, \mathbf{u}^{(k)\mathrm{T}}, \mathbf{u}^{(o)\mathrm{T}}, \mathbf{u}^{(p)\mathrm{T}}, \mathbf{u}^{(f)\mathrm{T}}, \mathbf{u}^{(e)\mathrm{T}}, \mathbf{u}^{(c)\mathrm{T}}, \mathbf{u}^{(v)\mathrm{T}} \right\|^{\mathrm{T}}$ is a vector of the generalized control, $\mathbf{h}_0$, $\mathbf{h}_1$ are known vector functions that are used for the state $\mathbf{x}$ end conditions at the time points $t = T_0$ and $t = T_f$, and the vector functions $\mathbf{q}^{(1)}, \mathbf{q}^{(2)}$ define the main spatio–temporal, technical and technological conditions for the SC execution.

On the whole the constructed model M (Eq. 10.21) is a deterministic non-linear non-stationary finite-dimensional differential system with a reconfigurable structure. Figure 10.2 shows the interconnection of models M_g, M_k, M_o, M_p, M_f, M_e, M_c and M_v *embedded* in the generalized model.

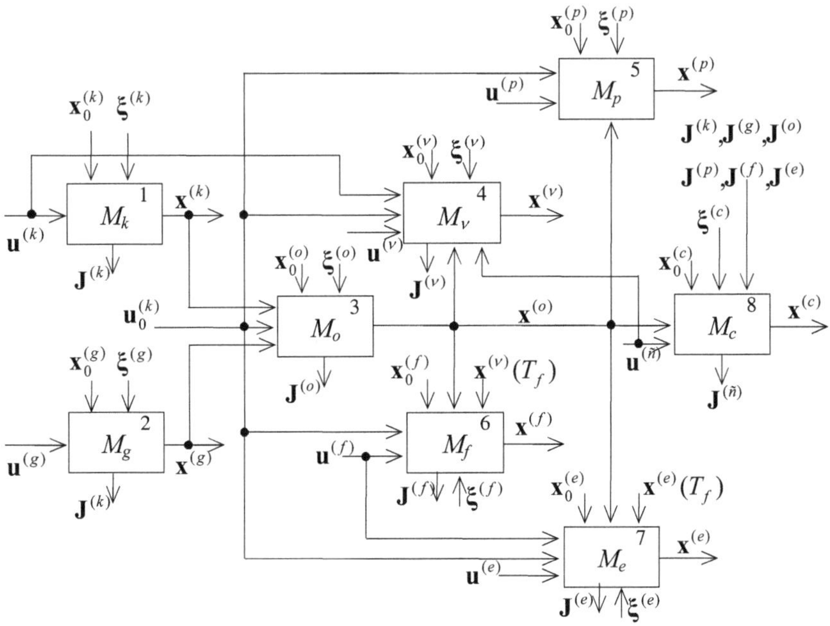

Fig.10.2 The scheme of model interconnection

In Fig. 10.2 the additional vector function of perturbation influences $\boldsymbol{\xi} = \left\| \boldsymbol{\xi}^{(g)\mathrm{T}}, \boldsymbol{\xi}^{(k)\mathrm{T}}, \boldsymbol{\xi}^{(o)\mathrm{T}}, \boldsymbol{\xi}^{(p)\mathrm{T}}, \boldsymbol{\xi}^{(f)\mathrm{T}}, \boldsymbol{\xi}^{(e)\mathrm{T}}, \boldsymbol{\xi}^{(c)\mathrm{T}}, \boldsymbol{\xi}^{(v)\mathrm{T}} \right\|^{\mathrm{T}}$ is introduced. This function describes the impact of the environment upon the SC functioning. In Chap. 12, selected models of this generalized framework will be considered in detail.

The solutions obtained in the multi-model complex presented are coordinated by the control inputs vector $\mathbf{u}^{(o)}(t)$ of the model M_o. This vector determines the sequence of interaction operations and fixes the SC resources allocation. The applied procedure of solution adjustment refers to resource coordination.

The model complex M evolves and generalizes the dynamic models of scheduling theory. The main distinctive feature of the complex is that non-linear technological constraints are actualized in the convex domain of allowable control inputs rather than in differential equations (see also Chap. 12).

10.6 Main Phases and Steps of Optimal Structure Dynamics Control Modelling

It is assumed that there are some perturbation influences (objective, subjective, internal and external) upon an SC. The current SC multi-structural macro-state is known. In this case, there are two groups of problems to be solved.

In the *first phase*, the formation of allowable multi-structural macro-states is performed. In other words, a structural–functional synthesis of a new SC should be fulfilled in accordance with an actual or forecasted situation.

In the *second phase*, a single multi-structural macro-state is selected, and adaptive plans (programmes) of SC transition to the selected macro-state are constructed. These plans should specify transition programmes, as well as programmes of stable SC operation in intermediate multi-structural macro-states.

10.6.1 Supply Chain Synthesis Algorithm

The general algorithm of the SC structural–functional (re)synthesis includes the following main steps.

Step 1. The gathering, analysis and interrelation of input data for the resynthesis of SC multi-structural macro-states. The construction or correction of the appropriate models.

Step 2. The planning of a solving process for the problem of the SC macro-states' resynthesis. The estimation of the time and other resources needed for the problem.

Step 3. The construction and approximation of an attainable set (AS) for a dynamic system. This set contains an indirect description of different variants of SC make-up (variants of SC multi-structural macro-states).

Step 4. The orthogonal projection of a set defining macro-state requirements for AS.

Step 5. The interpretation of the output results and their transformation to a convenient form for future use (for example, the output data can be used for the construction of adaptive plans for SC development).

10.6.2 Method of Attainable Sets for the Evaluation of Supply Chain Goal Abilities

Here, we introduce some comments on steps 3 and 4 of the previous section. Let us introduce the definition of an AS. The AS of a controllable dynamic system (Eq. 10.13) at $t_1 \in (T_0, T_f]$ includes points of all the system's state trajectories at time t_1 under the following conditions: each trajectory begins at time $t = T_0$ at the point $\mathbf{x}(T_0)$ and is formed through some allowable control $\mathbf{u}(t)$ over a time interval $(T_0, T_1]$. In other words, the AS is the set of all points to which the system can be steered at the instant of a given time (Chernousko 1994, Clarke *et al.* 1995). Herein, AS is denoted as $\mathbf{D}(t, T_0, \mathbf{x}(T_0), \mathbf{U}(\mathbf{x}(T_0)))$, where $\mathbf{U}(\mathbf{x}(t)) = \mathbf{Q}(\mathbf{x}(t), t) \times \mathbf{V}(\mathbf{x}(t), t)$ is a set of allowable control inputs.

AS is a very useful tool in the study of various problems of optimization, dynamical systems and differential game theory. Numerous papers have been devoted to the study of various properties of the AS of the control systems with geometric constraints on control (Chernousko 1994, Clarke *et al.* 1995, Motta and Sartori 2000, Sirotin and Formalskii 2003, Lou 2004, Guseinov 2009). In this study, we propose to apply attainable areas to the SCM domain. AS allows the explicit interconnecting of problems of estimation of SC goal abilities and stability. This opens up constructive ways to solve optimal control problems simultaneously with SC stability analysis.

The economic sense of the AS consists of the following: the AS characterizes a set of SC plans and the values of the SC's potential goals corresponding to them. The data on the AS and its basic characteristics in essence replace with themselves all the information necessary for the decision on problems of the evaluation of an SC' potential performance, the stability of its functioning and the synthesis of SCs and their development.

Let us explain the meaning of AS with an example. According to the classic scheme, the process of decision-making can be divided into two stages: (1) proving the existence of permissible solutions (i.e. proving SC manageability) and (2) searching for these solutions or an optimal solution. The AS allows the the first question to be answered. This defines the area where permissible solutions exist. For example, we are given the resources and time to achieve certain goals (e.g. the SC service level) with a forecasted demand. The AS analysis may provide the area in which permissible solutions (SC plans) are included. On the other hand, this may show that, with the given resource and at the given time horizon, it is impossible to achieve the required service level; hence, we should introduce additional resources or expand the SC cycle.

If the dimensionality of vectors $\mathbf{x}$ and $\mathbf{u}$ is high, the construction of an AS is a rather complicated problem. That is why an attainable set is usually approximated in different forms (Guseinov 2009). A possible approach to the AS approximation will be considered in Chapter 14.

The points of the attainable set are not of equal interest. It is important to construct a subset of macro-states favorable to the system's mission. To generate useful multi-structural macro-states, we need to set requirements for the characteristics of SCs in these macro-states. This can be done on the basis of the extended state vector

$$\mathbf{x}_p(t) = \left\| \mathbf{x}^{\mathrm{T}}(t), \mathbf{x}_{\tilde{a}}^{\mathrm{T}}(t) \right\|^{\mathrm{T}}, \qquad (10.23)$$

where $\mathbf{x}_{\tilde{a}}(t)$ are additional elements of SC state vector. These elements correspond to elements of the vector $\mathbf{J}$ transformed from an integral to a differential form via standard formulas (Gubarev *et al.* 1988). Now the variants of the required macro-states can be defined as

$$\tilde{\mathbf{x}}_{<\tilde{p},\tilde{\mu}>}(T_f) = \left\| \tilde{\mathbf{x}}^{\mathrm{T}}_{<\tilde{\mu}>}(T_f), \tilde{\mathbf{x}}^{\mathrm{T}}_{<\tilde{a},\tilde{\mu}>}(T_f) \right\|^{\mathrm{T}}, \tilde{\mu} = 1,\ldots,\tilde{e}\,, \tag{10.24}$$

where $\tilde{\mathbf{x}}^{\mathrm{T}}_{<\tilde{\mu}>}(T_f)$, $\tilde{\mathbf{x}}^{\mathrm{T}}_{<\tilde{a},\tilde{\mu}>}(T_f)$ are given values characterizing the preferable multi-structural macro-states. However the contradictory requirements for SC design can lead to violation of one or more constraints:

$$\mathbf{x}_{<\tilde{p}>}(T_f) = \mathbf{x}_{<\tilde{p},\tilde{\mu}>}(T_f),\ \tilde{\mu} = 1,\ldots,\tilde{e} \tag{10.25}$$

(for some fixed $\tilde{\mu}$). Therefore, a compromise solution should be found. It is desirable to construct a state vector $\mathbf{x}_{<\tilde{p}>}(T_f)$ that is the nearest for a given metric to $\mathbf{x}_{<\tilde{p},\tilde{\mu}>}(T_f)$. To receive such a solution, we use the points of $\mathbf{x}_{<\tilde{p},\tilde{\mu}>}(T_f)$ to form the convex capsule $\overline{E} = conv\{E_{\tilde{\mu}}, \tilde{\mu} = 1,\ldots,\tilde{e}\}$, where $E_{\tilde{\mu}}$ is a given point of an extended state space $\mathbf{x}_{<\tilde{p},\tilde{\mu}>}(T_f)$. It can be proved that the orthogonal projection of the set $\overline{E}$ to the AS $\mathbf{D}(t,T_0,\mathbf{x}(T_0))$ (in our case, to the approximation of AS) produces within $\mathbf{D}(t,T_0,\mathbf{x}(T_0))$ a set of non-dominated alternatives (Pareto's set). The attainable multi-structural macro-states of a stable SC execution are to be sought in Pareto's set.

There can be distinguished several variants of $\overline{E}$ and $\mathbf{D}(t,T_0,\mathbf{x}(T_0))$ inter-location (see Fig. 10.3) (Gubarev *et al.* 1988, Petrosyan 1994).

Variant 1: $\overline{E} \cap \mathbf{D}(t,T_0,\mathbf{x}(T_0)) = \varnothing$. In this case, the convex capsule $\overline{E}$ of points $E_{\tilde{\mu}}$ $(\tilde{\mu} = 1,\ldots,\tilde{e})$ and AS $\mathbf{D}(t,T_0,\mathbf{x}(T_0))$ do not intersect (see Fig. 10.3a). Let $\pi_{\mathbf{D}(t,T_0,\mathbf{x}(T_0))}\overline{E}$ be an operator of orthogonal projection of $\overline{E}$ to AS. Pareto's set $\Delta^{(nd)}$ is defined as follows:

$$\Delta^{(nd)}(t) = \left\{ \mathbf{x}_p(t) \mid \mathbf{x}_p(t) \in \pi_{\mathbf{D}(t,T_0,\mathbf{x}(T_0))}\overline{E} \right\}. \tag{10.26}$$

Therefore, for the first variant of $\overline{E}$ and $\mathbf{D}(t,T_0,\mathbf{x}(T_0))$ inter-location, Pareto-optimal points $\mathbf{x}_{\tilde{p}}(t)$ (characterizing preferable SC multi-structural macro-states) are projections of $\overline{E}$ to the bound of AS. In Fig. 10.3 a Pareto's set in the extended state space is marked with a solid line.

Variant 2: $\overline{E} \subset \mathbf{D}(t,T_0,\mathbf{x}(T_0))$. Then

$$\Delta^{(nd)}(t)=\left\{\mathbf{x}_{\tilde{p}}(t)\mid \mathbf{x}_{\tilde{p}}(t)\in \pi_{\mathbf{D}(t,T_0,\mathbf{x}(T_0))}\overline{E}=\overline{E}\right\}. \quad (10.27)$$

This is a rather rare case. Here, all the given points $E_{\tilde{\mu}}$ $(\tilde{\mu}=1,\ldots,\tilde{e})$ are attainable (see Fig. 10.3b). In this case, all Pareto-optimal points $\mathbf{x}_p$ are located in the convex capsule $\overline{E}$.

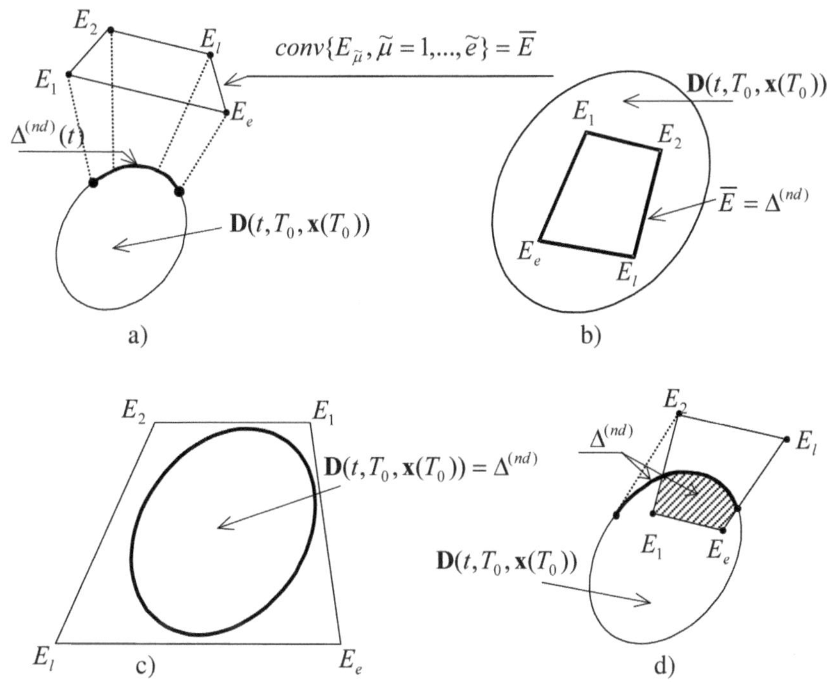

Fig. 10.3 Distinguished several variants of $\overline{E}$ and $\mathbf{D}(t,T_0,\mathbf{x}(T_0))$ inter-location

Variant 3: Let $\mathbf{D}(t,T_0,\mathbf{x}(T_0))\subset\overline{E}$. Then

$$\Delta^{(nd)}(t)=\left\{\mathbf{x}_{\tilde{p}}(t)\mid \mathbf{x}_{\tilde{p}}(t)\in \pi_{\mathbf{D}(t,T_0,\mathbf{x}(T_0))}\overline{E}=\mathbf{D}(t,T_0,\mathbf{x}(T_0))\right\}. \quad (10.28)$$

Therefore, in this case, all the boundary points of AS are Pareto-optimal (see Fig. 10.3c).

Variant 4: This is the most common case of $\overline{E}$ and $\mathbf{D}(t,T_0,\mathbf{x}(T_0))$ inter-location (see Fig. 10.3d): $\tilde{E}=\overline{E}\cap\mathbf{D}(t,T_0,\mathbf{x}(T_0))\neq\varnothing$. Then

$$\Delta^{(nd)}(t)=\left\{\mathbf{x}_{\tilde{p}}(t)\mid \mathbf{x}_{\tilde{p}}(t)\in \pi_{\mathbf{D}(t,T_0,\mathbf{x}(T_0))}\overline{E}\cup\tilde{E}\right\}. \quad (10.29)$$

Therefore, the first phase of the SC planning resulted in a set of non-dominated alternatives (Pareto's set). The elements of this set are possible SC plans. Further, it is necessary to analyze the influence of perturbation impacts to an SC.

The proposed original description of SC dynamics establishes dependence relations between control technology and the goals of the SC control system. For example, the methods of optimal control theory applied to the models M_o, M_e and M_n help to estimate the degree of interdependency between the quality of SC execution according to the specified goals and such aspects of SCM as SC resource allocation, routing, etc. Various combinations and interactions of particular control models in the general model M form the basis for a detailed multi-criteria analysis of the factors influencing the objective results of SC execution.

10.6.3 Supply Chain Resynthesis Algorithm

The phase of the SC reconfiguration is aimed at a solution of multi-level multi-stage optimization problems. The general algorithm of problem solving should include the following steps (mathematical algorithm description will be presented in Chap. 13).

Step 1. During this step, a structural and parametric adaptation of models, algorithms, and special software tools of the SS is fulfilled for the past and the current states of the environment, for the object-in-service and for the control subsystems embodied in the existing and developing SC. For missed data simulation, experiments with SS models or expert inquest can be used. SS is an important part of the proposed framework DIMA (see Chap. 9).

Step 2. Planning of the integrated modelling of adaptive SC control and development for the current and forecasted situation; planning of simulation experiments in SS; the selection of models; the selection of the model structure; the determination of methods and algorithms for particular modelling problems; the selection of models and the model structure for these problems; the estimation of the necessary time.

Step 3. Generating, via integrated modelling, feasible variants of SC functioning in the initial, intermediate, and required multi-structural macro-states; introducing the results to a decision maker; preliminary interactive structural–functional analysis of the modelling results; producing equivalent classes of SC multi-structural macro-states.

Step 4. Automatic putting into operation of the data of SC functioning variants; analysis of constraints' correctness; final selection of the aggregation level for SC SDC models and for computation experiments aimed at SC SDC programme construction.

Step 5. Search for optimal SC SDC programmes for the transition from a given multi-structural macro-state to a synthesized one and for stable SC operation in intermediate multi-structural macro-states.

Step 6. Simulation of programme execution under perturbation impacts for different variants of compensation control inputs received via methods and algorithms of real-time control.
Step 7. Structural and parametric adaptation of the plan and of SS software to possible (forecasted through simulation models) states of the SC and of the environment. Here, SC structural redundancy should be provided for the compensation of extra perturbation impacts. After reiterative computation experiments, the stability of the constructed SC SDC plan is estimated.
Step 8. The transfer of adaptive planning results to a decision maker; the interpretation and correction of these results.

One of the main opportunities of the proposed method of SC SDC is that, besides the vector $\mathbf{u}_{pl}(t)$, we receive a preferable multi-structural macro-state $\mathbf{x}_{\tilde{p}}(T_f) \in \Delta^{(nd)}(T_f)$ at time, $t = T_f$. This is the state of stable operating of the SC in the current (forecasted) situation.

The application of the proposed algorithm to the SC reconfiguration will be considered in Chap. 13. In Chap. 14, we will apply the AS to the SC stability analysis.

References

Aida-Zade R (2005) Solving Optimal Control Problems on the Class of Step Functions. Cybern Syst Anal 41(4):604–611

Anderson Jr EG, Morrice DJ, Lundeen G (2006) Stochastic optimal control for staffing and backlog policies in a two-stage customized service supply chain. Prod Oper Manag 15(2):262–278

Braun MW, Rivera DE, Flores ME, Carlyle WM, Kempf KG (2003) A model predictive control framework for robust management of multi-product, multi-echelon demand networks. Annu Rev Contr 27:229–245

Bryson AE, Ho Y-C (1975) Applied Optimal Control. Hemisphere, Washington

Chernousko FL (1994) State Estimation of Dynamic Systems. SRC Press, Boca Raton, Florida

Clarke FH, Ledyaev Yu S, Stern RJ, Wolenskii PR (1995) Qualitative properties of trajectories of control systems: a survey. J Dyn Control Syst 1:1–48

Disney SM, Towill DR (2002) A discrete linear control theory model to determine the dynamic stability of vendor managed inventory supply chains. Int J Prod Res 40(1):179–204

Disney SM, Towill DR, Warburton RDH (2006) On the equivalence of control theoretic, differential, and difference equation approaches to modeling supply chains. Int J Prod Econ 101: 194–208

Gubarev VA, Zakharov VV, Kovalenko AN (1988) Introduction in systems analysis. Study book. LGU, Leningrad

Guseinov KG (2009) Approximation of the attainable sets of the nonlinear control systems with integral constraint on controls. Nonlinear Analysis 71:622–645

Ivanov D (2009) Supply chain multi-structural (re)design. Int J Integrated Supply Manag 5(1):19–37

Ivanov D, Arkhipov A, Sokolov B (2007) Intelligent planning and control of manufacturing supply chains in virtual enterprises. Int J Manuf Tech Manag 11(2):209–227

Ivanov D, Sokolov B, Kaeschel J (2010) A multi-structural framework for adaptive supply chain planning and operations control with structure dynamics considerations. Eur J Oper Res 200: 409–420

Kalinin VN, Sokolov BV (1985) Optimal planning of the process of interaction of moving operating objects. Differ Equat 21(5):502–506

Kalinin VN, Sokolov BV (1987) A dynamic model and an optimal scheduling algorithm for activities with bans of interrupts. Autom Rem Contr 48(1-2):88–94

Kalinin VN, Sokolov BV (1996) Multi-model description of control processes for airspace crafts. J Comput Syst Sci Int 6:192–199

Kreipl S and Pinedo M (2004) Planning and scheduling in supply chains: an overview of issues in practice. Prod Oper Manag 13(1):77–92

Lalwani CS, Disney S, Towill DR (2006) Controllable, observable and stable state space representations of a generalized order-up-to policy. Int J Prod Econ 101:172–184

Lee EB, Markus L (1967) Foundations of optimal control theory. John Wiley & Sons, New York

Lou HW (2004) On the attainable sets of control systems with p-integrable controls. J Optim Theory Appl 123:123–147

Moiseev NN (1974) Element of the optimal systems theory. Nauka, Moscow (in Russian)

Motta M, Rampazzo F (2000) Multivalued dynamics on a closed domain with absorbing boundary, applications to optimal control problems with integral constraints. Nonlinear Anal TMA 41:631–647

Okhtilev M, Sokolov B, Yusupov R (2006) Intelligent technologies of complex systems monitoring and structure dynamics control. Nauka, Moskau (in Russian)

Petrosjan LA (1994) Strongly time consistent optimality principles in the games with discount payoffs. Lect Notes Contr Inform Sci. 197: 513–520

Pontryagin LS (1961) Theory of optimal process. Fizmatgiz, Moscow (in Russian)

Sethi SP, Thompson GL (2006) Optimal Control Theory: Applications to Management Science and Economics, Second Edition. Springer, Berlin

Sirotin AN, Formalskii AM (2003) Reachability and controllability of discrete-time systems under control actions bounded in magnitude and norm. Autom Rem Contr 64:1844–1857

Skurikhin VI, Zabrodsky VA and Kopeychenko YV (1989) Adaptive control systems in machinery industry. Mashinostroenie, Moscow (in Russian)

Sokolov BV, Yusupov RM (2004) Conceptual foundations of quality estimation and analysis for models and multi-model systems. J Comput Syst Sci Int 6:5–16

Van Houtum GY, Scheller-Wolf A, Yi J (2007) Optimal control of serial inventory systems with fixed replenishment intervals. Oper Res 55(4):674–687

Wang W, Rivera D, Kempf KG (2007) Model predictive control strategies for supply chain management in semiconductor industry. Int J Prod Econ 107:56–77

Zvirkun AD, Akinfeev VK(1993) Structure of the multi-level systems (synthesis and development). Nauka, Moscow (in Russian)

Chapter 11
Adaptive Planning of Supply Chains

It is a mistake to try to look too far ahead.
The chain of destiny can only be grasped one link at a time.

Winston Churchill

11.1 Planning

Planning, in the broad sense, is a purposeful, organized and continuous process including the synthesis of SC structures and elements, the analysis of their current state and interaction, the forecasting of their development for some period, the forming of mission-oriented programmes and schedules, and the development of SC SDC programmes for the transition to a required (optimal) structural macro-state.

Planning, as a phase of decision-making, has several peculiarities:

- Planning is a preliminary design of the organization design and functioning mechanism providing goal achievement by a given time.
- The result of planning is a system of interrelated distributed time-phased decisions, while the function of planning is directly connected to the function of the regulation control, since designing and keeping programme trajectories use common resources.
- The process of planning permanently approaches the end but never reaches it for two reasons: first, revising decisions lasts until concrete actions are performed; second, the system and the environment can change during the planning process; therefore, it is necessary to correct plans periodically.
- Planning is aimed at the prevention of erroneous operations and at a decrease in unimproved opportunities.

The main requirements for a plan are as follows:

- analysability (the plan execution should be subject to comprehensive analysis);
- controllability (the plan execution should be subject to control);
- adaptability (the plan should be able to be adapted in the planned and unplanned modes);
- goal approachability (the plan should ensure the fulfilment of management goals); and
- synchronizability (the plan should be coordinated in the horizontal mode under the SC partners and in the hierarchical mode with the plans of superordinated and underordinated levels).

In a general case, planning is concerned with the following groups of tasks (see Fig. 11.1):

- the formation of SC goals and objectives, i.e. the evaluation of preferable states and the time for achievement of goals and objectives;
- the determination of means and instruments for the achievement of goals and objectives;
- the determination of resources and their sources for the implementation of plans, as well as the development of principles and methods for resource allocation; and
- the design of the SC (first of all, the development of the SC's main structures) and SC functioning processes, providing continuity of comprehensive planning and control for the system structure-dynamics.

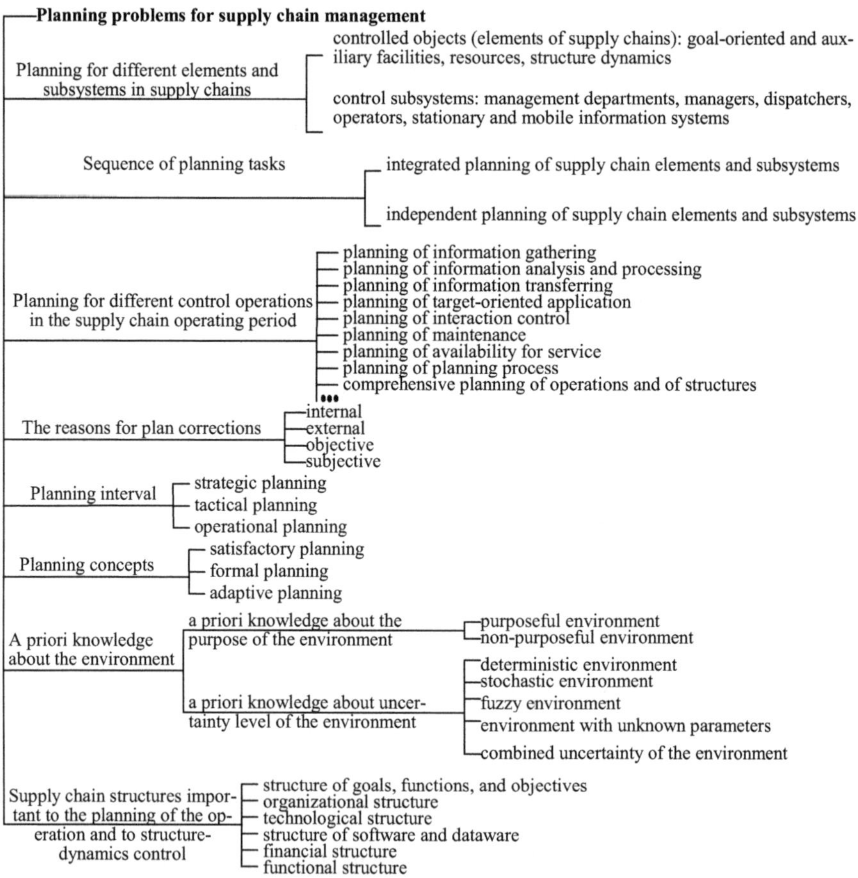

Fig. 11.1 Classification of planning problems

Planning approaches can be divided into *incremental planning*, which concentrates on situation predictions in terms of mathematical models, *satisfactory (formal) planning*, which considers SC reactions to external impacts and *adaptive planning*, which supports SC interaction with the environment.

Actually, the widespread techniques of OR mostly support incremental planning, which does not include dynamic feedback. Such an approach can be justified for problems for which a single plan computation should be fulfilled. These problems may be either of a very strategic nature or a very operative nature. In most tactical–operational problems that refer to the SC dynamics being under control, gathering current information about the SC execution and adapting SC operations, plans, and configurations as well as updating related models are mandatory.

The variety of *planning problems* is presented in Fig. 11.1. In practice, there are situations where planning as a goal approachability system should be given up and the desired goals approached empirically. The first ground for such a situation is system complexity, which makes planning insensible because of the following:

- too long-lasting and complex a planning process, during which the reality can change many times;
- the generated plan is neither analysable nor controllable or adaptable; and
- a plan is unstable and fails in a real execution environment.

Here, we can refer to the issue of complexity as considered in Chap. 5. The main source of system complexity is uncertainty, which is the primary cause of precise planning impossibility. In the context of adaptive planning, the following procedure is proposed. If there is not sufficiently full information about the system and operation dynamics, certain key operations and events must be determined, of which the uncertainty of states does not allow operation plans to be formed. As a result, a number of alternative operation dynamics scenarios can be elaborated. Further, with time, the uncertain (at the time of planning) operations' states become certain and possess certain characteristics (parameters). Hence, a systematic possibility to select a plan quickly from a fixed number of alternatives can be provided.

This analysis provides evidence that the planning and execution problems in SCs are tightly interlinked. In general, this interrelation is as follows: SC efficiency at the planning stage depends on two factors: (1) control actions that are planned and (2) future control actions to compensate for a possible deviation from the plan. On the other hand, adjustment actions' efficiency also depends on two factors: (1) control actions that are taken in operations' execution dynamics and (2) control actions that have been taken at the planning stage. Hence, the planning and execution models are to be inter-reflected, which means, in both of the models, that the decision-making principles of the other model are to be reflected.

11.2 Adaptive Planning

Adaptive planning is a continuous, event-driven, real-time (re)planning and control process. It uses not only simple open time slots (in contrast to incremental planning) but employs conflict-driven plan changes during the system execution (Andreev *et al.* 2007). In Fig. 11.2, the adaptive planning logic is presented.

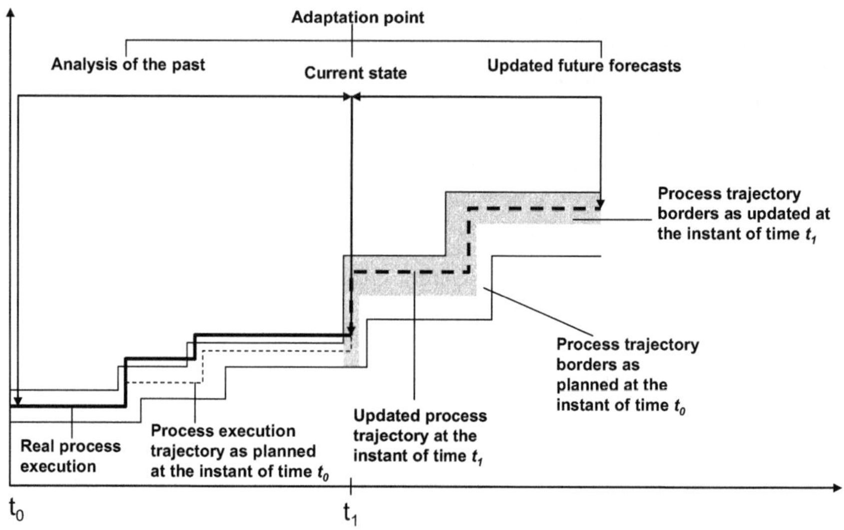

Fig. 11.2 Adaptive planning logic

Adaptive planning is a method of planning in which a plan is modified periodically by a change in the system parameters or the characteristics of control influences on the basis of information feedback about a current system state, the past and the updated forecasts of the future.

In adaptive planning, the precision of planning decisions decreases while moving away from the decision point (see Fig. 11.3).

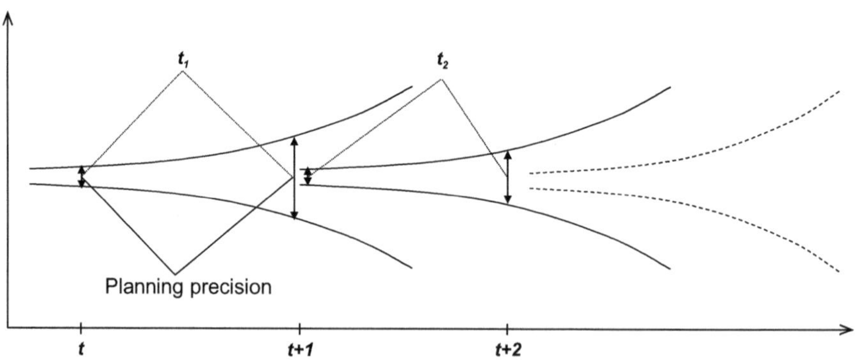

Fig. 11.3 Precision of planning decisions

As shown in Fig. 11.3, planning decisions become fuzzier with increasing distance from the decision-making point. This may be interpreted as a sequence of inter-inserted funnels. At certain intervals, the plans are updated, the "fuzzy" part of the plans becomes precise and a new funnel for the further process course appears. Such an approach provides the required flexibility with regard to dynamics and uncertainty. Hence, adaptive planning implies problem resolution and redefinition through the learning process, rather than problem solving. This allows us to interpret planning and scheduling not as discrete operations but as a continuous adaptive process. The adaptive planning procedure can be interconnected with the AC procedure. This allows reaction in the real-time mode to necessities for replanning due to disruptions in SCs (unplanned regulations).

11.3 Adaptation Framework

The main purpose of the adaptation framework is to ensure dynamic scheduling model parameter tuning with regard to changes in the execution environment. In the proposed framework, the plan adaptation is connected to the model adaptation. The parametric adaptation is enhanced by a structural adaptation. In Fig. 11.4, the general conceptual framework of the adaptive planning and scheduling is presented.

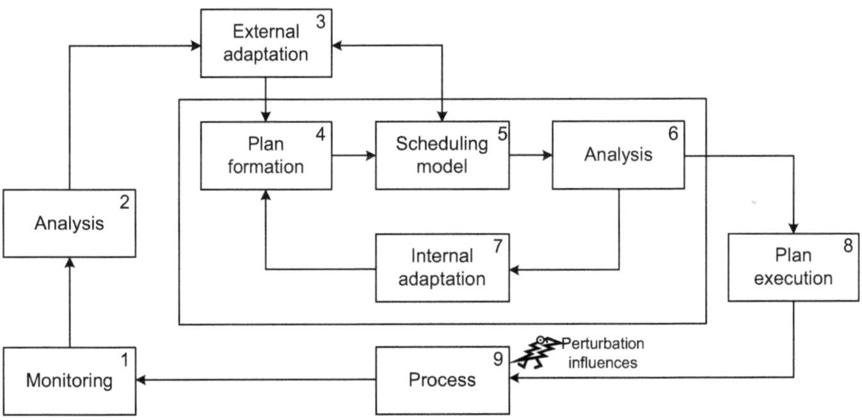

Fig. 11.4 General conceptual framework of the adaptive planning and scheduling (based on the principles described in Skurikhin *et al.* 1989).

An SC, when regarded as an object of control, is a non-stationary, non-linear and high-dimensional system with difficult-to-formalize aspects. The SC is characterized by the lack of *a priori* status information and non-strict criteria of decision making.

The proposed framework of adaptive planning and control includes the following main phases:

- parametric and structural adaptation of SDC models and algorithms for previous and current states of the SC (blocks 1, 2, 3);
- integrated planning and scheduling of SC operations (blocks 4, 5);
- simulation and analysis of SC execution according to different plans, schedules, and variants of control decisions in real situations (blocks 6);
- structural and parametric adaptation of the plans and schedules, control inputs, models, algorithms, which is based on results of simulations experiments (blocks 7); and
- plan and scheduling realization in real time, correction of the control action (blocks 8, 9).

To implement the proposed concept of adaptive planning and control, let us consider two groups of parameters for SDC models and algorithms:

- parameters that can be evaluated on the basis of real data available in the SC; and
- parameters that can be evaluated via simulation models for different scenarios of future events.

The *adaptation procedures* can be organized in two control loops:

- external adapter; and
- internal adapter.

The following parameters belong to the *first group* and can be evaluated through the external adapter:

- the values of end conditions of the SDC models (values of SC performance indicators);
- economic and technological characteristics of SC elements and subsystems; and
- probabilistic characteristics and values of real and observed random processes.

The *second group* of parameters being evaluated through the internal adapter includes such characteristics as:

- a redundancy rate for reserving of different types (functional, temporal, and informational reserving);
- priority of SC performance indicators; and
- parameters that define the variants of compensation for trajectory deviations (violations of the schedule) in the simulation models.

When the parametric adaptation of a SC does not provide simulation adequacy then structural transformations can be necessary. Two main approaches to structural model adaptation can usually be distinguished. The first approach lies in the selection of a model from a given set. The model must be the most adequate to supply the chain state and environment. The second approach stands for the SC SDC model construction of elementary models (modules) in compliance with the given requirements. The second approach provides more flexible adjustment of

SCs. However, the first one is faster and can be effective if the application knowledge base is sufficiently large.

Both approaches need the active participation of system analysts and decision makers in SCs and consider hard-formalizing factors and dependences. The structural adaptation of an SC takes a certain period of time, when the following main activities should be performed (Skurikhin *et al.* 1989):

- selection or construction (synthesis) of SC SDC models meeting the given requirements;
- selection or construction (synthesis) of SC SDC algorithms for the given conditions and control problems;
- synthesis of software and dataware for the given control problems; and
- adjustment of SC parameters for the current and predicted states of the SC (parametric adaptation).

Sometimes it is useful to adjust models and algorithms that are not currently used in SC control processes; this will provide fast utilization of additional models when they are needed. The considered adaptation should be based on the results of the SC SDC simulation.

11.4 Controller Concept

By designing the controller, we will reflect two particular features of SCs, namely:

- the delays between the deviations' identification and the adjustment decision making are handled within the SDC approach; and
- a combined people–machine adjustment system for the SC adaptation in the case of different disruptions will be used.

The perturbation impacts initiate SC structure dynamics. A hierarchy of adjustment actions is brought into correspondence with different deviations in the SC execution. For small perturbations, a state correction takes place (time horizon — minutes or hours). For greater perturbations, structural states or MSMS are adjusted (time horizon — weeks or months). In the paper by Ivanov *et al.* (2010), it has been proved that structure dynamics considerations may allow the establishment of adaptive feedback between the SC design, tactics, and operations.

The model of decision-making for the controller is based on the direct connection of business processes' stability estimation and analysis with problems of estimation and the analysis of their economic efficiency. The model is based on the dynamic interpretation of the SC's functioning process. The results of the stability analysis are brought into correspondence with different SC adjustment measures.

The stability analysis allows the definition of the admissible borders of deviation from SC execution parameters proceeding from the criteria of influence of these deviations on performance metrics and on the possibility of returning the SC to a planned (or wished for) state after disturbance. As a result of such an analysis, some zones of stability are defined, to each of which there corresponds a certain level of necessary control influences (see Chap. 7).

11.5 Supply Chain Planning Levels and Their Reflections

SCP is composed of setting management goals and defining measures for their achievement (Kreipl and Pinedo 2004). On the basis of the goals of the superordinated level of an SC, plans of a current level are formed. E.g. strategic goals can be referred to the service level and costs. The measures are in this case the realization of customers' orders. To fulfil these orders, schedules are to be constructed.

Planning decisions in SCM can be divided into strategic (SCD), tactics (SCP), and operations (see also the classification given in Sect. 9.3). As the term implies, strategic decisions are typically made over a longer time horizon. Over the last decade, a wealth of valuable approaches for SC strategic planning have been extensively developed (Simchi-Levi *et al.* 2004, de Kok and Graves 2004, Chopra and Meindl 2007). SC design deals with strategic issues of distribution networks and supplier/customer integration (Cohen and Lee 1988, Beamon 1999, Tayur *et al.* 1999, de Kok and Graves 2004, Simchi-Levi *et al.* 2004).

SCD is a critical source of competitive advantage and consists of SC structuring in accordance with the given competitive strategy, SC strategy, product programme, coordination strategy, distribution strategy, and financial plans (Chopra and Meindl 2007) as well as with demand and supply uncertainty (Lee *et al.* 1997, Santoso *et al.* 2005). Conventionally, the SCD's central question is to determine which suppliers, parts, processes, and transportation modes to select at each stage in the SC. The literature on SCM indicates a need in multi-structural SC treatment to take into account the product, business processes, technological, organizational, technical, topological, informational, and financial structures (Lambert and Cooper 2000, Bowersox *et al.* 2002).

SCP considers a shorter time horizon (months, weeks) and deals with demand forecasting, master production planning, supply planning, replenishment planning, inventory management and transport planning. Coordination in SC tactical planning plays a fundamental role to mitigate uncertainty (Holweg and Pil 2008) with the help of synchronizing information flows from a point-of-sale up to the raw material suppliers and material flows in the reverse way. One key problem in nearly all SCs is the so-called bullwhip effect (Lee *et al.* 1997). Different concepts of coordination have been developed over the last two decades, such as ECR, CPFR, JIT and VMI. Enablers of the coordination are information technologies, such as ERP, APS, EDI and RFID.

Operational decisions are short-term, and focus on activities on a day-to-day basis. While the SC is running, problems of operative order planning, SCMo, and reconfiguration in the case of operative disruptions (i.e. machine failures, human errors, information systems failure, cash-flow disruption or simply catastrophic events) as well as tactical/strategic changes (i.e. new products, new OPP, etc.) are to be solved (Graves and Willems 2005, Ijounu *et al.* 2007, Ivanov and Ivanova 2008).

However, conventionally the planning decisions at each of these levels have been considered in isolation from the other levels. In practice, the interrelation of these three management levels is very important. Moreover, decisions on SC strategy, design, planning, and operations are interlinked and dispersed over different SC structures (functional, organizational, informational, financial, etc.). The efficiency and applicability of the decisions decrease if decision-supporting models are considered in isolation for different SC managerial levels and structures (Ivanov *et al.* 2010).

In practice, partial SC strategy, design, planning and operations decisions are highly interlinked (Ivanov 2009b). The issues of aligning the SC strategy, design, tactics and operations have been highlighted in literature episodically. Harrison (2005) and Chopra and Meindl (2007) emphasized that SCD decisions are closely linked to the corporate and SC strategy. Chen (2007) presented a survey on integrated models of SCD. He pays particular attention to the strategic and tactical levels, considering three types of integrated problem: location-inventory, inventory-routing and location-routing. Sabri and Beamon (2000) developed an integrated multi-objective SC model for use in simultaneous strategic and operational SCP. Multiple objective decision analysis is adopted to allow the use of a performance measurement system that includes cost, customer service levels and flexibility (volume or delivery). This model incorporates production, delivery and demand uncertainty, and provides a multi-objective performance vector for the entire SC.

Guille´n *et al.* (2006) addressed the integrated planning/scheduling of SCs with multi-product, multi-echelon distribution networks, taking into account financial management issues. In order to tackle this problem, a mathematical formulation is derived, combining a scheduling/planning model with a cash flow and budgeting formulation. They also enhance the model by considering not only the insertion of financial aspects within an SCP formulation, but also the choice of a financial performance indicator, i.e. the change in equity, as the objective to be optimized in the integrated model (Guille´n *et al.* 2007). Moon *et al.* (2008) dealt with the integration of process planning and scheduling. They formulate an MIP model to solve this problem of integration. This model considers alternative resources: sequences and precedence constraints, and is solved with an evolutionary search approach. Ivanov *et al.* (2010) presented a conceptual framework and a mathematical model of multi-structural SC planning. Ivanov (2009a) developed a multi-disciplinary approach for integrated modelling SCs. Chandra and Grabis (2007) reported on simulation tool Flextronics which is used to transform SCs in accordance with the changes in market environment.

Let us consider the SC strategy, design, tactics and operations as a whole system in details (see Fig. 11.5).

Based on demand and supply forecasts (input $\mathbf{U}(t)$), SC plans are generated within the designed structures. If concrete orders penetrate a SC, SC operations plans are generated according to the orders' parameters (price, delivery place, batch size, etc.). While running a SC, different disturbances can affect the SC and cause deviations and disruptions.

SCMo is meant for the maintenance of output parameter values of an SC in accordance with the required ones of input signals from the SC plans, design and strategy. The results of the SC monitoring are reflected in the SC performance block.

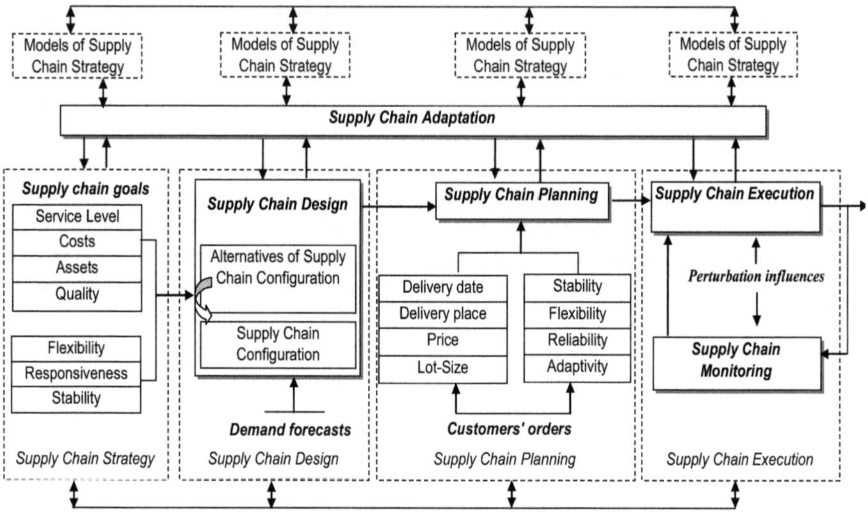

Fig. 11.5 Conceptual model of linking SC strategy, design, tactics, and operations (from Ivanov 2009b)

The SC adaptation serves for implicating monitoring results while the SC is running and for corrective control actions and to SC operations, plans, design and strategy. The adaptation is meant not only for processes but also for SC models, which should cohere with the current execution environment.

References

Beamon BM (1998) Supply chain design and analysis: models and methods. Int J Prod Econ 55(3):281–294

Bowersox D, Closs D and Cooper M (2002) Supply chain logistics management. McGraw-Hill/Irwin, New York

Chandra C, Grabis J (2007) Supply chain configuration. Springer, New York

Chen F, Drezner Z, Ryan JK and Simchi-Levi D (2000) Quantifying the bullwhip effect in a simple supply chain: the impact of forecasting, lead times, and information. Manag Sci 46(3):436–443

Chen ZI (2007) Integrated supply chain design models: a survey and future research directions. Journal of Industrial and Management Optimisation 3(1): 1–27

Chopra S, Meindl P (2007) Supply chain management. Strategy, planning and operation. Pearson Prentice Hall, New Jersey

Cohen MA and Lee HL (1988) Strategic analysis of integrated production-distribution systems: models and methods. Oper Res 36(2):216–228

Graves SC and Willems SP (2005) Optimizing the supply chain configuration for new products. Manag Sci 51(8):1165-1180

Guill´en G, Badell M, Espu˜na A, Puigjaner L (2006) Simultaneous optimization of process operations and financial decisions to enhance the integrated planning/scheduling of chemical supply chains. Comput Chem Eng 30:421–436

Guill´en G, Badell M, Puigjaner L (2007) A holistic framework for short-term supply chain management integrating production and corporate financial planning. Int J Prod Econ 106:288–306

Harrison TP (2005) Principles for the strategic design of supply chains. In: Harrison TP, Lee H L, Neale JJ (Ed) The practice of supply chain management. Kluwer academic publishers, Boston

Holweg M, Pil FK (2008) Theoretical perspectives on the coordination of supply chains. J Oper Manag 26(3):389–406

Ijioui R, Emmerich H, Ceyp M (2007) Supply Chain Event Management: Konzepte, Prozesse, Erfolgsfaktoren und Praxisbeispiele. Springer, Berlin

Ivanov D (2009a) DIMA – a research methodology for comprehensive multi-disciplinary modelling of production and logistics networks. Int J Prod Res 47(5):133–1155

Ivanov D (2009b) An adaptive framework for aligning (re)planning decisions on supply chain strategy, design, tactics, and operations. Int J Prod Res, DOI: 10.1080/00207540902893417, in press

Ivanov D, Ivanova M (2008) A framework of adaptive control for production and logistics networks. In: Haasis HD, Scholz-Reiter B, Kreowski J (Ed) Dynamics in logistics. Springer, Berlin

Ivanov D, Sokolov B, Kaeschel J (2010) A multi-structural framework for adaptive supply chain planning and operations with structure dynamics considerations. Eur J Oper Res 200:409–420

Kok de AG and Graves SC (2004) Supply chain management: design, coordination and operation. Elsevier, Amsterdam

Kreipl S and Pinedo M (2004) Planning and scheduling in supply chains: an overview of issues in practice. Prod Oper Manag 13(1):77–92

Lambert DM and Cooper MC (2000) Issues in supply chain management. International Marketing Management 29(1):65–83

Lee HL, Padmanabhan V and Whang S (1997) Information distortion in a supply chain: the bullwhip effect. Manag Sci 43(4):546–558

Moon C, Lee YH, Jeong CS, Yun YS (2008) Integrated process planning and scheduling in a supply chain. Comput Ind Eng 54(4):1048–1061

Muriel A and Simchi-Levi D (2004) Supply chain design and planning – application of optimization techniques for strategic and tactical models. In: Kok de AG, Graves SC (Ed) Supply chain management: design, coordination and operation. Elsevier, Amsterdam

Okhtilev M, Sokolov B, Yusupov R (2006) Intelligent technologies of complex systems monitoring and structure dynamics control. Nauka, Moscow (in Russian)

Sabri EH, Beamon BM (2000) A multi-objective approach to simultaneous strategic and operational planning in supply chain design. Omega 28:581–598

Santoso T, Ahmed S, Goetschalckx M and Shapiro A (2005) A stochastic programming approach for supply network design under uncertainty. Eur J Oper Res 167(1):96–115

Simchi-Levi D, Wu SD and Zuo-Yun S (2004) Handbook of quantitative supply chain analysis. Springer, New York

Skurikhin VI, Zabrodsky VA and Kopeychenko YV (1989) Adaptive control systems in machinery industry. Mashinostroenie, Moscow (in Russian)

Sokolov BV, Yusupov RM (2004) Conceptual foundations of quality estimation and analysis for models and multiple-model systems. J Comput Syst Sci Int 6:5–16

Tayur S, Ganeshan R and Magazine M (1999) Quantitative models for supply chain management. Kluwer Academic Publishers, Boston

Chapter 12
Modelling Operations Dynamics, Planning and Scheduling

The greatest challenge to any thinker is stating the problem
in a way that will allow a solution.
Bertrand Russell

12.1 Research Approach

The proposed approach is based on fundamental scientific results of the modern optimal control theory (Okhtilev *et al.* 2006, Sethi and Thompson 2006) in combination with the optimization methods of OR. This mathematical model is the extended application to the SCM domain of the scheduling model for complex technical systems (Kalinin and Sokolov 1985, 1987, 1996, Sokolov and Yusupov 2004, Okhtilev *et al.* 2006) and reflects the conceptual cybernetic framework of SC planning and execution presented for different SCM domains in Ivanov *et al.* (2007, 2009, 2010).

The proposed approach has the following particular features. *First*, we consider *planning and scheduling as an integrated function* within an adaptive framework. Recent studies (Moon *et al.* 2008, Shao *et al.* 2009) have provided evidence that the performance of SCs can be improved greatly if planning and scheduling are not performed in a sequential way but are integrated and considered simultaneously.

Second, we formulate the planning and scheduling models as optimal control problems, taking into account the *discreteness of decision making* and the *continuity of flows*. By special techniques for process dynamics models and constraint formulation, we will show how to transform the non-linear operations dynamics model into a linear one. In doing so, the dimensionality of problems can be reduced and discrete optimization methods of linear programming can be applied for solution within the general dynamic non-linear model.

In the model, a multi-step procedure for solving a multiple objective task of adaptive planning and scheduling is implemented. In doing so, at each instant of time while calculating solutions in the dynamic model with the help of the maximum principle, the linear programming problems to allocate jobs to resources and integer programming problems for (re)distributing material and time resources solved. The process control model will be presented as a dynamic linear system while the non-linearity and non-stationarity will be transferred to the model constraints. This allows us to ensure convexity and to use the interval constraints. As such, the constructive possibility of discrete problem solving in a continuous manner occurs.

Third, the modelling procedure is based on an essential reduction of a *problem dimensionality* that is under solution at each instant of time due to connectivity decreases. The problem dimensionality is determined by the number of independent paths in a network diagram of SC operations and by current economic, technical, and technological constraints. In its turn, the degree of algorithmic connectivity depends on a dimensionality of the main and the conjugate state vectors at a point when the solving process is being interrupted. If the vectors are known, then the schedule calculation may be resumed after the removal of the appropriate constraints. As such, the problem under solution can be presented with a polynomial complexity rather than with an exponential one. In contrast, traditional exact scheduling techniques work almost with the compete list of all the operations and constraints in SCs.

Fourth, for solving the problem, *Pontryagin's maximum principle* is applied (Pontryagin 1961, Shell 1968, Day and Taylor 2000). The algorithm of optimal control is based on a transformation of the optimal control problem to the boundary problem. Besides, the *Lagrange multipliers* will be presented as dynamic parameters of the conjunctive vector (eq. 10.25).

Fifth, the *multi-objective optimization* for taking into account the individual preferences of decision-makers is applied. The construction and narrowing of Pareto's sets is performed in the interaction mode providing decision-maker participation. The model basis in such a case is represented by discrete models of mathematical programming, queuing model, simulation models, and development control. Further considerations on the Pareto optimization and AS are included in Chaps. 10, 13 and 14.

12.2 Dynamic Models of Operations Dynamics Control

Let $\overline{B} = \{\overline{B}^{(i)}, i \in \overline{N}\}$, $\overline{N} = \{1,...,\overline{n}\}$ be a set of customers' orders that can be realized in an SC. Each order is characterized by operations $D_{\mu}^{(i)}$ $(\mu = 1,...,s_i;\ i = 1,...,\overline{n})$. Let $B = \{B^{(1)},...,B^{(n)}\}$ be a set of resources (enterprises) in an SC. The jobs' realization with these resources is connected to the flows (material, financial, etc.) $P^{(i,j)} = \{P^{(i,j)}_{<\mu,\rho>}, \mu = 1,...,s_i, \rho = 1,..., p_i\}$ The blocking of certain arcs in the dynamical graph resulting from the set-theoretical SC structure description given in (Ivanov *et al.* 2010) is possible to reflect non-cooperative relations between certain enterprises in an SC. The duration of operations and the resource productivity together with low and upper borders of perturbation impacts on resource availability and productivity are known. Enterprises can block the availability of their resources for certain periods of time (in the form of time–spatial constraints).

The goal is to (re)configure, (re)plan and (re)schedule the customer orders within the whole planning period subject to the maximization of the service level (the general volume of fulfiled orders in accordance with the delivery plan), the minimization of penalties for breaking delivery terms and the maximization of equal resource charge in the SC (the requirement for SC collaboration).

The formal statement of the scheduling problem will be produced, as noted above, via a *dynamic interpretation* of the operations' execution processes. In the further course of this section, we will consider the partial dynamic models of SC control in details. These models correspond to the models listed in Sect. 10.2.

12.2.1 Dynamic Model of Collaborative Operations Control

Mathematical model of operations control processes

Let us consider the mathematical model of the operation $D_\mu^{(i)}$ processing. The following notation can be introduced:

$x_{i\mu}^{(o)}$ is a variable characterizing the state of the operation $D_\mu^{(i)}$,

ε_{ij} is an element of the preset matrix time function of time-spatial constraints ($\varepsilon_{ij}=1$, if $t \geq t_k^i$, $\varepsilon_{ij}=0$, if $t < t_k^i$), (e.g. a constraint on the production shift from 7am to 4pm),

$u_{i\mu j}^{(o)}$ is a control action ($u_{i\mu j}^{(o)}=1$, if the resource $B^{(j)}$ processes the operation $D_\mu^{(i)}$, $u_{i\mu j}^{(o)}=0$ otherwise),

$x_{ij\eta\rho}^{(o)}$ is a variable characterizing the state of the operation to deliver the product flow $P_{<s_i,\rho>}^{(j,\eta)}$ to the customer $\overline{B}^{(\eta)}$ with the use of the resource $B^{(j)}$,

$u_{ij\eta\rho}^{(o)}(t)$ is control action ($u_{ij\eta\rho}^{(o)}(t)=1$, if the resource $B^{(j)}$ is used for the delivery of the product flow $P_{<s_i,\rho>}^{(j,\eta)}$ to the customer $\overline{B}^{(\eta)}$; $u_{ij\eta\rho}^{(o)}(t)=0$ otherwise,

t is current instant of time,

$t \in T=(T_0, T_f]$ is a planning horizon,

T_0 (T_f) is a start and end instants of time of the planning horizon.

The dynamics of the operation $D_\mu^{(i)}$ can be expressed as

$$\dot{x}_{i\mu}^{(o)} = \sum_{j=1}^{n} \varepsilon_{ij}(t) u_{i\mu j}^{(o)} . \tag{12.1}$$

Results of operations execution with regards to the flows consumption ρ is expressed through

$$\dot{x}_{ij\eta\rho}^{(o)} = u_{ij\eta\rho}^{(o)} . \tag{12.2}$$

The economic sense of eq. 12.1 consists of the operations dynamics representation in which process non-stationary and time resource consumption are reflected. Equation 12.2 describes the process of the delivery to the customer $\overline{B}^{(\eta)}$ the product flow $P^{(j,\eta)}_{<s_i,\rho>}$ with the use of the resource $B^{(j)}$.

Constraints on control actions

Let us introduce the following notation:

$a^{(o)}_{i\alpha}$, $a^{(o)}_{i\beta}$ are given quantities (end conditions), the values of which should have the corresponding variables $x^{(o)}_{i\alpha}, x^{(o)}_{i\beta}$ at the end of the planning interval at the instant of time $t = T_f$,

$\Gamma^-_{i\mu_1}$, $\Gamma^-_{i\mu_2}$ are the sets of operations which immediate precede the operation $D^{(i)}_\mu$,

$\sum_{j=1}^{n} u^{(o)}_{i\mu_j} \sum_{\alpha\in\Gamma^-_{i\mu_1}} (a^{(o)}_{i\alpha} - x_{i\alpha}(t)) = 0$ is a constraint "and" which means the condition of the total processing of all the predecessor operations,

$\sum_{j-1}^{n} u^{(o)}_{i\mu_j} \prod_{\beta\in\Gamma^-_{i\mu_2}} (a^{(o)}_{i\beta} - x^{(o)}_{i\beta}) = 0$ is a constraint "or", which means the condition of the processing of at least one of the predecessor operations,

$d^{(1)}_{ij\eta}$ is the maximal productivity of the resource $B^{(j)}$ subject to the collaboration operation $D^{(i)}_\mu$ with the customer $\overline{B}^{(i)}$, $\overline{B}^{(\eta)}$,

$d^{(2)}_{ij\rho}$ is the maximal productivity of the resource $B^{(j)}$ subject to the collaboration operations to deliver products ρ to customers $B^{(\eta)}$.

The control actions are subject to the following constraints:

$$\sum_{i=1}^{\overline{n}} \sum_{\mu=1}^{s_i} u^{(o)}_{i\mu j}(t) \le 1, \forall j; \ \sum_{j=1}^{n} u^{(o)}_{i\mu j}(t) \le 1, \forall i, \forall \mu, \tag{12.3}$$

$$\sum_{j=1}^{n} u^{(o)}_{i\mu j} [\sum_{\alpha\in\Gamma^-_{i\mu_1}} (a^{(o)}_{i\alpha} - x^{(o)}_{i\alpha}) + \prod_{\beta\in\Gamma^-_{i\mu_2}} (a^{(o)}_{i\beta} - x^{(o)}_{i\beta})] = 0, \tag{12.4}$$

$$u^{(o)}_{ij\eta\rho} [(a^{(o)}_{is_i} - x^{(o)}_{is_i}) + (a^{(f)}_{is_i} - x^{(f)}_{is_i})] = 0, \tag{12.5}$$

$$\sum_{\rho=1}^{p_i} u^{(o)}_{ij\eta\rho}(t) \le d^{(1)}_{ij\eta}; \ \forall i; \forall j; \forall \eta, \tag{12.6}$$

$$\sum_{\substack{\eta=1 \\ \eta \neq i}}^{\bar{n}} u_{ij\eta\rho}^{(o)}(t) \leq d_{ij\rho}^{(2)};\ \forall i; \forall j; \forall \rho\,, \tag{12.7}$$

$$u_{i\mu j}^{(o)}(t) \in \{0,1\};\ u_{ij\eta\rho}^{(o)}(t) \in \{0,1\}\,. \tag{12.8}$$

The left-hand part of the first constraint (Eq. 12.3) can be interpreted as the allocation problem. The value "1" reflects that only one operation can be processed by the resource. If we change this value to, e.g. "3", this will mean that three operations can be processed by the resource simultaneously. The right-hand part characterizes the non-interruptible operations. This part can be removed in the case when, e.g. the operation can be divided and processed simultaneously by different resources (e.g. in the case of operative outsourcing).

The second constraint (Eq. 12.4) brings the real process logic into the model. This determines which operations should be executed before the given operation. In particular, this constraint plays a significant role in problem dimensionality reduction. Equation 12.4 implies the blocking of operation $D_{\mu}^{(i)}$ until the previous operations $D_{\alpha}^{(i)}, D_{\beta}^{(i)}$ have been executed, i.e. the required amount of materials is received (processed, delivered). If $u_{i\mu j}^{(o)}(t) = 1$, this means that all the predecessor operations of the operation $D_{\mu}^{(i)}$ have been executed, and thus $[\sum_{\alpha \in \Gamma_{i\mu 1}^{-}} (a_{i\alpha}^{(o)} - x_{i\alpha}^{(o)}) + \prod_{\beta \in \Gamma_{i\mu 2}^{-}} (a_{i\beta}^{(o)} - x_{i\beta}^{(o)})] = 0$.

The presentation of these relations is implemented as DAMG (Kalinin and Sokolov 1985, 1996, Ivanov *et al.* 2010). In these graphs, different SC structures (organizational, functional, informational, financial, and technological) can be represented in their interrelations (Ivanov *et al.* 2010). In the case of the necessity to link customers' orders (e.g. one operation is used in different orders), we can implement these links by exception of the index *i* in the constraint (Eq. 12.4). In the constraint (Eq. 12.4), not only the requirement for timely operation fulfilment but also the requirement for resource and flow utilization can be introduced. This can be useful for the analysis of the operation processing quality.

The third constraint (Eq. 12.5) determines the possibility of the delivery to the customer $\overline{B}^{(\eta)}$ the product flow $P_{<s_i,\rho>}^{(j,\eta)}$ with the use of the resource $B^{(j)}$. The fourth and fifth constraints (Eqs. 12.6 and 12.7) reflect the intensity of resource consumption in the SC. According to Eq. 12.8, controls $u_{i\mu j}^{(o)}(t)$ and $u_{ij\eta\rho}^{(o)}(t)$ contain the values of the Boolean variables.

End conditions

$$\mathbf{h}_0^{(o)}(\mathbf{x}^{(o)}(T_0)) \le \mathbf{0};\ \mathbf{h}_1^{(o)}(\mathbf{x}^{(o)}(T_f)) \le \mathbf{0}, \tag{12.9}$$

where $\mathbf{h}_0^{(o)}$, $\mathbf{h}_1^{(o)}$ are known differentiated functions that determine the end conditions of the vector

$$\mathbf{x}^{(o)} = \left\| \mathbf{x}_{11}^{(o)\mathrm{T}}, \ldots, \mathbf{x}_{\bar{n}n\bar{n}p_i}^{(o)\mathrm{T}} \right\|^{\mathrm{T}}. \tag{12.10}$$

End conditions (Eqs. 12.11 and 12.12) specify the values of variables at the beginning and the end of the planning period:
At the moment $t = T_0$:

$$x_{i\mu}^{(o)}(T_0) = x_{ij\eta\rho}^{(o)}(T_0) = 0. \tag{12.11}$$

At the moment $t = T_f$:

$$x_{i\mu}^{(o)}(T_f) = a_{i\mu}^{(o)};\ x_{ij\eta\rho}^{(o)}(T_f) = a_{ij\eta\rho}^{(o)}. \tag{12.12}$$

Constraint (Eq. 12.11) reflects that, at the beginning, the volume of executed orders is equal to zero (in the case that a certain volume of orders is to be transferred from the previous planning period to the beginning of the current planning period, this should be reflected in Eq. 12.11. Condition (Eq. 12.12) reflects the SC's desired end state.

Goals
Let us introduce the following notation:

$$J_1^{(o)} = \frac{1}{2}\sum_{i=1}^{\bar{n}}\sum_{\mu=1}^{s_i}[(a_{i\mu}^{(o)} - x_{i\mu}^{(o)}(T_f))^2 + \sum_{j=1}^{n}\sum_{\substack{\eta=1 \\ \eta\ne i}}^{\bar{n}}\sum_{\rho=1}^{p_1}(a_{ij\eta\rho}^{(o)} - x_{ij\eta\rho}^{(o)}(T_f))^2], \tag{12.13}$$

$$J_2^{(o)} = \sum_{i=1}^{\bar{n}}\sum_{\mu=1}^{s_i}\sum_{j=1}^{n}\int_{T_0}^{T_f}\alpha_{i\mu}^{(o)}(\tau)u_{i\mu j}^{(o)}(\tau)d\tau. \tag{12.14}$$

The performance metric $J_1^{(o)}$ characterizes the accuracy of the end conditions' accomplishment. This can also express the extent of losses caused by non-fulfilment of the end conditions. In the settings of SC planning and scheduling, this means the service level of an SC. The goal function (Eq. 12.14) refers to the estimation of an operation's execution quality with regard to the supply terms ("ready for delivery"). The function $\alpha_{i\mu}^{(o)}(\tau)$ is assumed to be known for each operation. The goal function (Eq. 12.15) refers to the estimation of the penalties for breaking supply terms with regard to the product ρ and the end customer $B^{(\eta)}$:

$$J_3^{(o)} = \sum_{i=1}^{\bar{n}} \sum_{j=1}^{n} \sum_{\substack{\eta=1 \\ \eta \neq i}}^{\bar{n}} \sum_{\rho=1}^{p_i} \int_{T_0}^{T_f} \gamma_{i\eta\rho}^{(o)}(\tau) u_{ij\eta\rho}^{(o)}(\tau) d\tau \,. \tag{12.15}$$

In this case, the functions $\gamma_{i\eta\rho}^{(o)}(\tau)$ are assumed to be known. These functions define the time points of the penalties increase due to breaking supply terms.

12.2.2 Dynamic Model of Resource Control

Mathematical model of resource control
Let us introduce Eq. 12.16 to assess the total resource availability time:

$$\dot{x}_j^{(k)} = \sum_{i=1}^{\bar{n}} \sum_{\substack{\eta=1 \\ \eta \neq i}}^{\bar{n}} \sum_{\mu=1}^{s_i} \sum_{\rho=1}^{p_i} (u_{i\mu j}^{(o)} + u_{ij\eta\rho}^{(o)}) \,. \tag{12.16}$$

Equation 12.16 represents resource consumption in SC operations dynamics. Introduction of the time factor distinguishes this model from mathematical programming models.

End conditions

$$\mathbf{h}_0^{(k)}(\mathbf{x}^{(k)}(T_0)) \le \mathbf{0};\ \mathbf{h}_1^{(k)}(\mathbf{x}^{(k)}(T_f)) \le \mathbf{0} \,. \tag{12.17}$$

As in the previous models, $\mathbf{h}_0^{(k)}$, $\mathbf{h}_1^{(k)}$ are known differentiated functions that determine the end conditions of the vector

$$\mathbf{x}^{(k)} = \left\| \mathbf{x}_1^{(k)\mathrm{T}}, \ldots, \mathbf{x}_n^{(k)\mathrm{T}} \right\|^{\mathrm{T}} \,. \tag{12.18}$$

End conditions to specify the values of variables at the beginning and the end of planning period are similar to Eqs. 12.11 and 12.12.

Goals

$$J_1^{(k)} = \frac{1}{2}\sum_{j=1}^{n}(T - x_j^{(k)}(T_f))^2 . \tag{12.19}$$

The indicator $J_1^{(k)}$ helps to estimate the uniformity of channel use by the end point $t = T_f$ of the planning period. In our case, this characterizes the equality of resource charge in the SC (the requirement for SC collaboration).

12.2.3 Dynamic Model of Flow Control

In realizing processes and using resources in SC dynamics, different flows (material, informational, and financial) exist. Each flow is characterized by potential, planned and current intensity, flow capacity and parameters. The distribution of flows between the SC processes and resources depends on a great number of factors, such as structure dynamics, flow capacities and the speed of changes in flow characteristics and parameters.

Mathematical model of flow control
To take into account the flow dynamics, let us introduce Eqs. 12.20 and 12.21:

$$\dot{x}_{i\mu j}^{(f)} = u_{i\mu j}^{(f)} , \tag{12.20}$$

$$\dot{x}_{ij\eta\rho}^{(f)} = u_{ij\eta\rho}^{(f)} . \tag{12.21}$$

The economic sense of the Eq. 12.20 consists of the representation of flow consumption of the resource B_j. The dynamic model at Eq. 12.21 describes the delivery of the product ρ to the customer $B^{(\eta)}$.

Constraints on control actions
Let us introduce the following notations:

$c_{i\mu j}^{(f)}$ is a total potential productivity of the resource $B^{(j)}$ subject to the operation $D_\mu^{(i)}$,

$c_{i\eta\rho j}^{(f)}$ is a total potential productivity of the resource $B^{(j)}$ while delivering the product flow $P_{<s_i,\rho>}^{(j,\eta)}$ to the customer $\overline{B}^{(\eta)}$,

$\tilde{\tilde{R}}_{1j}^{(f)}$ is the maximal total productivity of the resource $B^{(j)}$ with regard to product flows,

$\tilde{\tilde{R}}_{1j\eta}^{(f)}$ is the maximal channel intensity to deliver products to the customer $B^{(\eta)}$ with the use of the resource $B^{(j)}$.

$$0 \le u_{i\mu j}^{(f)}(t) \le c_{i\mu j}^{(f)} \cdot u_{i\mu j}^{(o)}, \tag{12.22}$$

$$0 \le u_{ij\eta\rho}^{(f)}(t) \le c_{i\eta\rho j}^{(f)} \cdot u_{i\eta\rho j}^{(o)}, \tag{12.23}$$

$$\sum_{i=1}^{\bar{n}} \sum_{\mu=1}^{s_i} u_{i\mu j}^{(f)}(t) \le \tilde{\tilde{R}}_{1j}^{(f)} \cdot \xi_1^{(f)}(t), \tag{12.24}$$

$$\sum_{\rho=1}^{p_i} u_{ij\eta\rho}^{(f)}(t) \le \tilde{\tilde{R}}_{1j\eta}^{(f)} \xi_2^{(f)}(t). \tag{12.25}$$

Constraints 12.22 and 12.23 reflect the potential intensity of flows. Constraints 12.24 and 12.25 reflect the planned intensity of flows, taking into account the lower and upper boarders of perturbation impacts:

$$0 \le \xi_1^{(f)}(t) \le 1; \; 0 \le \xi_2^{(f)}(t) \le 1. \tag{12.26}$$

End conditions

$$0 \le \xi_1^{(f)}(t) \le 1; \; 0 \le \xi_2^{(f)}(t) \le 1. \tag{12.27}$$

As in the previous models, $\mathbf{h}_0^{(f)}, \mathbf{h}_1^{(f)}$ are known differentiated functions which determine the end conditions to the vector

$$\mathbf{x}^{(f)} = \left\| \mathbf{x}_{111}^{(f)\mathrm{T}}, \ldots, \mathbf{x}_{\bar{n}np_i}^{(f)\mathrm{T}} \right\|^{\mathrm{T}}. \tag{12.28}$$

End conditions to specify the values of variables at the beginning and the end of planning period are similar to Eqs. 12.10 and 12.11.

Goals

$$J_1^{(f)}=\frac{1}{2}\sum_{i=1}^{\bar{n}}\sum_{\mu=1}^{s_i}\sum_{j=1}^{n}[(a_{i\mu j}^{(f)}-x_{i\mu}^{(f)}(T_f))^2+\sum_{\substack{\eta=1\\ \eta\neq i}}^{\bar{n}}\sum_{\rho=1}^{p_i}(a_{ij\eta\rho}^{(f)}-x_{ij\eta\rho}^{(f)}(T_f))^2], \quad (12.29)$$

$$J_2^{(f)}=\frac{1}{2}\sum_{i=1}^{\bar{n}}\sum_{\mu=1}^{s_i}\sum_{j=1}^{n}\int_{T_0}^{T_f}\beta_{i\mu}^{(f)}(\tau)u_{i\mu j}^{(f)}(\tau)d\tau. \quad (12.30)$$

The performance metric $J_1^{(f)}$ characterizes the accuracy of the end conditions' accomplishment. This can also express the extent of losses caused by non-fulfilment of the end conditions. In the settings of SC planning and scheduling, this means the service level of an SC. The indicator $J_2^{(f)}$ refers to the estimation of an operation's execution quality with regard to the supply terms ("ready for delivery") and the penalties for breaking supply terms.

12.2.4 Integrated Dynamic Model of Supply Chain Operations Control

The above-mentioned models can be presented in the integrated form (model *M*) as shown in Sect. 10.5. As mentioned above, the model should provide the decision makers with alternatives to handle. The indicators may be weighted in dependence on the planning goals and SC strategies (e.g. a responsive or efficient SC). The performance metrics' preference calculation (minimax, maximin, etc.) form the Pareto space and allow the calculation of a general multi-model quality index (QI) (Eq. 12.31, see also Eq. 10.20).

$$\mathbf{J}(\mathbf{x}(t),\mathbf{u}(t),\xi(t),t)=\left\|J_1^{(o)},J_2^{(o)},J_3^{(o)},J_1^{(k)},J_1^{(f)},J_2^{(f)}\right\|^{\mathrm{T}}, \quad (12.31)$$

where $J_1^{(o)},J_2^{(o)},J_3^{(o)},J_1^{(k)},J_1^{(f)},J_2^{(f)}$ are values of the indicators characterizing the goals of SCM, for example, service level and profitability, within the corresponding plan $\mathbf{u}(t)$.

The aggregated general QI for all the above-mentioned models can be presented as

$$\mathbf{J}=\left\|\mathbf{J}^{(o)\mathrm{T}}\mathbf{J}^{(k)\mathrm{T}}\mathbf{J}^{(f)\mathrm{T}}\right\|^{\mathrm{T}}. \quad (12.32)$$

The performance metrics are tightly interlinked with the dynamic process model, constraints, and end conditions. E.g. if for conditions (Eqs. 12.12 and 12.27) the convergence is not reached during the scheduling model running, the adaptation of the initial variables of the plan (e.g. the introduction of additional resources, SC cycle extension etc.) is necessary. By often repeated deviations from the planned category, some systematic measures (e.g. safety stock volume increase) are mandatory.

12.2.5 Formal problem statement of complex operations dynamics control

The planning and scheduling problem can be formulated as the following problem of dynamic system control. This is necessary to find an allowable control $\mathbf{u}(t)$, $t \in (T_0, T_f]$ that ensures for the model (Eqs. 12.1, 12.2, 12.16, 12.20 and 12.21) meeting the requirements $\mathbf{q}^{(1)}(\mathbf{x},\mathbf{u}) = \mathbf{0}$, $\mathbf{q}^{(2)}(\mathbf{x},\mathbf{u}) \le \mathbf{0}$ (Eqs, 12.3–12.8 and 12.22–12.26), and guides the dynamic system (SC) $\dot{\mathbf{x}} = \mathbf{f}(\mathbf{x},\mathbf{u},t)$ from the initial state $\mathbf{h}_0$ to the specified final state $\mathbf{h}_1$. If there are several allowable controls (schedules), then the best one (optimal) should be selected in order to maximize (minimize) the components of vector (Eq. 12.32).

12.3 Algorithms of Operations Dynamics Planning and Scheduling

12.3.1 Transformation of the Optimal Control Problem to the Boundary Problem

It is assumed that standard methods (Kalinin and Sokolov 1985, 1987, Okhtilev *et al.* 2006) are used to transform the vector quality measure $\mathbf{J}$ to a scalar form J_G. According to Lee and Markus (1967), along with the initial class $\tilde{K}$ formed via constraints $\mathbf{q}^{(1)}$ and $\mathbf{q}^{(2)}$ describing the domain $\mathbf{Q}(\mathbf{x}(t))$, an extended class $\tilde{\tilde{K}}$ of control inputs can be considered. In the extended class $\tilde{\tilde{K}}$ the relay constraints $u^{(o)}_{i\mu j}(t) \in \{0;1\}$ are substituted for a less strict one $u^{(o)}_{i\mu j}(t) \in [0;1]$ ($\mathbf{u}$ is substituted for $\tilde{\tilde{\mathbf{u}}}$). In this case, an extended domain $\tilde{\tilde{\mathbf{Q}}}(\mathbf{x}(t))$ of allowable control inputs may be formed through special transformations ensuring the convexity and compactness of $\mathbf{Q}(\mathbf{x}(t))$ (Moiseev 1974, Kalinin and Sokolov 1985, 1987). An analysis of Lee and Markus (1967) and (Okhtilev *et al.* 2006) confirms that all the conditions of optimal control existence for the extended control class $\tilde{\tilde{K}}$ are valid. The work (Okhtilev *et al.* 2006) has shown that, if in a given class of permissible control actions $\tilde{\tilde{K}}$, the optimal control $\tilde{\tilde{\mathbf{u}}}(t)$ exists, then, as arises from the local section method, the control $\tilde{\tilde{\mathbf{u}}}(t)$ returns at each instant of time $t \in (T_0, T_f]$ at the set $\tilde{\tilde{\mathbf{Q}}}(\mathbf{x}(t))$ a maximum to the following Hamiltonian (Eqs. 12.33–12.37)

$$H(\mathbf{x}^*(t), \mathbf{u}^*(t), \boldsymbol{\psi}^*(t)) = \max_{\tilde{\tilde{\mathbf{u}}} \in \tilde{\tilde{\mathbf{Q}}}(\mathbf{x})} \sum_{\tilde{l}=1}^{4} H_l(\mathbf{x}(t), \mathbf{u}(t), \boldsymbol{\psi}(t)), \tag{12.33}$$

$$H_1 = \sum_{i=1}^{\bar{n}} \sum_{\mu=1}^{s_i} \sum_{j=1}^{n} [\psi_{i\mu}^{(o)} \cdot \varepsilon_{ij} + \psi_j^{(k)} + \lambda_2 \alpha_{i\mu j}^{(o)}] u_{i\mu j}^{(o)}, \tag{12.34}$$

$$H_2 = \sum_{i=1}^{\bar{n}} \sum_{\mu=1}^{s_i} \sum_{j=1}^{n} \sum_{\substack{\eta=1 \\ \eta \neq i}}^{\bar{n}} \sum_{\rho=1}^{p_i} [\psi_{ij\eta\rho}^{(o)} + \psi_j^{(k)} + \lambda_3 \gamma_{i\eta\rho}^{(o)}] u_{ij\eta\rho}^{(o)}, \tag{12.35}$$

$$H_3 = \sum_{i=1}^{\bar{n}} \sum_{\mu=1}^{s_i} \sum_{j=1}^{n} [\psi_{i\mu j}^{(f)} + \lambda_6 \beta_{i\mu}^{(f)}] u_{i\mu j}^{(f)}, \tag{12.36}$$

$$H_4 = \sum_{i=1}^{\bar{n}} \sum_{j=1}^{n} \sum_{\substack{\eta=1 \\ \eta \neq i}}^{\bar{n}} \sum_{\rho=1}^{p_i} \psi_{ij\eta\rho}^{(f)} u_{ij\eta\rho}^{(f)}. \tag{12.37}$$

Maximization of the Hamiltonians H_1 and H_2 solves the allocation problem. Maximization of the Hamiltonians H_3 and H_4 solves the linear programming problem.

The problem considered can be reduced to a two-point boundary problem via Pontryagin's maximum principle. Here, the conjugate system can be written as follows (Pontryagin 1961, Boltyanskiy 1966, Moiseev 1974):

$$\dot{\psi}_{\tilde{\tilde{l}}} = -\frac{\partial H}{\partial x_{\tilde{\tilde{l}}}} + \sum_{\tilde{\alpha}=1}^{\tilde{I}_1} \delta_{\tilde{\alpha}}(t) \frac{\partial \mathbf{q}_{\tilde{\alpha}}^{(1)}(\mathbf{x}(t), \mathbf{u}(t))}{\partial x_{\tilde{\tilde{l}}}} + \sum_{\tilde{\beta}=1}^{\tilde{I}_2} \rho_{\tilde{\beta}}(t) \frac{\partial \mathbf{q}_{\beta}^{(2)}(\mathbf{x}(t), \mathbf{u}(t))}{\partial x_{\tilde{\tilde{l}}}}. \quad (12.38)$$

The coefficients $\delta_{\tilde{\alpha}}(t)$, $\rho_{\tilde{\beta}}(t)$ can be determined through the following expressions:

$$\rho_{\tilde{\beta}}(t)\mathbf{q}_{\tilde{\beta}}^{(2)}(\mathbf{x}(t), \mathbf{u}(t)) \equiv 0, \ \tilde{\beta} \in \{1, \ldots, \tilde{I}_2\}, \quad (12.39)$$

$$\begin{aligned} &\operatorname{grad}_{\mathbf{u}} H(\mathbf{x}(t), \mathbf{u}(t), \boldsymbol{\psi}(t)) = \\ &= \sum_{\tilde{\alpha}=1}^{\tilde{I}_1} \delta_{\tilde{\alpha}}(t) \operatorname{grad}_{\mathbf{u}} \mathbf{q}_{\tilde{\alpha}}^{(1)}(\mathbf{x}(t), \mathbf{u}(t)) + \sum_{\tilde{\beta}=1}^{\tilde{I}_2} \rho_{\tilde{\beta}}(t) \operatorname{grad}_{\mathbf{u}} \mathbf{q}_{\tilde{\beta}}^{(2)}(\mathbf{x}(t), \mathbf{u}(t)). \end{aligned} \quad (12.40)$$

In Eqs. 12.38–12.40, $x_{\tilde{\tilde{l}}}$ are elements of a general state vector $\mathbf{x}(t)$ and $\psi_{\tilde{\tilde{l}}}$ are elements of a conjugate vector $\boldsymbol{\psi}(t)$. In accordance with the maximum principle, the following conjugate system can be presented (Eqs. 12.41–12.47):

$$\dot{\psi}_{i\mu}^{(o)} = -\sum_{j=1}^{n} [\psi_{i(\mu+1)}^{(o)} \varepsilon_{ij} + \psi_j^{(k)} + \lambda_2 \alpha_{i(\mu+1)j}^{(o)}] u_{i(\mu+1)j}^{(o)} \ , \quad (12.41)$$

$$\dot{\psi}_{is_i}^{(o)} = -\sum_{j=1}^{n} \sum_{\substack{\eta=1 \\ \eta \neq i}}^{\pi} \sum_{\rho=1}^{p_i} [\psi_{ij\eta\rho}^{(o)} + \psi_j^{(k)} + \lambda_3 \gamma_{is_i}^{(o)}] u_{ij\eta\rho}^{(o)} \ , \quad (12.42)$$

$$\dot{\psi}_{ij\eta\rho}^{(o)} = 0, \quad (12.43)$$

$$\dot{\psi}_j^{(k)} = 0, \quad (12.44)$$

$$\dot{\psi}_{i\mu j}^{(f)} = 0, \quad (12.45)$$

$$\dot{\psi}^{(f)}_{is_j j} = -\sum_{\substack{\eta=1\\ \eta\neq i}}^{\pi}\sum_{\rho=1}^{p_l}[\psi^{(o)}_{ij\eta\rho} + \psi^{(k)}_{j} + \lambda_3\gamma^{(o)}_{i\eta\rho}]u^{(o)}_{ij\eta\rho} \quad , \tag{12.46}$$

$$\dot{\psi}^{(f)}_{ij\eta\rho} = 0\,. \tag{12.47}$$

We assume, that the problem of multi-criteria vector optimization is solved for this particular task. In this case, in transversality conditions at the instant of time, the coefficients $\lambda_\varsigma,(\varsigma = 1,\ldots,\Im)$ are known (compare with Eq. 10.11). These coefficients will be used for multi-objective decision-making (see also Eqs. 12.31 and 12.32). The multiple-criteria optimization problem will be considered in Chap. 13 in detail. Hence, the transversality conditions can be formulated in the following way (Eqs. 12.48–12.54):

$$\psi^{(o)}_{i\mu}(T_f) = \lambda_1(a^{(o)}_{i\mu} - x^{(o)}_{i\mu}(T_f)); \tag{12.48}$$

$$\psi^{(o)}_{is_i}(T_f) = \lambda_1(a^{(o)}_{is_i} - x^{(o)}_{is_i}(T_f)); \tag{12.49}$$

$$\psi^{(o)}_{ij\eta\rho}(T_f) = \lambda_1(a^{(o)}_{ij\eta\rho} - x^{(o)}_{ij\eta\rho}(T_f)); \tag{12.50}$$

$$\psi^{(k)}_{j}(T_f) = \lambda_4(T - x^{(k)}_{j}(T_f)); \tag{12.51}$$

$$\psi^{(f)}_{i\mu j}(T_f) = \lambda_5(a^{(f)}_{i\mu j} - x^{(f)}_{i\mu j}(T_f)); \tag{12.52}$$

$$\psi^{(f)}_{is_i j}(T_f) = \lambda_5(a^{(f)}_{is_i j} - x^{(f)}_{is_i j}(T_f)); \tag{12.53}$$

$$\psi^{(f)}_{ij\eta\rho}(T_f) = \lambda_5(a^{(f)}_{ij\eta\rho} - x^{(f)}_{ij\eta\rho}(T_f)). \tag{12.54}$$

An analysis of Eqs. 12.33–12.37 shows that the Hamiltonian is linear in $\tilde{\tilde{\mathbf{u}}}$. Since $\tilde{\tilde{\mathbf{Q}}}(\mathbf{x}(t))$ is a linear capsule of $\mathbf{Q}(\mathbf{x}(t))$, the maximization of the Hamiltonian Eqs. 12.33–12.37 over the sets $\mathbf{Q}$ and $\tilde{\tilde{\mathbf{Q}}}$ leads to the same results. We come to the conclusion that the optimal control of the class $\tilde{K}$ belongs to the class $\tilde{\tilde{K}}$. Taking into account $\tilde{K} \subset \tilde{\tilde{K}}$, we see that the control is also optimal for the class $\tilde{K}$. Therefore the *relaxed problem* can be solved instead of the initial one to receive an optimal allowable control of the class $\tilde{K}$.

Equations 12.34 and 12.35 are a discrete form of an allocation problem at the each moment *t*, $t \in (T_0, T_f]$. Equations 12.36 and 12.37 are the linear programming problem. Equation 12.39 ensures the reduction of problem dimensionality at each instant of time in the calculation process due to recurrent operations description. At each instant of time, only those operations are considered that meet the requirements of constraints (the so-called active operations). Thus the problem dimensionality depends on the amount of active operations only. It was shown by Kalinin and Sokolov (1985, 1987) that the stated necessary conditions of optimality are also sufficient. Hence, a scheduling problem for SCs can be reduced to a boundary problem with the help of the local section method. It was shown by Okhtilev *et al.* (2006) that the stated necessary conditions of optimality are also the conditions of sufficiency.

12.3.2 Existing Methods for Optimal Control Based on the Maximum Principle

Let us consider the algorithmic realization of the maximum principle. In accordance with this principle, two systems of differential equations should be solved: the main system (Eq. 10.21) and the conjugate one (Eqs. 12.41–12.47). This will provide the optimal programme control vector $\mathbf{u}^*(t)$ and the state trajectory $\mathbf{x}^*(t)$. The vector $\mathbf{u}^*(t)$ at time $t = T_0$ under the conditions $\mathbf{h}_0(\mathbf{x}(T_0)) \leq \mathbf{O}$ and for the given value of $\psi(T_0)$ should return the maximum to the Hamilton's function.

The classification of methods and algorithms for optimal control problems is illustrated in Fig. 12.1 (Moiseev 1974). The most popular methods for the two-point boundary problems with fixed ends of the state trajectory $\mathbf{x}(t)$ and a fixed time interval $(T_0, T_f]$ are the following methods (Lee and Markus 1967, Moiseev 1974, Gigch 1978, Siliak 1990, Okhtilev *et al.* 2006):

- Newton's method and its modifications;
- methods of penalty functionals;
- gradient methods; and
- Krylov–Chernousko method.

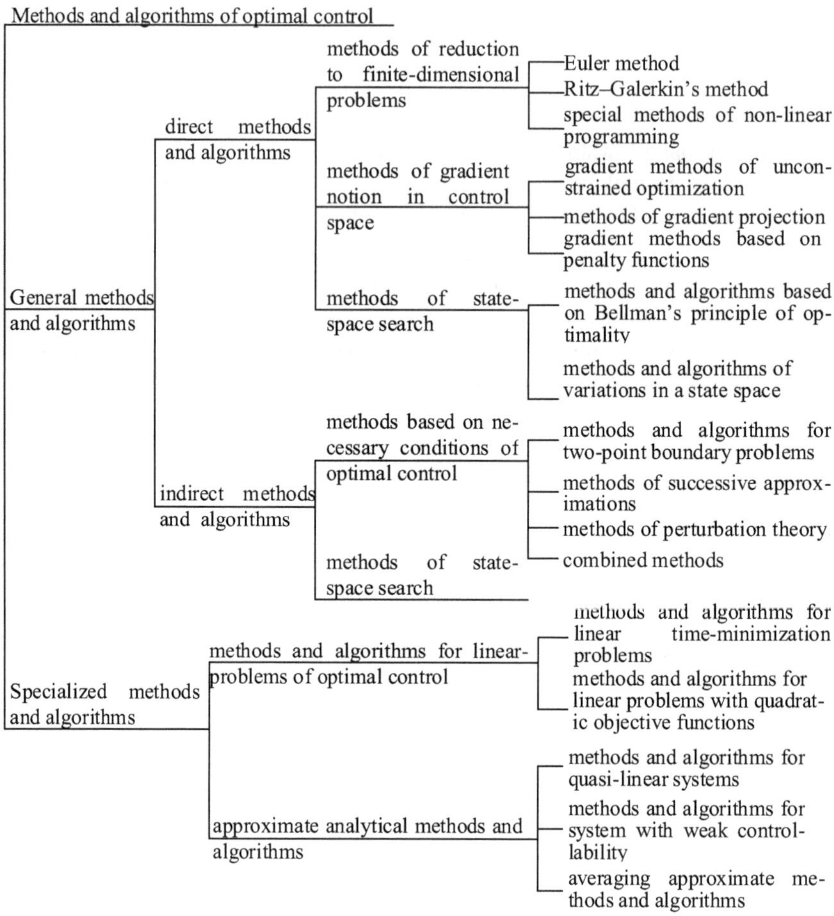

Fig. 12.1 Classification of methods and algorithms for optimal control problems

The main advantages $\psi(T_0)$ of Newton's method and its modifications are a simple realization (there is no need to integrate the conjugate system), a fast convergence (if the initial choice of $\mathbf{u}$ is good) and a high accuracy of the solution. The main disadvantage is the dependency of a convergence upon the choice of $\psi(T_0)$. In the case of the absence of a good heuristic plan these methods can be divergent. The method of penalty functionals is rather simple, but it does not provide an exact solution. Therefore, it is advisable to combine it with other methods. The main advantage of these algorithms over classical gradient algorithms is a simpler calculation of the direction vector at all iterations. However, this results in a slower convergence (sometimes in divergence). The convergence of all the gradient methods depends upon the initial approximation $\psi_{(0)}(T_0)$. For the problem of SC scheduling, we have chosen the Krylov-Chernousko method (Chernousko and Zak 1985, Chernousko 1994).

12.3.3 Proposed Algorithm of Supply Chain Scheduling

Let us consider the algorithm for the indicated boundary problem. As mentioned above, the "first approach" for launching the optimization procedure is a heuristics plan that can be generated either by a simple priority rule (e.g. FIFO – first-in-first-out) or by a high-level heuristic such as a genetic algorithm. Then, the scheme of computation can be stated as follows:

Step 1. An initial solution $\overline{\mathbf{u}}(t)$, $t \in (T_0, T_f]$ (an arbitrary allowable control, in other words, allowable schedule) is selected and $r = 0$.

Step 2. As a result of the dynamic model run, $\mathbf{x}^{(r)}(t)$ is received. Besides, if $t = T_f$ then the record value $J_G = J_G^{(r)}$ can be calculated. Then the transversality conditions are evaluated.

Step 3. The conjugate system (Eqs. 12.41–12.47) is integrated subject to $\mathbf{u}(t) = \overline{\mathbf{u}}(t)$ and over the interval from $t = T_f$ to $t = T_0$. For time $t = T_0$, the first approximation $\psi_i^{(r)}(T_0)$ is received as a result. Here, the iteration number $r = 0$ is completed.

Step 4. From the point $t = T_0$ onwards, the control $\mathbf{u}^{(r+1)}(t)$ is determined ($r = 0,1,2,\ldots$ is the number of iterations) through the conditions (Eqs. 12.48–12.54). In parallel with maximization of the Hamiltonian, the main system of equations and the conjugate one are integrated. The maximization involves solving several mathematical programming problems at each time point.

The iterative process of the optimal schedule search is terminated under the following circumstances: either the allowable solution to the problem is determined during the solving of a relaxed problem, or at the fourth step of the algorithm after the integration we receive

$$\begin{aligned} &\left| J_{\mathrm{G}}^{(r+1)} - J_{\mathrm{G}}^{(r)} \right| < \varepsilon_1, \\ &\left\| \mathbf{u}^{(r+1)} - \mathbf{u}^{(r)} \right\| < \varepsilon_2, \end{aligned} \tag{12.55}$$

where $\varepsilon_1, \varepsilon_2$ are given small values, $r = 0,1,2,\ldots$. If the condition (Eq. 12.55) is not satisfied, then the third step is repeated, etc.

The problem with the maximum principle-based algorithms is that they may calculate a solution with interruptions to operations. This may be inadmissible in real problems. This issue has been tackled by Kalinin and Sokolov (1987). In this study, a special algorithm for non-interruptible operations was proposed. We will not consider this approach here in detail.

The advantage of the proposed method is that it is not so sensitive to the selection of the "first approach" to optimal control compared with the other methods described in Sect. 12.3.2. Besides, the usage of this method allows the application of gradual approaches of control actions in the class of interrupted functions. This has great practical importance because of the relayed nature of resource distribution problems.

12.4 Concluding Remarks

The realization of the dynamic control approaches to SC planning and scheduling produces algorithmic and computational difficulties caused by high dimensionality, non-linearity, non-stationarity and uncertainty of the models. We proposed to modify the dynamic interpretation of the operations control processes. The main idea of the model simplification is to implement non-linear technological constraints in the sets of allowable control inputs rather than in the right parts of differential equations. In this case, Lagrange coefficients, keeping the information about economic and technological constraints, are defined via the local-sections method (Pontryagin 1961).

Computational investigations have shown that the use of the SC dynamics models entails a considerable dimensionality decrease for control problems to be solved in a real-time operation mode. The recurrence description of models allows parallel computations, accelerating problem solving. Furthermore, we proposed to use interval constraints instead of relay ones. Nevertheless, the control inputs take on Boolean values as caused by the linearity of differential equations and the convexity of the set of alternatives.

The proposed models are based on the structure dynamic control approach (see Chap. 10) of the modern control theory in combination with the optimization methods of OR and have some specific features in comparison with classic optimal control problems. The first feature is that the right parts of the differential equations undergo discontinuity at the beginning of interaction zones. The considered problems can be regarded as control problems with intermediate conditions. The second feature is the multi-criteria nature of the problems. The third feature is concerned with the influence of uncertainty factors. The fourth feature is the form of time-spatial, technical, and technological non-linear conditions that are mainly considered in control constraints and boundary conditions.

On the whole the constructed model is a non-linear non-stationary finite-dimensional differential system with a reconfigurable structure. The modelling procedure is based on an essential reduction of a problem dimensionality that is under solution at each instant of time due to connectivity decreases. Different variants of the model aggregation can be proposed. These variants produce a problem of model quality selection that is the problem of the model complexity reduction. Decision-makers can select an appropriate level of the model thoroughness in the interactive mode. The level of thoroughness depends on: input data, on external conditions, on required level of solution validity.

Finally, we would like to draw attention to the proposed model and algorithm allowing the achievement of better results in many cases in comparison with heuristics algorithms (see Chap. 15). However, this point is not the most important. The most important point is that this approach allows the interlinking of planning and scheduling models within an adaptation framework. Hence, the proposed modelling complex does not exist as a "thing in itself" but is embedded into the IDSS and guides the planning and scheduling decisions in dynamics on the principles of optimization and adaptation.

References

Boltyanskiy V (1966) Mathematical methos of optimal control. Nauka, Moscow (in Russian)

Chernousko FL (1994) State Estimation of Dynamic Systems. SRC Press, Boca Raton, Florida

Chernousko FL, Zak VL (1985) On Differential Games of Evasion from Many Pursuers. J Optim Theor Appl 46(4):461–470

Day T, Taylor PD (2000) A generalization of Pontryagin's maximum principle for dynamic evolutionary games among relatives. Theor Popul Biol 57(4):339–356

Gigch J (1978) Applied general systems theory. Harper and Row Publishers, New York, San Francisco, London

Ivanov D, Arkhipov A, Sokolov B (2007) Intelligent planning and control of manufacturing supply chains in virtual enterprises. Int J Manuf Tech Manag, Special issue on modelling and optimisation of supplier-based manufacturing and management 11(2):209–227

Ivanov D, Kaeschel J, Sokolov B (2009) Structure dynamics control-based framework for adaptive reconfiguration of collaborative enterprise networks. Int J Manuf Tech Manag 17(1-2):23–41

Ivanov D, Sokolov B, Kaeschel J (2010) A multi-structural framework for adaptive supply chain planning and operations control with structure dynamics considerations. Eur J Oper Res 200:409–420

Kalinin VN, Sokolov BV (1985) Optimal planning of the process of interaction of moving operating objects. Differ Equat 21(5):502–506

Kalinin VN, Sokolov BV (1987) A dynamic model and an optimal scheduling algorithm for activities with bans of interrupts. Autom Rem Contr 48(1-2):88–94

Kalinin VN, Sokolov BV (1996) Multiple-model description of control processes for airspace crafts. J Comput Syst Sci Int 6:192–199

Lee EB, Markus L (1967) Foundations of optimal control theory. John Wiley & Sons, New York

Moiseev NN (1974) Element of the optimal systems theory. Nauka, Moscow (in Russian)

Moon C, Lee YH, Jeong CS, Yun YS (2008) Integrated process planning and scheduling in an supply chain. Comput Ind Eng 54(4):1048–1061

Okhtilev M, Sokolov B, Yusupov R (2006) Intelligent technologies of complex systems monitoring and structure dynamics control. Nauka, Moscow (in Russian)

Pontryagin LS (1961) Theory of optimal process. Fizmatgiz, Moscow (in Russian)

Sethi SP, Thompson GL (2006) Optimal Control Theory: Applications to Management Science and Economics, Second Edition. Springer, Berlin

Shao X, Li X, Gao L, Zhang C (2009) Integration of process planning and scheduling. A modified genetic algorithm-based approach. Comput Ind 36(6):2082–2096

Shell K (1968) Applications of Pontryagin's maximum principle of economics. Working papers 16, Massachusetts Institute of Technology (MIT), Department of Economics

Siliak DD (1990) Decentralized control of complex systems. Academic Press, New York

Sokolov B (1992) Comprehensive operations planning and control in structures with active moving objects. Mozaiskay Academia, Saint Petersburg (in Russian)

Sokolov BV, Yusupov RM (2004) Conceptual foundations of quality estimation and analysis for models and multiple-model systems. J Comput Syst Sci Int 6:5–16

Chapter 13
Supply Chain Reconfiguration and Models' Adaptation

Who controls the past controls the future:
who controls the present controls the past.
George Orwell

13.1 Variety of Supply Chain Reconfiguration Issues

Figure 13.1 shows possible variants of SC reconfiguration (Ivanov *et al.* 2009).

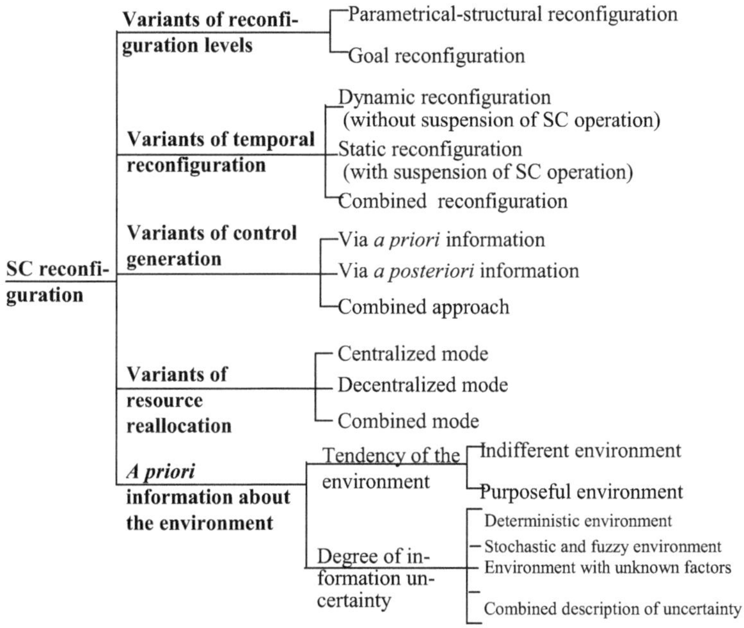

Fig. 13.1 Classification of the SC reconfiguration tasks

The components of this classification scheme are used in the SC reconfiguration control loop. This loop has also some additional functions:

- A function of diagnostics (FD) includes the following operations: SC state determination, localization of state changes, estimation of state changes' depth.

- A function of SC structure reconfiguration includes the following operations: SC state estimation in accordance with FD data, searching for MSMS, purposeful modification of SC structures, and SC structural impacts by means of FD.
- A function of protection consists of the failure type (simple or stop-all) estimation in accordance with FD data, failure consequence localization in the case of stop-all failure, SC transition to an operability state or to a simple failure state by means of structure reconfiguration.
- A function of reserve control includes the following operations: determination of SC element non-operability in accordance with FD data, reserve element switching on, inspection of element replacement results.
- A function of maintenance and function of repair include the following operations: determination of the recovery execution processes by means of FD and determination of the SC operation mode during adaptation.

The considerations presented led us to a wide interpretation within a new applied theory of SDC. In the next section, the formal model and algorithms of the SC reconfiguration will be presented.

13.2 Mathematical Model of the Supply Chain Reconfiguration

It is assumed that there are several variants of SC models inscribed in the set $\overline{\overline{M}} = \{M_1, M_2, ..., M_\theta\} = \{M_\Theta, \Theta \in \hat{I}\}$, $\hat{I} = \{1,...,\theta\}$; moreover, the vector $\boldsymbol{\beta}$ of the SC parameters includes the subvector $\boldsymbol{\beta}_0$ of fixed SC characteristics and beside it the subvector $\mathbf{w} = \left\| \mathbf{w}^{(1)\mathrm{T}}, \mathbf{w}^{(2)\mathrm{T}}, \mathbf{w}^{(3)\mathrm{T}} \right\|^{\mathrm{T}}$ of the parameters being adjusted through the SC external/internal adapter or defined within the structural adaptation.

These parameters are divided into the following groups (Skurikhin *et al.* 1989):

- $\mathbf{w}^{(1)}$ is a vector of the parameters being adjusted through the internal adapter. This vector consists of two subvectors. The first one, $\mathbf{w}^{(1,f)}$, belongs to the scheduling model, and the second one, $\mathbf{w}^{(1,p)}$, belongs to the model of control at the phase of plan execution.
- $\mathbf{w}^{(2)}$ is a vector of the parameters being adjusted through the external adapter. This vector consists of the subvector $\mathbf{w}^{(2,f)}$ belonging to the scheduling model and the subvector $\mathbf{w}^{(u)}$ including the parameters of the simulation model for SC functioning under perturbation impacts. In its turn, $\mathbf{w}^{(u)} = \left\| \mathbf{w}^{(2,o)\mathrm{T}}, \mathbf{w}^{(2,b)\mathrm{T}}, \mathbf{w}^{(2,p)\mathrm{T}} \right\|^{\mathrm{T}}$, where $\mathbf{w}^{(2,o)}$ is a vector of the parameters characterizing customers' orders in service; $\mathbf{w}^{(2,b)}$ is a vector of the parameters characterizing the environment; and $\mathbf{w}^{(2,p)}$ belongs to the model of control at the phase of plan execution.

- $\mathbf{w}^{(3)}$ is a vector of the parameters being adjusted within the structural adaptation of SC SDC models.

Now, we have the modified multiple model multi-criteria description of SC:

$$\mathbf{J}_\Theta\left(\mathbf{x}(t),\mathbf{u}(t),\boldsymbol{\beta},\xi(t),t\right)\to \underset{\mathbf{u}(t)\in\Delta_\Theta}{\text{extr}}\,, \tag{13.1}$$

$$\Delta_\Theta=\left\{\mathbf{u}(t)\,|\,\mathbf{x}(t)=\tilde{\boldsymbol{\varphi}}_\Theta\left(T_0,\mathbf{x}(T_0),\mathbf{x}(t),\mathbf{u}(t),\xi(t),\boldsymbol{\beta}_\Theta,t\right),\right. \tag{13.2}$$

$$\mathbf{y}(t)=\tilde{\boldsymbol{\psi}}_\Theta\left(\mathbf{x}(t),\mathbf{u}(t),\xi(t),\boldsymbol{\beta}_\Theta,t\right), \tag{13.3}$$

$$\mathbf{x}(T_0)\in X_0(\boldsymbol{\beta}_\Theta),\quad \mathbf{x}(T_f)\in X_f(\boldsymbol{\beta}_\Theta), \tag{13.4}$$

$$\mathbf{u}(t)=\left\|\mathbf{u}_{pl}^{\mathrm{T}}(t),\mathbf{v}^{\mathrm{T}}\left(\mathbf{x}(t),t\right)\right\|^{\mathrm{T}};\ \mathbf{u}_{pl}(t)\in\mathbf{Q}_\Theta\left(\mathbf{x}(t),t\right);\ \ \mathbf{v}(t)\left(\mathbf{x}(t),t\right)\in\mathbf{V}_\Theta\left(\mathbf{x}(t),t\right), \tag{13.5}$$

$$\xi(t)\in\boldsymbol{\Xi}_\Theta\left(\mathbf{x}(t),t\right);\ \ \boldsymbol{\beta}_\Theta\in\mathrm{B}\,, \tag{13.6}$$

$$\mathbf{x}(t)\in\mathbf{X}\left(\xi(t),t\right), \tag{13.7}$$

$$\begin{aligned}
\mathbf{x}(t)&=\left\|\mathbf{x}^{(g)\mathrm{T}}(t),\mathbf{x}^{(o)\mathrm{T}}(t),\mathbf{x}^{(k)\mathrm{T}}(t),\mathbf{x}^{(p)\mathrm{T}}(t),\mathbf{x}^{(f)\mathrm{T}}(t),\mathbf{x}^{(e)\mathrm{T}}(t),\mathbf{x}^{(c)\mathrm{T}}(t),\mathbf{x}^{(v)\mathrm{T}}(t)\right\|^{\mathrm{T}};\\
\mathbf{y}(t)&=\left\|\mathbf{y}^{(g)\mathrm{T}}(t),\mathbf{y}^{(o)\mathrm{T}}(t),\mathbf{y}^{(k)\mathrm{T}}(t),\mathbf{y}^{(p)\mathrm{T}}(t),\mathbf{y}^{(f)\mathrm{T}}(t),\mathbf{y}^{(e)\mathrm{T}}(t),\mathbf{y}^{(c)\mathrm{T}}(t),\mathbf{y}^{(v)\mathrm{T}}(t)\right\|^{\mathrm{T}};\\
\mathbf{u}(t)&=\left\|\mathbf{u}^{(g)\mathrm{T}}(t),\mathbf{u}^{(o)\mathrm{T}}(t),\mathbf{u}^{(k)\mathrm{T}}(t),\mathbf{u}^{(p)\mathrm{T}}(t),\mathbf{u}^{(f)\mathrm{T}}(t),\mathbf{u}^{(e)\mathrm{T}}(t),\mathbf{u}^{(c)\mathrm{T}}(t),\mathbf{u}^{(v)\mathrm{T}}(t)\right\|^{\mathrm{T}};\\
\xi(t)&=\left\|\xi^{(g)\mathrm{T}}(t),\xi^{(o)\mathrm{T}}(t),\xi^{(k)\mathrm{T}}(t),\xi^{(p)\mathrm{T}}(t),\xi^{(f)\mathrm{T}}(t),\xi^{(e)\mathrm{T}}(t),\xi^{(c)\mathrm{T}}(t),\xi^{(v)\mathrm{T}}(t)\right\|^{\mathrm{T}};\\
\boldsymbol{\beta}_\Theta&=\left\|\boldsymbol{\beta}_\Theta^{(g)\mathrm{T}},\boldsymbol{\beta}_\Theta^{(o)\mathrm{T}},\boldsymbol{\beta}_\Theta^{(k)\mathrm{T}},\boldsymbol{\beta}_\Theta^{(p)\mathrm{T}},\boldsymbol{\beta}_\Theta^{(f)\mathrm{T}},\boldsymbol{\beta}_\Theta^{(e)\mathrm{T}},\boldsymbol{\beta}_\Theta^{(c)\mathrm{T}},\boldsymbol{\beta}_\Theta^{(v)\mathrm{T}}\right\|^{\mathrm{T}};
\end{aligned} \tag{13.8}$$

$$\begin{aligned}
\boldsymbol{\beta}_\Theta&=\left\|\boldsymbol{\beta}_0^{\mathrm{T}}\ \mathbf{w}^{\mathrm{T}}\right\|^{\mathrm{T}};\ \ \mathbf{w}=\left\|\mathbf{w}^{(1)\mathrm{T}}\ \mathbf{w}^{(2)\mathrm{T}}\ \mathbf{w}^{(3)\mathrm{T}}\right\|^{\mathrm{T}};\\
\mathbf{w}^{(1)}&=\left\|\mathbf{w}^{(1,f)\mathrm{T}}\ \mathbf{w}^{(1,p)\mathrm{T}}\right\|^{\mathrm{T}};\\
\mathbf{w}^{(2)}&=\left\|\mathbf{w}^{(2,f)\mathrm{T}}\ \mathbf{w}^{(2,\xi)\mathrm{T}}\right\|^{\mathrm{T}},\\
\mathbf{w}^{(\xi)}&=\left\|\mathbf{w}^{(2,o)\mathrm{T}}\ \mathbf{w}^{(2,b)\mathrm{T}}\ \mathbf{w}^{(2,p)\mathrm{T}}\right\|^{\mathrm{T}}\ \Big\}.
\end{aligned} \tag{13.9}$$

The formulas define a dynamic system describing SC reconfiguration control processes. As stated in Chap. 10, the problem of SC structure dynamics control includes tasks of three main classes. Here, we consider a formal description and solving algorithms for the class C problems. In a general case, the formal statement of the SC SDC problem can be written as follows.

We are given: space-time, technical, and technological constraints (Eqs. 13.2–13.7) and (13.8) determining the variants of SC SDC at the operation phase; vector (Eq. 13.1) of SC performance metrics.

We should determine: $\mathbf{u}_{pl}(t)$, $\mathbf{x}(\mathbf{x}(t),t)$, and $\boldsymbol{\beta}_{\Theta}$ meeting the constraints (Eqs. 13.2–13.7) and returning optimal value to the general performance metric $J_{<G,\Theta>} = J_{<G,\Theta>}(\mathbf{x}(t),\mathbf{y}(t),\boldsymbol{\beta}_{\Theta},\mathbf{u}(t),\xi(t))$.

The problem described is too complex to be solved as a whole. That is why we propose a decomposition by splitting the initial problem into tasks of five main subclasses (Okhtilev *et al.* 2006):

- *Subclass C1 problems* (problems of real-time structural–functional synthesis of the SC design);
- *Subclass C2 problems* (problems of optimal control programme selection for SC SDC);
- *Subclass C3 problems* (problems of control input generation for optimal conditions of control programme execution);
- *Subclass C4 problems* (problems of SC parameter optimization); and
- *Subclass C5 problems* (problems of the structural and parametric adaptation of SC models and algorithms).

Let us consider the main features of the subclasses C1–C5. *The main peculiarity* of the enumerated problems is the coordination of heterogeneous multilevel models of SC elements and subsystems. The coordination of models should be accompanied by the inter-model and intra-model coordination of performance metrics (goal functions) used for the comparison of alternative decisions.

Let us complete this description with the classification of coordination algorithms. The following variants of coordination are distinguished for the control system: goal functions of the $\tilde{\gamma}$ level are constructed via the decomposition of the $(\tilde{\gamma}+1)$-level coordinator's goal function.

The goal function of the $(\tilde{\gamma}+1)$-level coordinator is constructed via the composition of lower-level goal functions in compliance with their priorities. In this case, the coordinator adjusts the plans of the lower subsystems and takes into account the limited common resources.

The elements and subsystems of the $\tilde{\gamma}$ level of the SC have their own interests and appropriate goal functions. The coordinator of the $(\tilde{\gamma}+1)$ level considers these interests and uses his goal function to produce coordinating control inputs.

There are two approaches to the organization of iterative coordination:

- The coordinator and subsystems communicate during the coordination process.
- The coordinator analyses the information received from the subsystems and then accomplishes the iterative process of coordination and produces control signals for the subsystems.

The following features of *C1 problems* can be noted: optimal control programmes for the SC's can be implemented only when the list of functions and algorithms for control in the SC elements is known. In its turn, the distribution of the functions and algorithms among the SC elements depends upon the actual control laws for these elements. The described contradictory situation is complicated by the changes in SC parameters and structures as caused by different reasons during the SC life cycle.

Problems of the subclass C3 are related to the following functions:

- the production of control inputs for a customer order according to a constructed schedule (programme);
- the monitoring of the programme execution and of the control inputs; and
- the production (if necessary) of auxiliary control inputs to compensate for perturbations disturbing the execution of the programme.

It is obvious that the *subclass C3* problems are different from the typical automatic control problems. First, the complexity and inertness of the SC are to be reflected in the special technologies of control inputs produced for the realization of the SC's operation plans. Second, the *function of regulation* (real-time control) is directly connected with the *function of plan monitoring*. The latter function is quite different from the function of output evaluation in classic automatic systems.

According to the classification given in Chap. 7, we shall distinguish the following types of regulative inputs (types of real-time control) in the SC:

- *Process correction* or selectable control (dispatching) is a selection of control inputs appropriate for the current situation. It is based on reserves of different types and on the alternation of functioning modes.
- *Plan correction* (control inputs produced as a function of the difference $\Delta\mathbf{x}(t)$ between the planned state trajectory and the real one, under the assumption that $\Delta\mathbf{x}(t) < \varepsilon_1$ (ε_1 is a given value) and that perturbation inputs $\xi(t)$ are stochastic processes with known or evaluated characteristics).
- *Real-time replanning* is the construction of a new plan and the production of appropriate correcting inputs for the transition from the actual SC state trajectory $\mathbf{x}(t)$ to the planned one $\mathbf{x}_{pl}(t)$ at the time interval $(t',t''] \in (T_0,T_f]$ or by the final time: $\mathbf{x}(T_f) = \mathbf{x}_{pl}(T_f)$. A modification of this task is also possible when the difference is minimized:

$$\Delta\mathbf{x}(t) = \int_{t'}^{t''} \left(\mathbf{x}(\tau) - \mathbf{x}_{pl}(\tau)\right)^{\mathrm{T}} \left(\mathbf{x}(\tau) - \mathbf{x}_{pl}(\tau)\right) d\tau , \tag{13.10}$$

$$\Delta\mathbf{x}(T_f) = \left(\mathbf{x}(T_f) - \mathbf{x}_{pl}(T_f)\right)^{\mathrm{T}} \left(\mathbf{x}(T_f) - \mathbf{x}_{pl}(T_f)\right). \tag{13.11}$$

The *subproblems C2 and C3* are interrelated at the conceptual level as follows. The planning performance depends upon two factors: the current plan and future compensations for disruptions. On the other hand, the performance of the control inputs depends upon the current input and the future correcting inputs eliminating deviation from the proper trajectory. Therefore, the subsystems of regulation and planning should implement reciprocal reflection, i.e. consider the decision procedures of the other subsystem.

The *subclass C4* problems are particular classes of the subclass C2 problems. That is why the main models and algorithms constructed for the subclass C2 can also be used for the subclass C4.

Let us consider a more detailed statement of the *subclass C5* problems. We shall use the following modification of Eqs. 13.2 and 13.3:

$$\begin{aligned}
&\mathbf{x}(t_{<k>}) = \tilde{\boldsymbol{\varphi}}_{\Theta}\left(T_{<0,k>}, \mathbf{x}(T_{<0,k>}), \mathbf{x}(t_{<k>} - 1), \mathbf{u}(t_{<k>}), \xi(t_{<k>}), \boldsymbol{\beta}_{\Theta}, t_{<k>}\right);\\
&\mathbf{y}(t_{<k>}) = \tilde{\boldsymbol{\psi}}_{\Theta}\left(\mathbf{x}(t_{<k>} - 1), \mathbf{u}(t_{<k>}), \xi(t_{<k>}), \boldsymbol{\beta}_{\Theta}, t_{<k>}\right);\\
&t_{<k>} \in (T_{<0,k>}, T_{<f,k>}],\ k = 1,2,\ldots,K.
\end{aligned} \tag{13.12}$$

In Eq. 13.12, as compared with Eqs. 13.2 and 13.3, discrete time points are used, and k time intervals are substituted for the single interval $(T_0, T_f]$ of the SC control. Discrete time is typical of hierarchical control systems, where plans are worked out for fixed periods of time, such as an 1 h, 24 h, a week, a ten-day period, a month, a quarter, etc. It has already been mentioned that the following information is needed for a plan correction at the *k*-th control cycle: current SC state information, SC *a posteriori* state information for the previous $< k-1 >$ control cycle, and prognoses information for the next $< k+1 >$ control period. If these data are received, then the problems of parametric adaptation for the SCP models can be written as follows.

Problem of subclass C 5.1. The vector $\mathbf{w}^{(u)}_{<k-1>}$ (vector of parameters of the simulation model of SC functioning under perturbation impacts) should be determined, such that the simulation output data are proximate of the results of the SC SDC programme execution. The closeness of the data can have different meanings: stochastic, fuzzy, minimum of maximal distance, etc. In other words, the simulation model is adapted to the "past". The external adapter should solve this problem in the SS.

The problem can be written as follows (Skurikhin *et al.* 1989):

$$F_1\left(\{\hat{\mathbf{y}}(t_{<k-1>})\}, \{\mathbf{y}(t_{<k-1>})\}\right) \to \underset{\mathbf{w}^{(u)}_{<k-1>} \in \mathbf{W}^{(u)}_{<k-1>}}{\operatorname{extr}}, \tag{13.13}$$

where $\{\hat{\mathbf{y}}(t_{<k-1>})\}$ is a set of values of simulation output characteristics for the SC SDC at the plan cycle $<k-1>$; $\{\mathbf{y}(t_{<k-1>})\}$ is set of real SC SDC characteristics at the control cycle $<k-1>$, and $\mathbf{W}^{(u)}_{<k-1>}$ is a set of the vector $\mathbf{w}^{(u)}_{<k-1>}$ values. Let us recall that the vector $\mathbf{w}^{(u)}_{<k-1>}$ of model parameters is being adjusted by the external adapter according to the results of the SC functioning at the control cycle $<k-1>$. In Eq. 13.13, F_1 is a possibility measure. For example, it can be used for distribution fitting in standard statistical procedures (for the estimation of the closeness of real and model distributions).

Problem of subclass C 5.2. The vector $\mathbf{w}^{(2,f)}_{<k-1>}$ (vector of parameters for the model of SC SDC programme construction at control cycle $<k-1>$) should be determined, such that

$$F_2\left(J^{(G)}_{<\Theta,k-1>}\left(\mathbf{w}^{(2,f)}_{<k-1>},\mathbf{w}^{(\xi)}_{<k-1>}\right),\tilde{\Pi}_{<k-1>}\right)\rightarrow \underset{\mathbf{w}^{(2,f)}_{<k-1>}\in \mathbf{W}^{(2,f)}_{<k-1>}}{\text{extr}}, \tag{13.14}$$

where F_2 is a given functional characterizing the adequacy of the planning model. For example, this functional can be expressed as an expectation $M\left(J^{(G)}_{<\Theta,k-1>}-\tilde{\Pi}_{<k-1>}\right)$. Here, $J^{(G)}_{<\Theta,k-1>}$ is a planned value of SC SDC performance at cycle $<k-1>$, $\Pi_{<k-1>}$ is an index of the total losses caused by the necessity for correction inputs at cycle $<k-1>$, and $\mathbf{W}^{(2,f)}_{<k-1>}$ is a set of allowable values of the vector $\mathbf{w}^{(2,f)}_{<k-1>}$.

Problem of subclass C 5.3. The vector $\mathbf{w}^{(1)}_{<k>}$ (a parameter vector to be adjusted by the internal adapter of the SS) is to be determined, such that

$$F_3\left(J^{(G)}_{<\Theta,k>}\left(\mathbf{w}^{(1)}_{<k>},\mathbf{w}^{(2)}_{<k-1>}\right),\tilde{\Pi}^{(\xi)}_{<k>}\right)\rightarrow \underset{\mathbf{w}^{(1)}_{<k>}\in \mathbf{W}^{(1)}_{<k>}}{\text{extr}}, \tag{13.15}$$

where F_3 is a given functional that establishes an interrelation for the planned value of SC SDC performance at the $<k>$- cycle and the expected (estimated via simulation) losses caused by the necessity for correction inputs at the cycle $<k>$.

General formal statements for structure adaptation of SC SDC modules can be written as problems of two subclasses.

Problem of subclass C 5.4.

$$AD\left(M^{(rr)}_{\Theta},\bar{P}_{cs}\right)\rightarrow \min, \tag{13.16}$$

$$t_{st}\left(\mathbf{w}^{(3)},M^{(rr)}_{\Theta}\right)\leq \bar{t}_{st}, \tag{13.17}$$

$$M_{\Theta}^{(rr)} \in \overline{\overline{M}}, \mathbf{w}^{(3)} \in W^{(3)},\ M_{\Theta}^{(rr)} = \overline{\overline{\Phi}}\left(M_{\Theta}^{(rr-1)}, \mathbf{w}^{(3)}, \overline{P}_{cs}\right),\ rr = 1,2,\ldots, \tag{13.18}$$

where $AD\left(M_{\Theta}^{(rr)}, \overline{P}_{cs}\right)$ is a functional characterizing the adequacy of the model $M_{\Theta}^{(rr)}$ for the SC. The latter is described, in its turn, with a set $\overline{P}_{cs}(t) = \left\{\overline{P}_{\overline{g}}^{(cs)}, \overline{g} = 1,\ldots,\overline{G}\right\}$ of characteristics, t_{st} is a total time of SC SDC models' structure adaptation, $\bar{t}_{st}$ is a maximal allowable time of structural adaptation; $\overline{\overline{\Phi}}$ is an operator of iterative construction (selection) of the model $M_{\Theta}^{(rr)}$, rr is the current iteration number and $W^{(3)}$ is a set of allowable values for the vectors of structure-adaptation parameters.

Problem of subclass C 5.5.

$$t_{st}\left(\mathbf{w}^{(3)}, M_{\Theta}^{(rr)}\right) \rightarrow \min, \tag{13.19}$$

$$AD\left(M_{\Theta}^{(rr)}, \overline{P}_{cs}\right) \leq \varepsilon_2, \tag{13.20}$$

$$M_{\Theta}^{(rr)} \in \overline{\overline{M}}, \mathbf{w}^{(3)} \in W^{(3)},\ M_{\Theta}^{(rr)} = \overline{\overline{\Phi}}\left(M_{\Theta}^{(rr-1)}, \mathbf{w}^{(3)}, \overline{P}_{cs}\right), \tag{13.21}$$

where ε_2 is a given constant establishing an allowable level of the SC SDC model $M_{\Theta}^{(rr)}$ adequacy, and $\Theta \in \hat{I}$, $\overline{\overline{M}}$ is a set of SC SDC models.

The analysis of Eqs. 13.16–13.21 shows that the structural adaptation starts and stops according to a criterion characterizing the similarity of a real object and an object described via models (a condition of model adequacy is applied). The adequacy of SC models does not mean a description of all the "details". It means that the simulation results meet the changes and relations observed in reality.

The main purpose of the quantitative estimation of the model M_{Θ} adequacy at time t is to raise the decision maker's confidence in the conclusions made for the real situation. Therefore, the utility and correctness of SC SDC simulation results can be measured via the adequacy degree of models and objects.

The adequacy functional should meet the following requirements:

$$AD\left(M_{\Theta}^{(rr)}, \overline{P}_{cs}\right) > 0,\quad \forall\, M_{\Theta}^{(rr)} \in \overline{\overline{M}}, \overline{P}_{cs} \in \overline{\overline{P}}_{cs}, \tag{13.22}$$

where $\overline{\overline{M}}$ is a set of SC models and $\overline{\overline{P}}_{cs}$ is a set of possible values of SC characteristics.

$$AD\left(M_{\Theta}^{(rr)}, \overline{P}_{cs}^{(1)}\right) > AD\left(M_{\Theta}^{(rr)}, \overline{P}_{cs}^{(2)}\right), \quad (13.23)$$

where the model $M_{\Theta}^{(rr)}$ is more adequate for the SC with the characteristics set $\overline{P}_{cs}^{(2)}$ than for the SC with the characteristics set $\overline{P}_{cs}^{(1)}$.

$$AD\left(M_{\Theta_1}^{(rr)}, \overline{P}_{cs}^{(1)}\right) > AD\left(M_{\Theta_2}^{(rr)}, \overline{P}_{cs}^{(1)}\right), \quad (13.24)$$

where the model $M_{\Theta_2}^{(rr)}$ is more adequate than the model $M_{\Theta_1}^{(rr)}$ for the SC with the characteristics set $\overline{P}_{cs}^{(1)}$.

It is assumed that the parameters of the models are adjusted for the particular SC. It is important that the changes in SC characteristics should be observed and forecasted so that corrections of the models' structure and parameters can be carried out in time. The time of corrections can be determined as a compromise between an aspiration for receiving proper values of $\overline{P}_{cs}$ and the necessity for the construction, adjustment, and preparation for use of a new model.

13.3 Algorithms of Supply Chain Parametrical and Structural Model Adaptation

There are two main groups of methods to be used for the stated tasks (Bellmann 1972, Moiseev 1974, Rastrigin 1980, 1981, Skurikhin *et al.* 1989) :

- identification methods of parametric adaptation; and
- simulation methods of parametric adaptation.

The identification methods for model (Eqs. 13.9 and 13.12) (for SC with a single output) are based on a solution to the following optimization problem (it can be solved via gradient procedures and their modifications). The vector $\boldsymbol{\beta}_{\Theta}$ of the model parameters should be adjusted to minimize the quadratic residual of the estimation $y(t_{<k>})$ received for the real SC output $\widetilde{y}(t_{<k>})$. Thus, the main expressions of the problem can be written as follows:

$$q_{<k>}^{2}(\boldsymbol{\beta}_{\Theta}) = [\widetilde{\psi}_{\Theta}(\mathbf{x}(t_{<k>}-1), \mathbf{u}(t_{<k>}), \boldsymbol{\xi}(t_{<k>}), \boldsymbol{\beta}_{\Theta}, t_{<k>}) - \widetilde{y}(t_{<k>})]^2 \rightarrow \min_{\boldsymbol{\beta}_{\Theta} \in \mathrm{B}}. \quad (13.25)$$

Standard formulas of the gradient method are

$$\boldsymbol{\beta}_{\Theta}^{(rr)} = \boldsymbol{\beta}_{\Theta}^{(rr-1)} - \widetilde{\widetilde{\alpha}}\, \mathrm{grad}_{\boldsymbol{\beta}_{\Theta}} q_{<k>}^{2}(\boldsymbol{\beta}_{\Theta})\Big|_{\boldsymbol{\beta}_{\theta} = \boldsymbol{\beta}_{\Theta}^{(rr-1)}}, \quad (13.26)$$

where the elements of the gradient vector can be estimated as

$$\frac{\partial \psi_\Theta}{\partial \boldsymbol{\beta}_{<\Theta,\hat{\vartheta}>}} = \frac{1}{2g_{<\hat{\vartheta}>}} \Big[\tilde{\psi}_\Theta \Big(\mathbf{x}(t_{<k>} - 1), \mathbf{u}(t_{<k>}), \xi(t_{<k>}), \boldsymbol{\beta}_\Theta^{(rr-1)} + \mathbf{e}_{<\Theta,\hat{\vartheta}>} g_\vartheta, t_{<k>} \Big) - \\ - \tilde{\psi}_\Theta \Big(\mathbf{x}(t_{<k>} - 1), \mathbf{u}(t_{<k>}), \xi(t_{<k>}), \boldsymbol{\beta}_\Theta^{(rr-1)} - \mathbf{e}_{<\Theta,\hat{\vartheta}>} g_\vartheta, t_{<k>} \Big) \Big] \\ \hat{\vartheta} \in \{g, k, o, f, p, e, v, c\}. \tag{13.27}$$

Here, the parameter $\tilde{\tilde{\alpha}}$ regulates the speed of convergence, the vector $\mathbf{e}_{<\Theta,\hat{\vartheta}>}$ is a unit vector in the space of parameters $\boldsymbol{\beta}_\Theta$ and $g_{\hat{\vartheta}}$ is a base of estimation.

The most complicated step in the procedure of parameter correction is the estimation of the gradient of the local residual function. The elements of the gradient are the partial derivatives of the local residual function with respect to the parameters being adjusted.

The input data are received at the previous control cycle $< k+1 >$ of the SC or via SC SDC simulation for the control cycle $< k >$. If a new control is needed at the next control cycle $< k+1 >$, in response to newly received SC characteristics (this usually occurs in practice), then the SC SDC model will not be "ready".

Therefore, the following procedure for producing control is useful here. The "trial" increment $\delta\mathbf{u}(t_{<k>})$ is produced for the optimal control $\mathbf{u}^*(t_{<k>})$. This correction does not distort the optimal control and gives a more precise adjustment of the SC SDC model. In other words, we propose to plan experiments during SC control process. The increment $\delta\mathbf{u}(t_{<k>})$ should be selected in order to synthesize a more accurate SC SDC model via parameter correction $\Delta\boldsymbol{\beta}_\Theta$. In this case, the total control $\tilde{\mathbf{u}}^*(t_{<k>}) = \mathbf{u}^*(t_{<k>}) + \delta\mathbf{u}(t_{<k>})$ has two purposes: to fulfil the SC mission and to construct an adequate model of the SC SDC. The control $\tilde{\mathbf{u}}^*(t_{<k>})$ is optimal with respect to the dual objective. It is called a dual control. A new goal function should be constructed for the dual control production. This function expresses the total losses including the cost of unachieved control objectives (as caused by the increment $\delta\mathbf{u}(t_{<k>})$) and the cost of model inaccuracy for SC SDC.

13.3.1 Parametric Adaptation

The input data for the SC adaptation is gathered during the SC functioning at the state $S_{\delta'-1}$ and is received for the state $S_{\delta'}$. Thus, we obtain the formulas

$$J_G = J_G\big(\mathbf{x}(t), \mathbf{u}_{pl}(t), \mathbf{v}(\mathbf{x}(t), \xi), \xi\big) \to \max_{\mathbf{u} \in \Delta}, \tag{13.28}$$

$$\Delta = \mathbf{Q}(\mathbf{x}(t)) \times \mathbf{V}(\mathbf{x}(t), \xi, t), \quad \mathbf{u} = \mathbf{u}_{pl}(t) \times \mathbf{v}(\mathbf{x}(t), \xi), \tag{13.29}$$

where $\mathbf{x}(t)$ is a state vector and $\mathbf{u}_{pl}(t)$ is the main vector of control inputs, in other words it is a control programme for the SC dynamics, $\mathbf{v}$ is a vector of control inputs compensating for perturbation impacts over the control programme, $\mathbf{Q}(\mathbf{x}(t))$ and $\mathbf{V}(\mathbf{x}(t), \xi, t)$ are the sets of allowable controls $\mathbf{u}_{pl}(t)$ and $\mathbf{v}(\mathbf{x}(t), \xi)$, respectively and $\xi(t)$ is a vector of perturbation impacts, where $\xi(t) \in \Xi(\mathbf{x}(t), t)$.

The general performance metric of the forecasted states can be evaluated as a functional of the enumerated values via simulation experiments with the SC operation model. In this case, the following group of tasks is substituted for the initial problem of the SC control:

$$J_G = J_G(\mathbf{x}(t, \boldsymbol{\lambda}'), \mathbf{u}_{pl}(t, \boldsymbol{\lambda}'), \mathbf{v}(\mathbf{x}(t, \boldsymbol{\lambda}'), \xi), \xi) \to \max_{\boldsymbol{\lambda}' \in \Delta'}, \tag{13.30}$$

$$\Delta' = \{\boldsymbol{\lambda}' \mid \mathbf{u}_{pl}(t, \boldsymbol{\lambda}') \times \mathbf{v}(\mathbf{x}(t, \boldsymbol{\lambda}'), \xi) \in \mathbf{Q}(\mathbf{x}(\boldsymbol{\lambda}')) \times \mathbf{V}(\mathbf{x}(\boldsymbol{\lambda}'), \xi)\}, \tag{13.31}$$

$$\sum_{\bar{\vartheta} \in \Gamma'} \lambda'_{\bar{\vartheta}} J_{\bar{\vartheta}}(\mathbf{x}_{\bar{\vartheta}}) \to \underset{\mathbf{x}_{\bar{\vartheta}} \in D_{\bar{\vartheta}}(T_f, T_0, \mathbf{x}_{\bar{\vartheta}}(T_0))}{extr}, \tag{13.32}$$

$$\sum_{\bar{\vartheta} \in \bar{L}} \lambda'_{\bar{\vartheta}} = 1, \; \lambda'_{\bar{\vartheta}} \geq 0, \; \mathbf{x}_{\bar{\vartheta}} = \left\| \mathbf{x}^{(\bar{\vartheta})\mathrm{T}} \mathbf{x}^{(o)\mathrm{T}} \right\|^{\mathrm{T}}, \bar{\vartheta} \in \bar{L} = \{g, k, f, p, e, v, c\}. \tag{13.33}$$

Here, the vectors $\mathbf{x}_{\bar{\vartheta}}(T_f)$ returning optimal values are sought, while the vector $\boldsymbol{\lambda}'_{(rr)}$ is fixed ($rr = 0,1,2,\ldots$ is the number of the current iteration). The received problems of mathematical programming have several important features. The search for components of the vector $\mathbf{x}_{\bar{\vartheta}}^{(rr)}$ can be fulfilled over subsets of the AS $D_{\bar{\vartheta}}(T_f, T_0, \mathbf{x}_{\bar{\vartheta}}(T_0))$ rather than over the whole sets of allowable alternatives (see Chap. 10). The subsets include non-dominated alternatives of the enumerated models. The non-dominated alternatives can be received via the orthogonal projection of the goal sets to the AS $D_{\bar{\vartheta}}(T_f, T_0, \mathbf{x}_{\bar{\vartheta}}(T_0))$. Each particular model includes the state vector $\mathbf{x}^{(o)}$ of the operation model $M_{<o>}$ besides its own state vectors $\mathbf{x}^{(g)}$, $\mathbf{x}^{(k)}$,..., $\mathbf{x}^{(c)}$. The above-mentioned structural features of Eqs. 13.30–13.33 allow us to use decomposition and overcome the problem of high dimensionality.

When the vector $\mathbf{x}_{\bar{\vartheta}}^{(rr)}(T_f)$ is known, the optimal programmes $\mathbf{u}_{pl}^{(rr)}(t, \boldsymbol{\lambda}'_{(rr)})$ for the SC control can be defined within each model $M_{\bar{\vartheta}}$ via numerical methods (for example, via Krylov and Chernousko's method) (Chernousko and Zak 1985, Chernousko 1994). The programmes $\mathbf{u}_{pl}^{(rr)}(t, \boldsymbol{\lambda}'_{(rr)})$ are used for the evaluation of a new approximation of the vector $\boldsymbol{\lambda}'_{(rr+1)}$ in the simulation model $M_{\bar{g}}$ describing the SC functioning under perturbation impacts.

The problem of the $\boldsymbol{\lambda}^*$ search is similar to the problem of optimal experiment design. Here, elements of the vector $\boldsymbol{\lambda}'$ are endogenous variables and the performance metric (Eq. 13.30) is an exogenous one. Different methods can be used for the optimal design of experiments, for example the method of quickest ascent, the methods of random search, the method of ψ-transformation (Chichinadze 1980). In conclusion, we note that components of the vector $\boldsymbol{\lambda}'$ can define the preferable control inputs $\mathbf{v}(\mathbf{x}(t, \boldsymbol{\lambda}'), \xi)$ for the compensation of the mismatch of the planned trajectory of SC dynamics with the predictable (via simulation) trajectory.

The matching of simulation and analytical models is based on the Pareto principle by means of iterations during the process of information exchange. This approach is based on the assumption that the global extremum of the generalized quality level of the performance of SC execution is situated in one of the points of Pareto's set, defined by certain quality levels, obtained by means of non-formal decomposition of the problem (and corresponding simulation model).

As a matter of fact, this assumption is fulfilled in any case when we deal with a monotone utility function as a function of particular performance metrics. Besides, multi-criteria optimization is performed by means of different types of models. The narrowing of Pareto's set in discrete models is performed in the interactive mode by means of eliminating elements from this set. The elimination is based on the mathematical investigations of Pareto's set features and the consideration of decision-makers' opinions (evaluation of the set's power, of the performance metrics' range, and of the performance metrics' contradictoriness).

If the power of Pareto's set becomes acceptable, then the SC structures selected on the base of static models can be checked by means of queuing theory models and then by means of simulation models. If the constraints characterizing the models are not fulfilled, then corresponding structure variants are no longer considered.

The SC synthesis process is finalized with the search for appropriate programmes for the development of the SC structures (that is, programmes promoting the existing SC structures to advanced ones).

13.3.2 Structural Adaptation

Let us consider two groups of algorithms for the structural adaptation of SC SDC models. All the algorithms are based on the choice of model structure for a given set of possible models. The algorithms of the first group use the procedures of fuzzy clusterization.

The second group of algorithms for the structural adaptation of SC SDC models is based on the evolutionary (genetic) approach. Let us exemplify these algorithms in the structural adaptation of a model describing the structure dynamics of one SC output characteristic (of one element of the vector $\mathbf{y}(t_{<k>})$).

The residual of its estimation via the model M_{Θ}, as compared with the observed value of the characteristic, can be expressed in Eq. 13.34 (Rastrigin 1980, 1981):

$$Q_{<k>}^{<\Theta>} = \left[\widetilde{\psi}_{<\Theta,k>}\left(\mathbf{x}(t_{<k>}-1), \mathbf{u}(t_{<k>}), \boldsymbol{\xi}(t_{<k>}), \boldsymbol{\beta}_{\Theta}, t_{<k>}\right) - \widetilde{y}(t_{<k>})\right]. \tag{13.34}$$

To simplify the formulas, we assume that the perturbation influences $\boldsymbol{\xi}(t)$ are described via stochastic models. Thus, the following performance metric can be introduced for the model M_{Θ}:

$$\overline{Q}_{<K>}^{<\Theta>} = \sum_{k=1}^{K} g^{(K-k)} Q_{<k>}^{<\Theta>}, \tag{13.35}$$

where $0 \le g^{(K-k)} \le 1$ is a "forgetting" coefficient that "depreciates" the information received in the previous steps (control cycles). If $g^{(K-k)} = 0$ then $\overline{Q}_{<K>}^{<\Theta>} = Q_{<K>}^{<\Theta>}$, i.e. the weighted residual is equal to one received in the last step, as the prehistory has been "forgotten". The coefficient $g^{(K-k)}$ was substituted for the function $f(K)$:

$$\overline{Q}_{<K>}^{<\Theta>} = \sum_{k=1}^{K} f(K-k) Q_{<k>}^{<\Theta>}, \quad \Theta = 1,\ldots,\theta. \tag{13.36}$$

Here, $f()$ is a monotone decreasing function of "forgetting". It has the following properties:

$$f(\alpha) > 0,\ f(0) = 1,\ \lim_{\alpha \to \infty} f(\alpha) = 0,\ f(\alpha) \ge f(\alpha+1),\ \alpha = 0,1,.. \tag{13.37}$$

Now, the structural adaptation algorithm is reduced to a search for the structure $M_{\Theta'}$, such that

$$\overline{Q}_{<K>}^{<\Theta'>} = \min_{\Theta=1,\ldots,\theta} \overline{Q}_{<K>}^{<\Theta>}. \tag{13.38}$$

Thus, it is necessary to calculate the performance metrics (Eq. 13.38) for all the competitive structures M_{Θ}, $\Theta = 1,...,\theta$ of SC SDC models at each control cycle $k = 1,...,K$. All the performance metrics should be compared, and the structure $M_{\theta'}$ with the best measure (minimal residual) should be chosen.

The parametric adaptation of the model $M_{\Theta'}$ should follow the structural one. It is important to determine a proper "forgetting" function under the perturbation impacts $\xi(t)$. The higher the noise level in the SC, the slower the decrease of the function should be implemented. However, if the SC greatly changes its structure, then the function $f(\alpha)$ should decrease rapidly in order to "forget" the results of the previous steps. It can be shown that the structural-adaptation algorithms based on the model construction (synthesis) of atomic models (modules) are rather similar to the algorithms of the SC structural-functional synthesis. These algorithms only differ in the interpretation of the results. The examples of the above-mentioned algorithms, results of calculations are illustrated in Ivanov *et al.* (2009).

References

Bellmann R (1972) Adaptive control processes: a guided tour. Princeton Univ. Press, Princeton, New Jersey.

Chernousko FL (1994) State Estimation of Dynamic Systems. SRC Press, Boca Raton, Florida

Chernousko FL, Zak VL (1985) On Differential Games of Evasion from Many Pursuers. J Optim Theor Appl 46(4):461–470

Chichinadze VK (1980) Solving Dynamic Optimal Problems by the Y-Transformation Method. In: Szego GP, Dixon LCW (Eds) Numerical Optimisation of Dynamic Systems. Amsterdam, North-Holland

Ivanov D, Kaeschel J, Sokolov B (2009) Structure dynamics control-based framework for adaptive reconfiguration of collaborative enterprise networks. Int J Manuf Tech Manag 17(1–2):23-41

Moiseev NN (1974) Elements of the optimal systems theory. Nauka, Moscow (in Russian)

Okhtilev M, Sokolov B, Yusupov R (2006) Intelligent technologies of complex systems monitoring and structure dynamics control. Nauka, Moscow (in Russian)

Rastrigin LA (1980) Modern principles of control for complex objects. Sovetscoe Radio, Moskau (in Russian)

Rastrigin LA (1981) Adaptation of complex systems. Zinatne, Riga (in Russian).

Chapter 14
Models of Supply Chain Global Stability and Manageability

What we anticipate seldom occurs;
what we least expect generally happens.
Benjamin Disraeli

What is now proved was only once imagined
William Blake

14.1 General Remarks on the Evaluation of Supply Chain Goal Abilities

One of the important problems in SCM is the evaluation of goal abilities, i.e. the potential of the SC to perform its missions in different situations. Thus, the preliminary analysis of the functioning and goal abilities (FA and GA) of a SC can be used to obtain reasonable means of the SC execution under different uncertainty conditions.

In Chap. 7, we formulated the SC global stability as a dynamic SC property that emerges through controlled adaptability on the basis of feedback loops. In this chapter, the mathematical model complex of SC stability analysis is presented. These formal models present on the mathematical level the conceptual model of the global stability that has been considered in Chap. 7 in the STREAM concept.

The numerical estimations of FA and GA of a SC control system should be based on the system of measures. These measures can be regarded as characteristics of SC potential effectiveness and efficiency. The GA measures characterizing different levels of SC are interrelated and have a hierarchical structure (e.g. within the SCOR-model).

In parallel with the enumerated measures of FA the following measures of GA can be used: the total possible number of objects-in-service (OS), e.g. customers' orders, over the time period T and the total time that is necessary for the execution of all interaction operations with OS (SC cycle). If the uncertainty factors are considered (the stochastic, probabilistic, or fuzzy models can be applied) the measures of GA can be evaluated as the expectation (or the fuzzy expectation) of the number of serviced objects by a given time point and the probability (its statistical estimation) of successful service for the given objects. Similar measures can be proposed for FA estimations, for example the expectation of the number of objects in a given macro-state and the possibility of operations' fulfilment.

The problem of SC GA and FA evaluation and analysis can be solved on the basis of SDC models (the model M and its components M_o, M_k, M_f, M_e, M_g, M_v, M_c, M_p). These models have a form of non-stationary finite-dimensional differential dynamic systems with reconfigurable structures, so the problem of GA and FA evaluation can be regarded as a problem of controllability analysis. The latter problem, in its turn, can be solved by the AS $D(t,T_0,x(T_0))$ construction. If the AS is obtained, the solvability of the previously stated boundary problems for SDC can be checked in accordance with the sets of initial X_0 and final X_f states $(x(T_0)\in X_0, x(T_f)\in X_f)$, with the considered period of time, with time–spatial, technical, and technological constraints.

Moreover, the problems of SC SDC stated can be formulated as follows:

$$J'_{\mathrm{G}}(\mathbf{x}(\cdot)) \to \min_{\mathbf{x}(\cdot)\in D_{\hat{i}}(T_f,T_0,\mathbf{x}(T_0))}, \tag{14.1}$$

where $D_{\hat{i}}(T_f,T_0,\mathbf{x}(T_0))$ is the AS of the dynamic system (model M) and $J'_{\mathrm{G}}(\mathbf{x}(\cdot))$ is the initial functional (e.g. see Eq. 12.32) transformed to the form of Mayer's functional. It is important that the alteration of the objective functional does not imply the recalculation of the AS $D_{\hat{i}}(T_f,T_0,\mathbf{x}(T_0))$. If the dimensionality of the SC SDC problem is high, then the construction of the AS becomes a rather complicated problem. Therefore, the approximations of $D_{\hat{i}}(T_f,T_0,\mathbf{x}(T_0))$ is used.

14.2 Construction of the Attainable Sets

As has been mentioned in Chap. 10, AS is a very useful tool in the study of various problems of optimization, dynamic systems and differential game theory. Numerous papers have been devoted to the study of various properties of the AS of the control systems (Chernousko 1994, Clarke *et al.* 1995, Motta and Sartori 2000, Sirotin and Formalskii 2003, Lou 2004, Guseinov 2009). In this study, we propose to apply attainable areas to the SCM domain.

If the dimensionality of a problem is high, the construction of an AS is a rather complicated problem. That is why an AS is usually approximated in different forms (Guseinov 2009). Let us introduce the algorithm of $D(T_f,T_0,\mathbf{x}(T_0))$ construction. The boundary points of the set $D(T_f,T_0,\mathbf{x}(T_0))$ are obtained as the solutions to the optimal control problems (Okhtilev *et al.* 2006):

$$J''_{\mathrm{G}}(\mathbf{x}(\cdot)) = \mathbf{c}^{\mathrm{T}}\mathbf{x}(T_f) \to \min_{\tilde{\tilde{\mathbf{u}}}\in\tilde{\tilde{\mathbf{Q}}}(\mathbf{x})}, \tag{14.2}$$

where $\mathbf{c}$ is a vector such that $|\mathbf{c}|=1$. For vector $\boldsymbol{c}$ we obtain the optimal control $\mathbf{u}^*(t)$, the state vector $\mathbf{x}^*(T_f)$ that is equal to boundary point of $D(T_f,T_0,\mathbf{x}(T_0))$ and the hyperplane $\mathbf{c}^{\mathrm{T}}\mathbf{x}^*(T_f)$ to $D(T_f,T_0,\mathbf{x}(T_0))$ at the point $\mathbf{x}^*(T_f)$.

Let $\overline{\overline{\Delta}}$ be the number of different vectors $\mathbf{c}_{\overline{\beta}}$, $\overline{\beta}=1,\ldots,\overline{\overline{\Delta}}$, then the external approximation $D^+(T_f,T_0,\mathbf{x}(T_0))$ of the set $D(T_f,T_0,\mathbf{x}(T_0))$ is a polyhedron whose faces lie on the corresponding hyperplanes, and the internal approximation $D^-(T_f,T_0,\mathbf{x}(T_0))$ of $D(T_f,T_0,\mathbf{x}(T_0))$ is a polyhedron whose vertices are the points $\mathbf{x}^*_{\beta}(T_f)$, i.e. $D^-(T_f,T_0,\mathbf{x}(T_0))=C_0(\mathbf{x}_1(T_f),\ldots,\mathbf{x}_{\overline{\overline{\Delta}}}(T_f))$. The larger $\overline{\overline{\Delta}}$, the better the approximation of the AS $D(T_f,T_0,\mathbf{x}(T_0))$ can be obtained. It can be proved (Kalinin and Sokolov 1987) that the value $\overline{\overline{\Delta}}$ is defined by the total number of possible interruptions for SC interaction operations over a given time period (T_0,t). To obtain D^+,D^- Krylov and Chernousko's method was used (Moiseev 1974). Instead of the vector $\boldsymbol{c}$ the vector $\boldsymbol{\psi}(T\ )$ of conjugate variables varies.

Besides the general dynamic model of SC SDC (the model *M*) its aggregated variants can be used for the AS construction. Let us exemplify this approach via the models M_o, M_k. Besides, we prescribe and allow the interruptions to operations. So the aggregated models of objects' interactions and channels' use can be stated as follows:

$$\dot{\tilde{x}}_i^{(o)}=\sum_{j=1}^{n}\varepsilon_{ij}(t)\tilde{u}_{ij}^{(o)}\,, \tag{14.3}$$

$$\dot{\tilde{x}}_{ij}^{(k)}=\sum_{\substack{\eta=1\\ \eta\neq i}}^{n}\tilde{u}_{\eta j}^{(k)}\frac{h_{\eta i}^{(j)}-\tilde{x}_{ij}^{(k)}}{\tilde{x}_{\eta j}^{(k)}}\gamma_-\left(\tilde{x}_{\eta j}^{(k)}\right), \tag{14.4}$$

where $\tilde{x}_i^{(o)}=\sum_{\mu=1}^{s_i}x_{i\mu}^{(o)}$, $\tilde{u}_{ij}^{(o)}=\sum_{\mu=1}^{s_i}u_{i\mu j}^{(o)}$ are the aggregating functions. The classes $\tilde{\tilde{K}}_t^{(o)}$, $\tilde{\tilde{K}}_t^{(k)}$ of allowable control inputs are defined as follows (Eqs. 14.5 and 14.6):

$$\tilde{\tilde{K}}_t^{(o)}=\left\{\tilde{\tilde{U}}_t^{(o)}=\left\|\tilde{u}_{ij}^{(o)}\right\|\,\middle|\,\sum_{i=1}^{\bar{n}}\tilde{u}_{ij}^{(o)}(t)\le 1,\right.$$
$$\left.\sum_{j=1}^{n}\tilde{u}_{ij}^{(o)}(t)\le 1,\ \tilde{u}_{ij}^{(o)}\cdot\tilde{x}_{ij}^{(o)}=0,\ \tilde{u}_{ij}^{(o)}(t)\in[0,1];\ \tilde{s}_t^{(o)}\right\}, \tag{14.5}$$

$$\widetilde{\widetilde{K}}_t^{(k)} = \left\{ \widetilde{U}_t^{(k)} = \left\| \widetilde{u}_{ij}^{(k)} \right\| \left| \sum_{i=1}^{\bar{n}} \widetilde{u}_{ij}^{(k)}(t) \le 1, \sum_{j=1}^{n} \widetilde{u}_{ij}^{(k)}(t) \le 1, \widetilde{u}_{ij}^{(k)}(t) \in [0,1]; \widetilde{s}_t^{(k)} \right. \right\}, \quad (14.6)$$

where $\widetilde{s}_t^{(o)}$, $\widetilde{s}_t^{(k)}$ are function-theoretic constraints imposed on the classes of allowable controls.

We assume that the control inputs are piecewise continuous functions. We introduce vector $\widetilde{\mathbf{x}}^{(o)} = \left\| \widetilde{x}_1^{(o)}, \ldots, \widetilde{x}_{\bar{n}}^{(o)} \right\|^{\mathrm{T}}$ and vector $\widetilde{\mathbf{x}}^{(k)} = \left\| \widetilde{x}_{11}^{(k)}, \ldots, \widetilde{x}_{\bar{n}n}^{(k)} \right\|^{\mathrm{T}}$. Let $\widetilde{\mathbf{x}}^{(o)}(t_0) = 0$, $\widetilde{\mathbf{x}}^{(k)}(t_0) = \widetilde{\mathbf{x}}_0^{(k)}$. Then the AS in the state space of the dynamic system (Eqs. 14.3 and 14.4) can be obtained as follows:

$$\widetilde{D}_{(o,k)} = \left\{ \widetilde{\mathbf{x}} \left| \widetilde{x}_i^{(o)} = \int_{T_0}^{T_f} \sum_{j=1}^{n} \varepsilon_{ij}(\tau) \widetilde{u}_{ij}^{(o)}(\tau) d\tau, \ \widetilde{\widetilde{U}}_t^{(o)} \in \widetilde{\widetilde{K}}_t^{(o)}, \right. \right.$$
$$\left. \widetilde{x}_{ij}^{(k)} = \int_{T_0}^{T_f} \sum_{\substack{\eta=1 \\ \eta \ne l}}^{\bar{n}} \widetilde{q}_{\eta j}(\tau) \widetilde{u}_{\eta j}^{(k)}(\tau) d\tau, \quad \widetilde{\widetilde{U}}_t^{(k)} \in \widetilde{\widetilde{K}}_t^{(k)} \right\}, \quad (14.7)$$

where $\mathbf{x} = \left\| (\widetilde{x}^{(o)})^{\mathrm{T}} (\widetilde{x}^{(k)})^{\mathrm{T}} \right\|^{\mathrm{T}}$, $\widetilde{q}_{\eta j} = \dfrac{h_{\eta i}^{(j)} - \widetilde{x}_{ij}^{(k)}}{\widetilde{x}_{\eta j}^{(k)}} \gamma_-\left(\widetilde{x}_{\eta j}^{(k)}\right)$, $\gamma_-\left(\widetilde{x}_{\eta j}^{(k)}\right) = 1$, if $\widetilde{x}_{\eta j}^{(k)} > 0$, $\gamma_-\left(\widetilde{x}_{\eta j}^{(k)}\right) = 0$, if $\widetilde{x}_{\eta j}^{(k)} \le 0$.

The following theorem (Kalinin and Sokolov 1985, 1987, Okhtilev *et al.* 2006) expresses the characteristics of the AS.

Theorem. Let the functions $\varepsilon_{ij}(t)$ be nonnegative bounded functions having at most denumerable points of discontinuity, let the classes of allowable controls be defined by (14.5), (14.6), and let the AS $\widetilde{D}_{(o,k)}$ meet the following conditions:

1. It is bounded, closed, and convex. It lies in the nonnegative orthant of the space $\widetilde{X} = \mathsf{R}^{(\bar{n}+\bar{n}n)}$.
2. $\widetilde{D}_{(o,k)}^{-} \subseteq \widetilde{D}_{(o,k)} \subseteq \widetilde{D}_{(o,k)}^{+}$.

Here,

$$\widetilde{D}_{(o,k)}^{-} = \left\{ \widetilde{\mathbf{x}} \left| 0 \le \widetilde{x}_i^{(o)} \le \overline{\overline{\xi}}_i \widetilde{\widetilde{x}}_i^{(o)}, 0 \le \widetilde{x}_{ij}^{(k)} \le \overline{\overline{\chi}}_i \varphi_{ij}^{(k)}, \overline{\overline{\xi}}_i \ge 0, \right. \right.$$
$$\left. \sum_{i=1}^{n} \overline{\overline{\xi}}_i = 1; 0 \le \overline{\overline{\chi}}_i \le 1 \right\}, \quad (14.8)$$

$$\widetilde{D}^{+}_{(o,k)} = \left\{\mathbf{x} \middle| 0 \le \widetilde{x}_i^{(o)} \le \widetilde{\widetilde{x}}_i^{(o)}, 0 \le \widetilde{x}_{ij}^{(k)} \le \overline{\overline{\chi}}_i \, \varphi_{ij}^{(k)}, 0 \le \overline{\overline{\chi}}_i \le 1\right\}, \tag{14.9}$$

where $\widetilde{\widetilde{x}}_i^{(o)} = \int_{T_0}^{T_f} \left[\max_{j=1,\dots,n} \varepsilon_{ij}(\tau) d\tau \right]$ under the conditions $x_{ij}^{(k)} \equiv 0 \; \forall t, \forall i$, $\varphi_{ij}^{(k)} = \max_{j=1,\dots,n} \{h_{\eta i}^{(j)}\} \forall j$.

The theorem is of great importance to the preliminary analysis of SC performance, as the calculation of the values $\widetilde{\widetilde{x}}_i^{(o)}$, $\varphi_{ij}^{(k)}$ is rather simple, while the sets $D^{-}_{(o,k)}$, $D^{+}_{(o,k)}$ allow, in many cases, the verification of the end conditions and the calculation of the range of variation for the performance metrics of SCM.

14.3 Dynamic Multi-criteria Model of Supply Chain Global Stability Analysis

The traditional understanding of stability analysis (based on BIBO stability) consists of proving system stability with regard to small perturbation impacts (Lyapunov 1966). As discussed in Chap. 7, this approach has limitations regarding the SCM domain. *First*, the SCs as management systems evolve from state to state not only through perturbation influences (compare with Ashby 1956) but through control managerial actions of both a planned and regulative (as a reaction to mitigate negative perturbation influences) nature. *Second*, in management systems, unlike in mechanical systems, there is usually no need to ensure 100% stability. The nature of management systems lies in taking entrepreneurship risks. Additionally, these risks are perceived individually by different managers. Hence, the essence of SC stability is, in our opinion, to ensure such a SC functioning so that the management goals (e.g. service level) can be achieved at a level that would be acceptable to managers (or that these goals' values would lie within some predetermined intervals).

The SC stability analysis model presented in this chapter addresses the problem of the direct connection of business processes' stability estimation and analysis with problems of estimation and the analysis of their economic performance. This approach commits to principles that are laid down in the *global asymptotic stability* by Lyapunov, which allows uncertainty in dynamics, the system's parameters, and control actions (Casti 1979). In this approach, stability is considered as a *dynamic property* that emerges through feedback loops. Hence, stability can be considered as a system behaviour property that should be maintained despite perturbation influences by means of corresponding control actions in the feedback loops. As such, stability becomes interconnected with adaptivity within the so-called stabilizing feedback control (Casti 1979). This conceptual understanding of stability is very close to the global stability that we determined in the STREAM concept (see Chap. 7).

The model we will present in the further course of this chapter is of a generic methodological nature and needs localization for concrete applications. The model is based on the dynamic interpretation of the SC's functioning process and uses the method of AS that has been explained in Sects. 10.4 and 14.2.

In the case of multi-criteria problems, a stability estimation can be performed on the basis of the AS $D(t, T_0, X_0)$, where X_0 is a set of possible initial states of the system (Okhtilev *et al.* 2006). The model allows multi-criteria estimation and the analysis of SCs stability to be made, considering the combined variants of initial data about possible perturbation influences (the determined, indistinct, stochastic, interval data, and their combinations). The model allows one (1) to analyse the stability of various alternative SCs plans to be made concerning various kinds and scales of perturbation and control influences, (2) to calculate for each of the plans and possible scenarios a stability index, having given to the decision maker the possibility of a choice of that plan that corresponds to his/her individual risk perception. The essence of a stability index calculation is based on the construction and comparison of two sets (the area of admissible values of SC goal indicators and the approximated area of SC attainability under the influence of perturbation factors). The *stability index* is expressed as the area of intersection of these two rectangles.

The presented model is logically and technically interconnected with the models of operations dynamics control and the SC adaptation presented in Chaps. 10–13. Hence, the notation of the stability models is also based on the notations from the above-mentioned models.

In order to assess the economic performance and stability of SC plans, the models of SDC (Eq. 10.17) and vector (Eq. 12.31) of SC economic performance indicators under perturbation influences (see Chaps. 10 and 12) are used. The stability analysis is possible with regard to deterministic, stochastic and interval input data. These particular cases will be considered in the next section.

14.4 Stability Analysis Models for Different Input Data

14.4.1 Deterministic Data

We use polar diagrams which help to analyse both SC state vectors $\mathbf{x}(t)$ and SC performance metrics, like SC costs, lead-time, etc. $\mathbf{J} = \left\| J_1, J_2, \ldots, J_{I_M} \right\|^{\mathrm{T}}$. To simplify the pictures only some elements of the vector $\mathbf{x}$, $\mathbf{J}$ are depicted (see Fig. 14.1).

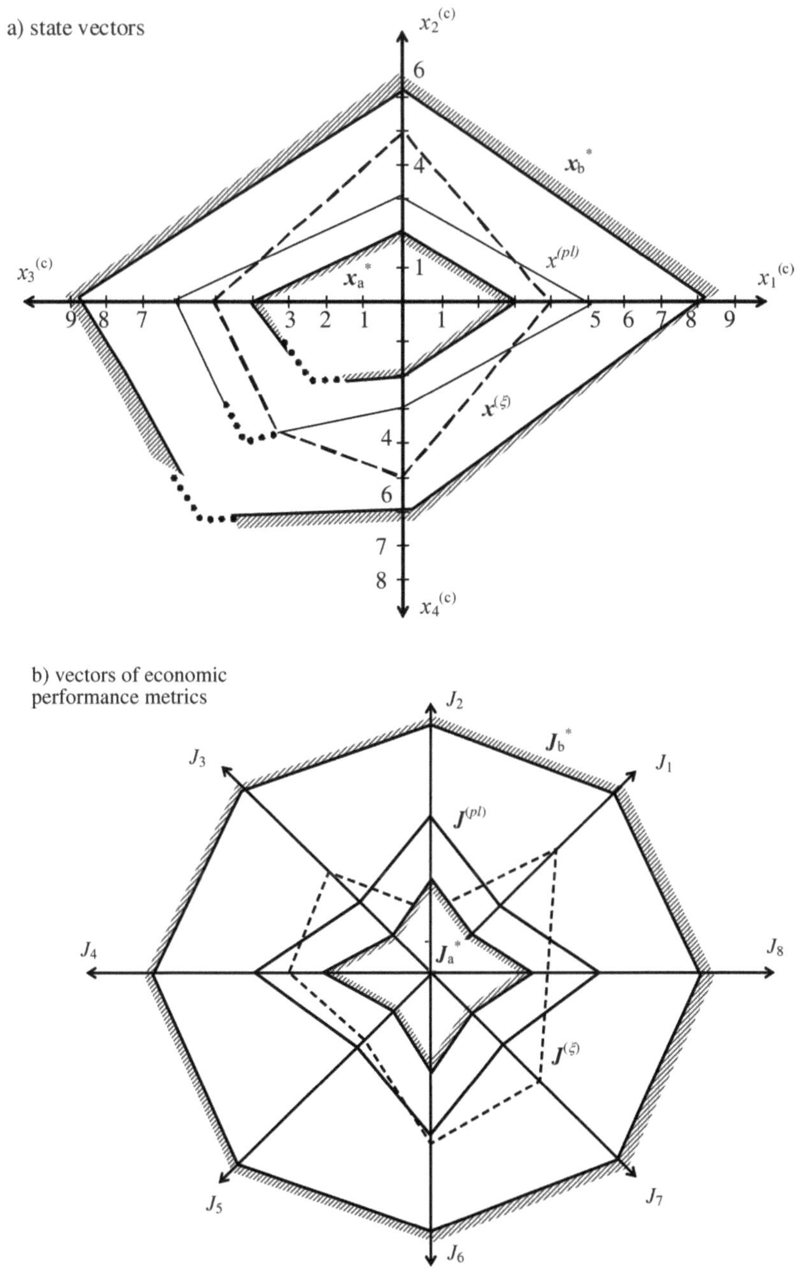

Fig. 14.1 Polar diagram for SC: (a) state vectors, and (b) SC performance metrics

Each element $x_i^{(c)}$ of the vector $\mathbf{x}^{(c,3)}$ is equal to the number of objects in the macro-state S_δ.

Let us introduce the following notation:

$\mathbf{x}^{(pl)}(T_f) = \left\| x_1^{(pl)}, x_2^{(pl)}, \ldots, x_{K_\sigma}^{(pl)} \right\|^{\mathrm{T}}\Big|_{t=T_f}$ is an SC state vector at the time point $t = T_f$;

$\mathbf{x}^{(\xi)}(T_f) = \left\| x_1^{(\xi)}, x_2^{(\xi)}, \ldots, x_{K_\sigma}^{(\xi)} \right\|^{\mathrm{T}}\Big|_{t=T_f}$ is a perturbed SC vector obtained as a result of simulation replicating the conditions of plan realization;

$\mathbf{x}_a^*(T_f) = \left\| x_{a1}^*, \ldots, x_{aK_\sigma}^* \right\|^{\mathrm{T}}\Big|_{t=T_f}$ and $\mathbf{x}_b^*(T_f) = \left\| x_{b1}^*, \ldots, x_{bK_\sigma}^* \right\|^{\mathrm{T}}\Big|_{t=T_f}$ are vectors defining respectively the lower and upper bounds for vectors $\mathbf{x}^{(pl)}(T_f)$, $\mathbf{x}^{(\xi)}(T_f)$;

$\mathbf{J}_{\tilde{h}}^{(pl)} = \left\| J_{\tilde{h}1}^{(pl)}, \ldots, J_{\tilde{h}I_M}^{(pl)} \right\|^{\mathrm{T}}\Big|_{t=T_f}$ is a vector of SC effectiveness measures (measures of SC goals) for the case of zero perturbation actions ($\tilde{h} = 1, \ldots, \tilde{H}$);

$\mathbf{J}_{\tilde{h}}^{(\xi)} = \left\| J_{\tilde{h}1}^{(\xi)}, \ldots, J_{\tilde{h}I_M}^{(\xi)} \right\|^{\mathrm{T}}\Big|_{t=T_f}$ is a vector obtained as a result of simulation replicating the conditions of plan realization ($\tilde{h} = 1, \ldots, \tilde{H}$);

$\mathbf{J}_a^* = \left\| J_{a1}^*, \ldots, J_{aI_M}^* \right\|^{\mathrm{T}}\Big|_{t=T_f}$, $\mathbf{J}_b^* = \left\| J_{b1}^*, \ldots, J_{bI_M}^* \right\|^{\mathrm{T}}\Big|_{t=T_f}$ a vectors defined respectively the lower and upper bounds of $\mathbf{J}^{(pl)}(T_f)$, $\mathbf{J}^{(\xi)}(T_f)$.

The following algorithm can be used to evaluate the stability of SC plans.

Step 1. Let $\tilde{h}$ be the number of a current plan $\tilde{h} = 1, \ldots, \tilde{H}$, then the following conditions are verified:

$$\mathbf{x}_a(T_f) \le \mathbf{x}_{\tilde{h}}^{(\xi)}(T_f) \le \mathbf{x}_b(T_f), \tag{14.10}$$

$$\left\| \mathbf{x}_{\tilde{h}}^{(pl)}(T_f) - \mathbf{x}_{\tilde{h}}^{(\xi)}(T_f) \right\| \le \varepsilon_1^{(pl)}, \tag{14.11}$$

where $\varepsilon_1^{(f)}$ is a given constant.

If a condition is not fulfilled the current plan is stated to be invalid, and its stability is not evaluated. In Fig. 14.1 the vector $\mathbf{x}^{(\xi)}(T_f)$ is shown by means of the dashed line. The shaded regions denote the sets of unstable plans. So as a result of step 1 the set $\tilde{H}^{(1)} = \tilde{H} / \overline{H}^{(1)}$ is constructed, where $\overline{H}^{(1)}$ is a set of subscripts enumerating the invalid plans.

Step 2. For every $\mathbf{x}_{\widetilde{h}}^{(\xi)}(t)$, $\widetilde{h} \in \widetilde{H}^{(1)}$ the following conditions are verified:

$$J^*_{a\widetilde{\delta}} \le J_{\widetilde{h}\widetilde{\delta}} \le J^*_{b\widetilde{\delta}}, \tag{14.12}$$

$$\left| J^{(f)}_{\widetilde{h}\widetilde{\delta}} - J^{(\xi)}_{\widetilde{h}\widetilde{\delta}} \right\|_{t=T_f} < \varepsilon_2^{(f)},\ \widetilde{\delta} = 1,\ldots,I_M, \tag{14.13}$$

where $\varepsilon_2^{(f)}$ is a given value.

If for some plan $\widetilde{h} \in \widetilde{H}^{(1)}$ at least one of conditions is not satisfied then the plan is stated to be unstable. So $\widetilde{H}^{(2)} = \widetilde{H}^{(1)} / \overline{H}^{(2)}$, where $\overline{H}^{(2)}$ is a set of subscripts for unstable plans.

Step 3. Now one or more plans are to be chosen from the set $\widetilde{H}^{(2)}$ of the stable plans. The choice can be performed in the interactive mode; here the polar diagrams may be useful. Another approach to the choice problem is to construct a general stability criterion as a convolution of particular measures or by means of metrics in the criteria space. In the latter case we should obtain a solution of the optimization problem

$$\rho(\mathbf{J}_{\widetilde{h}}^{(pl)}, \mathbf{J}_{\widetilde{h}}^{(\xi)}) \to \min . \tag{14.14}$$

14.4.2 Stochastic Data

For stochastic input data, uncertainty factors of the environment influence upon the SC are replicated in detail. To provide statistical significance of stability estimations the multiple simulation experiments should be fulfilled. Here the methods of variance lowering can be applied to reduce the number of experiments. If we use the stochastic input data then the stability can often be expressed as a probability of some event. The most appropriate event for this purpose is the completion of a given mission in accordance with the plan. For certain cases, the necessary level of stability can be defined as

$$P\{\hat{z}_h \ge z_\alpha\} = \alpha^{(\xi)}, \tag{14.15}$$

where z_α, α are given values, $\hat{z}_{\widetilde{h}} = \rho\left(\hat{\mathbf{x}}_{\widetilde{h}}^{(\xi)}(T_f),\ \hat{\mathbf{x}}_{\widetilde{h}}^{(pl)}(T_f)\right)$ is the estimation of the difference between the planned SC state and the perturbed one.

The stability of SC plan can be indirectly estimated by means of the following objective function:

$$\hat{M}_{es}(J_{\tilde{h}}^{(pl)} - J_{\tilde{h}}^{(\xi)}), \tag{14.16}$$

where $\hat{M}_{es}$ is the expectation sign, $J_{\tilde{h}}^{(pl)}$ is a general measure of the SC performance (a convolution $J_{1\tilde{h}}^{(pl)},...,J_{I_M\tilde{h}}^{(pl)}$) and $J_{\tilde{h}}^{(\xi)}$ measures losses caused by perturbation influences and resources consumption for adjustment managerial actions.

14.4.3 Interval Data

Let us suppose that the area of admissible disturbances $\Xi(\mathbf{x}(t),t)$ is defined as

$$\boldsymbol{\xi}_j^{(1)}(t) \le \xi_j(t) \le \xi_j^{(2)}(t),\ j = 1,...,n., \tag{14.17}$$

where $\boldsymbol{\xi}_j^{(1)}$, $\boldsymbol{\xi}_j^{(2)}$ are vectors functions for minimal and maximal disturbances. We propose to call this area as the AS of the SC under disturbances. We define it as

$$D_x^{(\xi)}(T_f, T_0, X_0, \Xi, \mathbf{u}_{\tilde{h}}). \tag{14.18}$$

The set $D_x^{(\xi)}(T_f, T_0, X_0, \Xi, \mathbf{u}_i)$ corresponds to the indicators values, which assess the SC's economic performance and stability. The latter we define as

$$D_J^{(\xi)}(T_f, T_0, X_0, \Xi, \mathbf{u}_{\tilde{h}}). \tag{14.19}$$

To make the further material more illustrative, we will examine only two components of vector index. These components correspond to the indicators of SC service level (J_1) and SC profit (J_2). In this case, while geometrically describing the AS, it becomes possible to use Descartes coordinate system.

Let the admissible limits of oscillations be as follows:

$$J_{a1} \le J_1 \le J_{b1}, \tag{14.20}$$

$$J_{a2} \le J_2 \le J_{b2}. \tag{14.21}$$

They construct special area P_J in the indices space. As explained in Chap. 10, we are not interested in all the AS but only in the subarea, in which the SC is capable of carrying out the planned processes. To perform the stability analysis, internal $D^{-}(t,T_0,X_0)$ and external $D^{+}(t,T_0,X_0)$ approximations of $D(t,T_0,X_0)$ should be constructed. The construction of the approximation is based on the following rule:

$$J_g = \mathbf{c}^{\mathrm{T}}\mathbf{J} \rightarrow \min_{\boldsymbol{\xi}_i \in \Xi}, \tag{14.22}$$

where $\mathbf{c} = \| c_1, c_2 \|^{\mathrm{T}}$ — given vector that fulfils the normal conditions

$$|\mathbf{c}| = \sqrt{c_1^2 + c_2^2} = 1 \tag{14.23}$$

and $\mathbf{J} = \|J_1, J_2\|$ — vector of particular indices of the SC effectiveness.

Within the framework of SC plan stability assessment, the task is to find the point $\mathbf{J}^* = \| J_1^*, J_2^* \|^{\mathrm{T}}$, which lies on the border of the set (Eq. 14.19) and some line of the following type:

$$\tilde{n}_1 J_1^*, + \tilde{n}_2 J_2^* = 0, \tag{14.24}$$

which is a tangent for the given set and includes the point J^*. After determining the multitude of points $J_{\overline{\beta}}^*$ and appropriate tangents for some variants of vector $\mathbf{c}_{\overline{\beta}}$ components $\overline{\beta} = 1,\ldots,\overline{\overline{\Delta}}$ ($\overline{\overline{\Delta}}$ — number of variants of indices $\mathbf{c}_{\overline{\beta}}$), we obtain the external approximation of the set (Eq. 14.19), which is defined as follows:

$$\overline{\overline{D}}_J^{(\xi)}(T_f, T_0, X_0, \Xi, \mathbf{u}_{\tilde{h}}). \tag{14.25}$$

This AS approximation is a geometrical figure that lies between the lines determined as $\mathbf{c}_{\overline{\beta}}^{\mathrm{T}}\mathbf{J}'^*, \overline{\beta} = 1,\ldots,\overline{\overline{\Delta}}$. Let us examine the case when $\overline{\overline{\Delta}} = 4$, and vectors $\mathbf{c}_{\overline{\beta}}$ are $\mathbf{c}_1 = \|0,1\|^{\mathrm{T}}$; $\mathbf{c}_2 = \|0,-1\|^{\mathrm{T}}$; $\mathbf{c}_3 = \|1,0\|^{\mathrm{T}}$; $\mathbf{c}_4 = \|-1,0\|^{\mathrm{T}}$ and fulfil the normal conditions. To approximate the AS, it is necessary to solve four tasks of the type (Eq. 14.18)

$$J_1' = J_2 \rightarrow \min_{\boldsymbol{\xi}_j \in \Xi} : J_2' = -J_2 \rightarrow \min_{\boldsymbol{\xi}_j \in \Xi} : J_3' = J_1 \rightarrow \min_{\boldsymbol{\xi}_j \in \Xi} : J_4' = -J_1 \rightarrow \min_{\boldsymbol{\xi}_j \in \Xi}. \tag{14.26}$$

The result of solving the task (Eq. 14.26) is the coordinates of the points $\mathbf{J}_1'^*$, $\mathbf{J}_2'^*$, $\mathbf{J}_3'^*$, $\mathbf{J}_4'^*$, which make it possible to construct an external approximation of the AS considered. This approximation is a rectangular area constructed as a result of four lines crossing.

14.5 Stability Index Calculation

A stability assessment of the SC plans comes down to the calculation and analysis of a general index of the SC stability for disturbance scenarios $\boldsymbol{\xi}(t) \in \boldsymbol{\Xi}(\mathbf{x}(t),t)$ and control influences $\mathbf{v}(\mathbf{x}(t),t) \in \mathbf{V}(\mathbf{x}(t),t)$ within each generated SC plan $\mathbf{u}_{\tilde{h}}(t) \in \mathbf{Q}(\mathbf{x}(t),t)$ ($\widetilde{h} = 1,...,\widetilde{H}$, where $\widetilde{h}$ is a number of alternative SC plans).

If, for a SC plan $\mathbf{u}_{\tilde{h}}(t)$, $(\widetilde{h} = 1,...,\widetilde{H})$ under disturbances $\xi_j(t)$, the requirement (Eq. 14.27) is fulfilled:

$$D_J^{(\xi)}(T_f, T_0, X_0, \Xi, \mathbf{u}_{\tilde{h}}) \subset P_J \,. \tag{14.27}$$

the $\mathbf{u}_{\tilde{h}}(t)$ SC is considered to be stable under disturbances $\xi_j(t)$. In other words, feasible J_1, J_2 deviations are considered to be acceptable.

The selection of the most stable SC plan is carried out according to condition

$$S_{\tilde{h}}^*(\mathbf{u}_{\tilde{h}}(t)) = \max_{1 \le \tilde{h} \le \tilde{H}} \min_{1 \le \tilde{j} \le \tilde{m}} S_{\tilde{h}}(\mathbf{u}_{\tilde{h}}(t)) \tag{14.28}$$

where $S_{\tilde{h}}(\mathbf{u}_{\tilde{h}}(t))$ is the area of sets $\overline{\overline{D}}_J^{(\xi)}(T_f, T_0, X_0, \Xi, \mathbf{u}_{\tilde{h}})$ and P_J intersection; $\widetilde{h}$ — the total amount of analysed SC plans; and $\widetilde{m}$ — the total amount of disturbance scenarios at the stage of SC plan realization. The square $S_{\tilde{h}}(\mathbf{u}_{\tilde{h}}(t))$ of the intersection of these two rectangles reflects the desirable result declared at the beginning of this subsection — the general stability index of the SC. The essence of a stability index calculation is based on the construction and comparison of two sets (the area of admissible values of SC goal indicators and the approximated area of SC goal attainability under the influence of perturbation factors). The smaller the square, the more stable the SC plan. The larger the square, the less stable the SC plan.

So, considering the above-mentioned positions, we conclude that the task of the SCP within the proposed dynamic interpretation comes to the search for an SC plan $\mathbf{u}_{\tilde{h}}(t)$, $t \in (T_0, T_f]$, for which all restrictions (Eqs. 10.14–10.19) are fulfilled, and all components of a general index of SC functioning quality (Eq. 10.20) are extremities.

It is possible to show that the search for the most stable SC functioning plan due to the statement at Eq. 14.28 is a realization of the multi-criteria selection under uncertainty, i.e. the principle of the guaranteed result. Figure 14.2 shows the most distinctive cases of areas P_J and $\overline{\overline{D}}_J^{(\xi)}$ arranged for different SC plans.

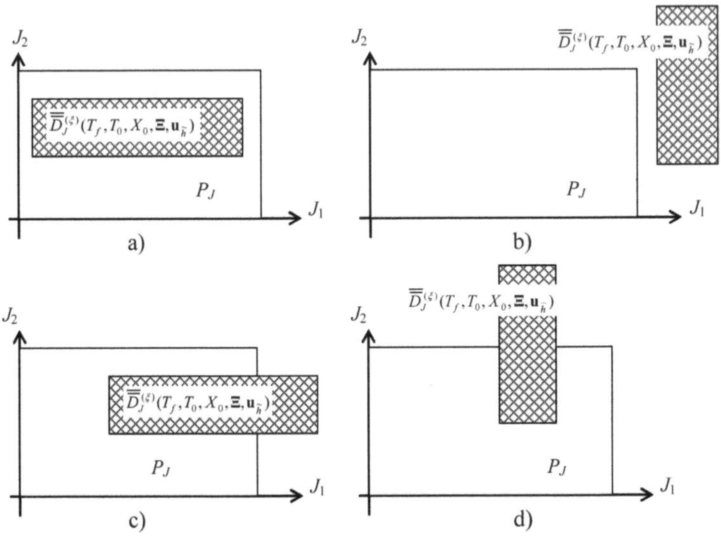

Fig. 14.2 The most distinctive cases of areas P_J and $\overline{\overline{D}}_J^{(\xi)}$ arrangement for different SC plans: (a) unstable plan with regard to both J_1 and J_2, (b) stable plan with regard to both J_1 and J_2, (c) unstable plan with regard to J_2, (d) unstable plan with regard to J_1

The following statements hold true:

- In case *a*, possible deviations of the SC operations' quality metrics, which are caused by disturbances, are non-acceptable, and the SC is unstable under these disturbances.
- In case *b*, possible deviations of the SC operations' quality metrics (metrics of economic performance and stability), which are caused by disturbances, are acceptable, and the SC plan is stable under these disturbances.
- In case *c*, the SC plan is unstable under disturbances that influence the metric of SC performance J_2 most of all.
- In case *d*, the SC plan is unstable under disturbances that influence the metric of SC performance J_1 most of all.

References

Ashby WR (1956) An introduction to cybernetics. Chapman & Hall, London
Casti JL (1979) Connectivity, complexity and catastrophe in large-scale systems. Wiley-Interscience, New York and London
Chernousko FL (1994) State Estimation of Dynamic Systems. SRC Press, Boca Raton, Florida
Clarke FH, Ledyaev Yu S, Stern RJ, Wolenskii PR (1995) Qualitative properties of trajectories of control systems: a survey. J Dyn Control Syst 1:1–48
Guseinov KG (2009) Approximation of the attainable sets of the nonlinear control systems with integral constraint on controls. Nonlinear Anal 71:622–645
Kalinin VN, Sokolov BV (1985) Optimal planning of the process of interaction of moving operating objects. Differ Equat 21(5):502–506
Kalinin VN, Sokolov BV (1987) A dynamic model and an optimal scheduling algorithm for activities with bans of interrupts. Autom Rem Contr 48(1-2):88–94
Lou HW (2004) On the attainable sets of control systems with p-integrable controls. J Optim Theory Appl 123:123–147
Lyapunov AM (1966) Stability of motion. Academic Press, New York and London
Moiseev NN (1974) Element of the optimal systems theory. Nauka, Moscow (in Russian)
Motta M, Rampazzo F (2000) Multivalued dynamics on a closed domain with absorbing boundary, applications to optimal control problems with integral constraints. Nonlinear Anal TMA 41:631–647
Okhtilev M, Sokolov B, Yusupov R (2006) Intelligent technologies of complex systems monitoring and structure dynamics control. Nauka, Moscow (in Russian)
Sirotin AN, Formalskii AM (2003) Reachability and controllability of discrete-time systems under control actions bounded in magnitude and norm. Autom Rem Contr 64:1844–1857

Chapter 15
Experimental Environment

A theory can be proved by experiment;
but no path leads from experiment to the birth of a theory.
Albert Einstein

No amount of experimentation can ever prove me right;
a single experiment can prove me wrong.
Albert Einstein

15.1 Concept of the United Experimental Environment

A vision of a special software environment, which contains a simulation and optimization "engine" of SCP, a Web platform, an ERP system, and a SC monitor are presented in Fig. 15.1.

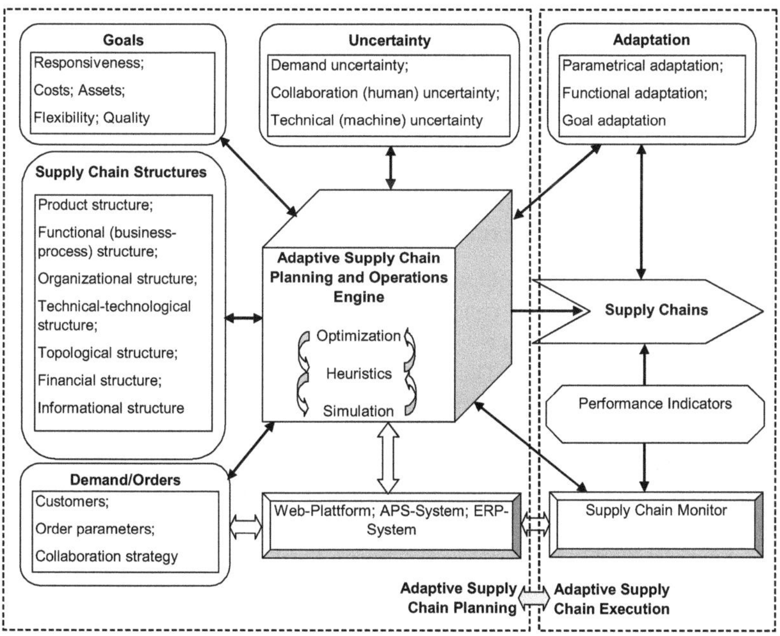

Fig. 15.1 The vision of software environment (from Ivanov *et al.* 2010)

For the experiments, we elaborated a software environment that is composed of two main software prototypes:

- SNDC – Supply Network Dynamics Control; and
- SCPSA – SC Planning and Stability Analysis.

15.2 Integrated Supply Chain Planning and Scheduling

For the experiments, we elaborated a software prototype "Supply network dynamics control". SC planning and execution are based on the simultaneous consideration of different structures in their interrelations and dynamics. The set of SC multi-structural alternatives is formed not on the basis of concrete values of structure-relevant parameters, but first upon data structures. Then, we specify one or several structures with deterministic or stochastic parameters as well as select solution procedures (optimization or heuristics) for partial planning sub-problems (i.e. demand forecasting, production planning, procurement planning etc.). Therefore, a number of SC multi-structural alternatives are identified, evaluated in regard to the goal criteria (costs, supply cycle time), and stability, and the best one is selected.

The software has three modes of operation. The first mode includes the interactive preparation of the data and data input subject to models in Chaps. 10 and 12. The second mode lies in the evaluation of heuristic and optimal SC schedules. The following operations can be executed in an interactive regime:

- multi-criteria rating, analysis, and the selection of SC plans and schedules;
- the evaluation of the influence that is exerted by time, economic, technical, and technological constraints upon SC structure dynamics control; and
- the evaluation of a general quality measure for SC plans and schedules, and the evaluation of particular performance indicators.

The third mode provides interactive selection and visualization of SC SDC and report generation. An end user can select the modes of program run, set and display data via a hierarchical menu.

Let us consider an example. The first step is to gain the heuristic solution. We programmed two simple priority rules — equal charge and FIFO. The next step is to optimize the schedule. While optimizing, the program addresses an external optimization library (MatLab or MS Excel Solver).

For the optimal schedule calculated with the proposed in Chap. 12 DYN algorithm, the general QI (Eqs. 12.31-12.32) usually possesses better values compared with the heuristic plans. Of course, this index is of a relative nature and needs concretizations in any particular SC environment. In our opinion, the most important result is even not the optimal solution but the evidence that complex problems of SC scheduling can be solved in an appropriate time (e.g. for the problems with more than ten resources, orders and operations, the calculating time did not exceed several minutes). The tool SNDC facilitates the simulation of the SC design, planning, scheduling and control processes (Ivanov *et al.* 2004, 2005). The results of the planning are the structural–functional–informational SC execution plans. The special feature of the approach implemented in the SDNC is the possibility of analysing the compiled plan on the basis of the MSMS (see Fig. 15.2).

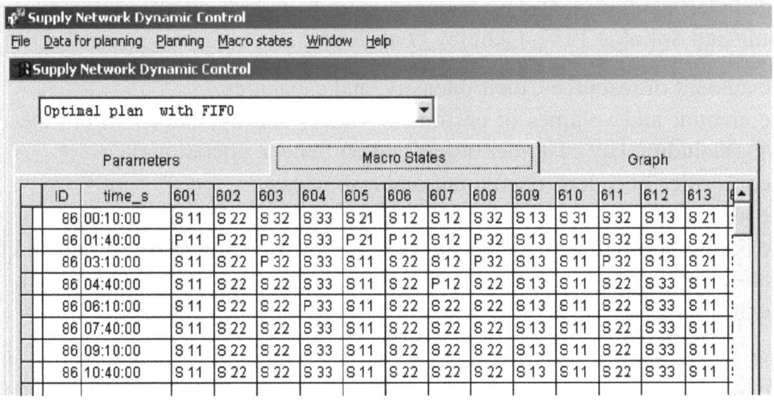

ID	time_s	601	602	603	604	605	606	607	608	609	610	611	612	613
86	00:10:00	S 11	S 22	S 32	S 33	S 21	S 12	S 12	S 32	S 13	S 31	S 32	S 13	S 21
86	01:40:00	P 11	P 22	P 32	S 33	P 21	P 12	S 12	P 32	S 13	S 11	S 32	S 13	S 21
86	03:10:00	S 11	S 22	P 32	S 33	S 11	S 22	S 12	P 32	S 13	S 11	P 32	S 13	S 21
86	04:40:00	S 11	S 22	S 22	S 33	S 11	S 22	P 12	S 22	S 13	S 11	S 22	S 33	S 11
86	06:10:00	S 11	S 22	S 22	P 33	S 11	S 22	S 22	S 22	S 13	S 11	S 22	S 33	S 11
86	07:40:00	S 11	S 22	S 22	S 33	S 11	S 22	S 22	S 22	S 13	S 11	S 22	S 33	S 11
86	09:10:00	S 11	S 22	S 22	S 33	S 11	S 22	S 22	S 22	S 13	S 11	S 22	S 33	S 11
86	10:40:00	S 11	S 22	S 22	S 33	S 11	S 22	S 22	S 22	S 13	S 11	S 22	S 33	S 11

Fig. 15.2 SC plan as macro-states

The application of the methodology of multi-structural macro-states makes it possible to represent and evaluate SC configurations as well as SC execution plans on the aggregated detailed level. This allows managers to gain a complex presentation of dynamical order execution in the SC. This dynamics may also be represented as a chart (see Fig. 15.3).

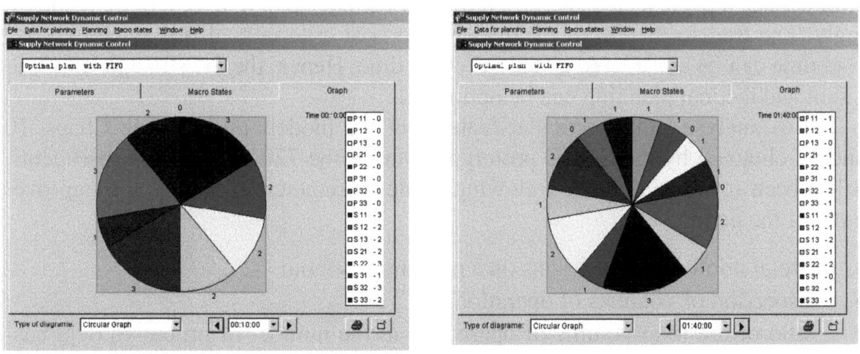

Fig. 15.3 Graphical representation of SC dynamics: start and end multi-structural macro-states

The developed prototype provides a wide range of the analysis possibilities from such points of view as SC structure content (initial data for scheduling) and the approachability of the tactical planning goals, taking into account individual managers' risk perception, the execution dynamics of customers' orders, and operations within the orders (including key customer orders and critical operations).

The results of the SC dynamics simulation make it possible to analyse the complex dynamics of supply cycles and form necessary management adjustments. The scheduling model is very flexible. More than 15 parameters can be changed to investigate different interrelations of schedule parameters and SC tactical goals (e.g. service level) achievement. For example, there is an explicit possibility to change (Kalinin and Sokolov 1987, Okhtilev *et al.* 2006):

- the amount of resources, their intensity, and capacities;
- the amount and volumes of customers' orders and operations within these orders (including key customer orders and bottleneck operations);
- the priorities of orders, operations, and resources;
- the lead times, supply cycles, and penalties for breaking delivery terms;
- the perturbation influences on resources and flows in the SC (e.g., demand fluctuations, technological failures, purposeful threats like thefts or terrorism); and
- the priorities of the goal criteria.

The Pareto optimality-based multi-criteria problem formulation allows us to take into account individual managers' preferences, SC strategies, etc. The general schedule QI enables us to compare alternative schedules that in turn can undergo a detailed analysis with regard to concrete orders and operations.

The proposed model interprets dynamic SC scheduling as a response to planning goal changes, demand fluctuations, and resource availability. In this interpretation, the problem is to schedule SCs in order to achieve the plan goals (e.g. SC service level).

The model is scaleable to other management levels of SCs, i.e., orders and operations can be presented as SC configuration elements and orders correspondingly. The transformation of parameters and goal criteria is also possible, i.e. the lead time can be considered as the SC cycle time. Hence, the SC strategic configuration and tactical planning can be optimized.

Let us analyse some particular features of the models presented in Chaps. 10 and 12 obtained by the SNDC prototype. During the 720 conducted experiments, it has been revealed that the following model parameters influence the improvement of the general QI:

- the total number of operations on a planning horizon;
- a dispersion of volumes of operations;
- a ratio of the total volume of operations to the number of processes (e.g. customers' orders); and
- a ratio of the amount of data of operation to the volume of the operation (relative operation density).

The total number of operations on a planning horizon and operation density render the greatest influence (see Fig. 15.4). In Fig. 15.4, the results of 720 experiments are visualized as the surface in space of the following partial QI: the total number of processes; operation density; and improvement of the general QI (see Eq. 10.20). Improvement of the QI is a ratio of a difference of the quality index of heuristic (FIFO) and the optimum plan to the quality index of the heuristic plan.

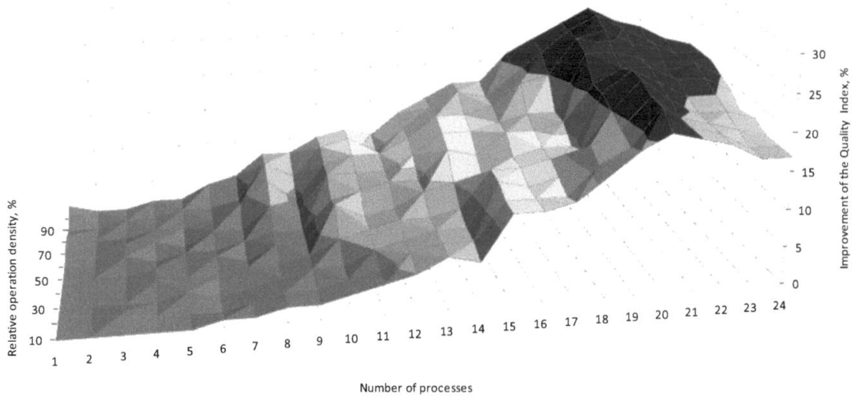

Fig. 15.4 Analysis of general quality index improvements

Fig. 15.4 depicts that the usage of the proposed DYN algorithm is especially sensible in the situation with the operation density of 80%. In this case, the improvement in the general QI amounts to up to 27%. In the case of low resource loading (low operation intensity), the heuristic algorithms are preferable.

Besides, the efficiency of the proposed algorithm is sensitive to the number of processes (e.g., customers' orders). The most benefits have been achieved by a relatively high number of processes to be executed. In the experiments conducted, we considered the maximal number of processes as equal to 24. The best results were achieved with regard to the number of processes equal to 20 or 21. When approaching the maximum number of processes, the improvements in the quality index decrease. This means that additional resource volumes should be introduced into the SC. Another explanation of the "hump" in Fig. 15.4 is connected with the non-stationary character of the ordering. The additional resources or increasing the resource productivity shifts the surface to the right side.

The conducted experiments showed that the application of the presented dynamic scheduling model is especially useful for the problems where a number of operations are arranged in a certain order (e.g. technological restrictions). This is the case in SC planning and scheduling.

As a problem, we found that the convergence of the iterative process decreases with increasing resource usage. In cases with seven and fewer processes, 98% of the experiments converged within one iteration. In cases with 7–13 processes, 85% of the experiments converged within 2 iterations. In cases with 14–21 processes, 70% of the experiments converged within 3 iterations. In cases with 22 and more processes, 87% of the experiments converged within 4 iterations; some divergence cases have been observed.

Based on the obtained optimal solutions, we can methodically justify the usage and quality of certain heuristics for certain variants of initial data (see Fig. 15.5).

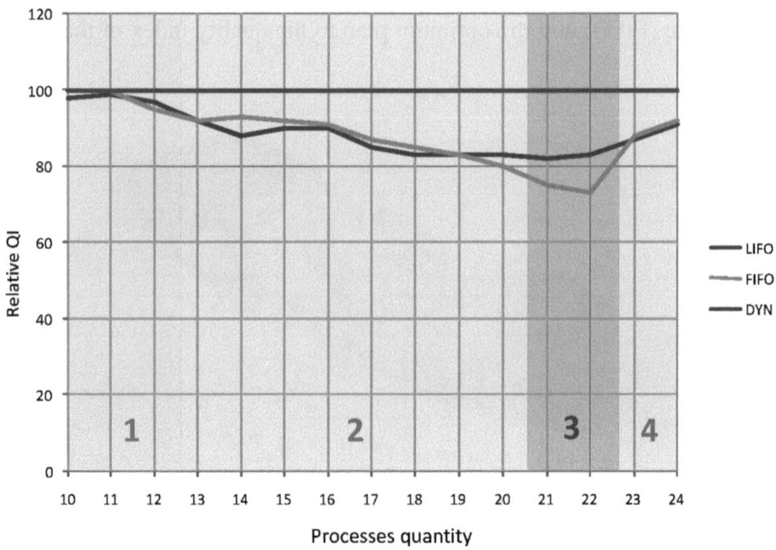

Fig. 15.5 Comparison of heuristic algorithms' quality

Having calculated optimal solutions for several points, it is possible to validate the decision to use either dynamic or one of the heuristic planning algorithms. In Fig. 15.5, the relative QI of the optimal solutions is assumed to be 100%. The relative QI of the heuristic solutions is calculated as a fraction of the optimal one, i.e. it can be observed that, in the case of a number of processes between 10 and 12, the quality of the heuristic and optimal solutions does not differ by more than 4%. In area 2, the DYN algorithm is preferable to the heuristics. If still using the heuristics, the FIFO algorithm is preferable to the LIFO (last-in-first-out) one. The most benefit from using the DYN algorithm is achieved in area 3. In this area, the LIFO algorithm is preferable to the FIFO algorithm.

Finally, we would like to draw attention to the proposed model and algorithm allowing the achievement of better results in many cases in comparison with heuristics algorithms. However, this point is not the most important. The most important point is that this approach allows the interlinking of planning and scheduling models within an adaptation framework. Hence, the proposed modelling complex does not exist as a "thing in itself" but works in the IDSS and guides the planning and scheduling decisions in dynamics on the principles of optimization and adaptation.

15.3 Supply Chain Stability Analysis

The second part of our experiments dealt with stability analysis. For experimental calculations on the basis of the model presented above, a software product has been developed on the basis of C++ and XML (Ivanov *et al.* 2009). Let us provide an example.

Example. The problem consists of the analysis of the stability of three alternative SC configurations (or plans) concerning three variants of perturbation influences $\xi(t)$ at the given area of perturbation influences $\Xi(\mathbf{x}(t),t)$ (see notation in Chaps. 10 and 14). The SC structures are characterized by the different areas of the planned admissible control influences $\mathbf{Q}(\mathbf{x}(t),t)$ with regard to the reliability and the given area of general admissible control influences $\mathbf{V}(\mathbf{x}(t),t)$ with regard to the flexibility. Structure 1 is characterized by lower profitability, a higher service level and higher excessiveness costs (with regard to the reliability and flexibility) compared with structures 2 and 3. In structure 1, the areas $\mathbf{Q},\mathbf{V}(\mathbf{x}(t),t)$ are balanced with the area $\Xi(\mathbf{x}(t),t)$. In structures 2 and 3, the areas $\mathbf{Q},\mathbf{V}(\mathbf{x}(t),t)$ are smaller than the area $\Xi(\mathbf{x}(t),t)$.

In the example given, the following perturbation impacts have been considered: a decrease in the availability of resources of 30% (scenario 1), a resource productivity decline of 5% (scenario 2), and the cumulative influence of these two perturbation impacts (scenario 3). In Fig. 15.6, the experimental results are presented.

		Supply chain structure 1			Supply chain structure 2			Supply chain structure 3		
		Service level	Profit	Stability index	Service level	Profit	Stability index	Service level	Profit	Stability index
Ideal scenario	No perturbations, $\xi = 1$	42	11	0	41	12	0	37	12	0
Scenario 1	Small perturbations, $\xi = 0{,}9$	40	10	0	38	12	0	35	9	0
Scenario 2	Middle perturbations, $\xi = 0{,}7$	38	7	0	35	6	0	29	7	1
Scenario 3	Strong perturbations, $\xi = 0{,}5$	27	5	9	21	4	36	18	5	36

Fig. 15.6 Results of the SC stability analysis

With regard to the structures and execution scenarios considered, the values of the performance metrics and stability index are calculated. The cases with zero value of the stability index mean that the *excessiveness* (see Chap. 7) of the corresponding SC structures is enough to protect the SC against the given perturbation influences and to consider the SC as stable. In other cases, additional reserves should be introduced or possibilities of dynamic SC adjustments in the case of temporary loss of stability taken into account.

For the evaluation of the results by managers, a special interface is elaborated for SC stability and economic performance analysis (see Fig. 15.7). In the example given, on an axis of abscises, the metric of the SC service level J_1, and on an axis of ordinates, a metric of the SC profitability J_2 are presented. The stability index is defined on the basis of the area of intersection of the two AS. In the worst case scenario (scenario 3), for case 1 (SC structure no. 1), it is equal to 9 (this case is presented on the interface in Fig. 15.7), for case *b* (SC structure no. 2) — 36, and for case *c* (SC structure no. 3) – 36. As such, the SC structure no. 1 is the most stable, i.e. the SC remains stable even in the case of the occurrence of the considered perturbation influences.

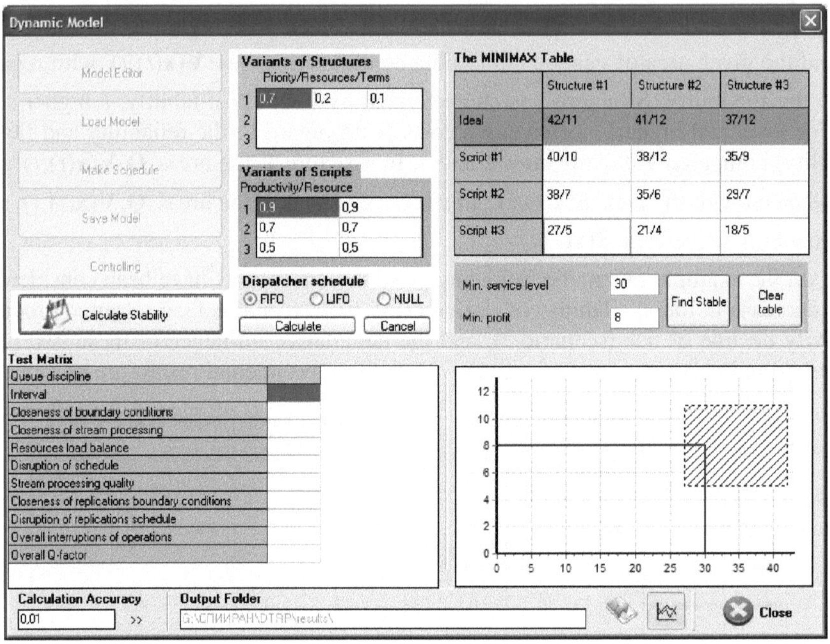

Fig. 15.7 Simultaneous analysis of SC economic performance and stability

The selection of the final SC configurations (plans) from the set of analysed scenarios and structures is based on the psychological type of the decision maker and his own *individual risk perception*. For each of the scenarios and structures considered, the values of performance metrics and the stability index are calculated and serve as a basis for decision-making, e.g. if the decision maker is a pessimistic psychological type and puts particular emphasis on stability, structure 1 would the preferable option because, in the worst case scenario, this structure would be the most stable. If the manager is an optimist and can take risk to a higher extent, structures 1 or 2 could be selected with regard to the ideal case or scenario 1 (small perturbations).

Let's analyse the results. As already mentioned, SC no. 1 was initially characterized by the lowest level of planned profit since it had been constructed with the greatest redundancy and requirements for the stability and service level. In comparison with SCs nos. 2 and 3, which were characterized initially by a higher profit level, SC no. 1 in the case of the negative scenario (a simultaneous decrease in the availability of a resource of 30% and a resource decline of productivity of

5%) has appeared even more profitable in comparison with SC no. 2, and on the service level is the best among the SC structures considered. Thus, the proposition on the global stability condition is proved because, in structure 1, area $\mathbf{Q}, \mathbf{V}(\mathbf{x}(t), t)$ is balanced with area $\Xi(\mathbf{x}(t), t)$.

During the experiments, we could confirm the proposition that the plan stability decreases with increasing number of processes to be operated. With regard to the situations with high operation density, we found out that the competencies of enterprises in an SC do affect the stability immediately. During the experiments, we fixed the total resource productivity and varied the number of resources (competencies) and their partial productivity. The plan stability increases by about 12% if there are substitute resources in the SC. Additionally, the stability can be increased by about 7% if the total productivity is broken down into a greater number of autonomous resources (competencies). Hence, configuring SCs with unique resources may potentially decrease their stability.

In the programme, the decision makers have a wide range of additional analytical possibilities with regard to the different SC structures and execution scenarios, i.e. they can change the admissible intervals of the goal parameters oscillations and scope and the scale of the perturbation impacts. Additionally, the priorities of the SC goal metrics can be changed. There is also a possibility for a detailed analysis of the order dynamics, operation dynamics, enterprise activities dynamics, and "bottlenecks". The developed prototype implements the above-mentioned theoretical models and makes a step towards designing stable and profitable SCs.

References

Ivanov D, Arkhipov A, Sokolov B (2004) Intelligent Supply Chain Planning in Virtual Enterprises. In: Camarihna-Matos LM (Ed) Virtual Enterprises and Collaborative Networks. Kluwer academic publishers, Boston

Ivanov D, Käschel J, Arkhipov A, Sokolov B and Zschorn L (2005) Quantitative models of collaborative networks. In: Camarihna-Matos LM, Afsarmanesh H, Ortiz A (Ed) Collaborative networks and their breeding environments. Springer, New York

Ivanov D, Sokolov V, Potryasaev S (2009) Issues of stability analysis in flexible supply networks and possible ways of their decisions. In: Proceedings of the 13th IFAC Symposium INCOM. Moskau, preprint: 574–579

Ivanov D, Sokolov B, Kaeschel J (2010) A multi-structural framework for adaptive supply chain planning and operations control with structure dynamics considerations. Eur J Oper Res 200:409–420

Kalinin VN, Sokolov BV (1987) A dynamic model and an optimal scheduling algorithm for activities with bans of interrupts. Autom Rem Contr 48(1-2):88–94

Okhtilev M, Sokolov B, Yusupov R (2006) Intelligent technologies of complex systems monitoring and structure dynamics control. Moskau, Nauka (in Russian)

Conclusion

Not the place where we are
but the direction we are following is important.
Leo Tolstoy

Discussion of Findings

In this book, we considered new viewpoints on decision-making in the SCM domain. *First*, we assumed that the changes in the global economic environment will inevitably cause changes in SCM. In these settings, we stated the new SCM paradigm as the maintenance of stability and harmonization of value chains with possibly full customer satisfaction and optimal resource consumption for ensuring the performance of production-ecological systems at the infinite time horizon.

Second, we conceptualized some cross-linked issues in SCM problem semantics. We considered SCs as integrated multi-structural systems. Planning and scheduling have also been considered as an integrated management function within an adaptation framework.

Third, as for these new conceptual frameworks, new tools for decision-making support are needed, and we further developed the corresponding mathematical models. We started from the presumption that SCs as complex systems are described by a number of different models. On the basis of the integrated consideration of control theory, systems analysis, OR and artificial intelligence, we developed a new viewpoint on quantitative decision-making support in SCM.

Fourth, to implement at the formal level the integrated consideration of SC effectiveness and efficiency as composed of SC economical performance and stability, we applied AS. For the integrated planning and scheduling, we proposed the optimization-based modelling complex for SC operation dynamics that is based on the combined application of control theory and OR. In these models and algorithms, we proposed new constructive ways to apply optimal control approaches to the SCM domain. Besides, these models exist not as "things in themselves" but are embedded into the adaptation framework that supports the SCM in dynamics. For multi-structural SC dynamics control, new mathematical tools such as DAMG and MSMS have been developed.

Based on the results gained in this study, the following contributions with regard to OR and control theory can be stated.

With regard to *operations research*:

- Problems of high dimensionality and complexity can be solved with optimization techniques.
- The dynamics of real processes can be reflected.

- Models of planning and execution can be explicitly interconnected in terms of uncertainty.
- The linear discrete world of operations research is enriched by the categories of dynamics, non-linearity, non-stationarity, adaptability and stability.

With regard to *control theory*:

- A novel controller concept is proposed to take into account managerial control actions (unlike technical systems with automated tuning).
- New principles of planning and scheduling problem formulation within the operation dynamics model allow us to apply optimal control techniques to the domain of complex management systems with the discreteness of decision-making.
- The optimization models of operation dynamics are interconnected with the adaptation control loops and make possible optimization-based tuning processes under control with regard to both the current execution environment and goal criteria.
- Problems of planning, monitoring and adjustment are explicitly interconnected with each other on the basis of the unified decision-making principles both in the planning and regulation (replanning) models.

The advancements gained will be discussed in detail below. The study introduced a new conceptual framework for the *multi-structural* planning and execution control of adaptive SCs with *structure dynamics* considerations. SCs are modelled in terms of dynamic multi-structural macro-states, based on the simultaneous consideration of management as a function of both states and structures. We proposed new tools for the multi-structural SCP – MSMS and DAMG. The results show that the multi-structural and inter-disciplinary treatment of SC planning allows integrated and realistic planning problem formulation and solution. The proposed multi-structural treatment also allows the establishment of links to uncertainty analysis and especially to SC execution and reconfiguration.

We conceptualized the subject domain of SCM under uncertainty from the uniform system-cybernetic and SCM points of view. The concept STREAM and corresponding engineering tools have been developed. We considered the SC property to approach the real SC performance with the planned one under the colliding SC processes in the real perturbed execution environment with regard to the variety of execution and goal criteria as the *SC global stability*. Hence, the essence of SC stability is, in our opinion, to ensure such a SC functioning that the management goals (e.g. service level) can be achieved at a level that would be acceptable to managers.

We formulated and proved the proposition that SC goals can be achieved only if there are enough control actions to guide the system to the achievement of its goals in a real execution environment with negative perturbation influences. Indeed, the achievement of the balance between the available control actions subject to a certain scale of perturbation influences makes the SC manageable and leads to the stabilization of SCs. For detailed considerations of this balancing, the business and formal SC properties of SC reliability, flexibility, security, vulnerability, BIBO stability, resilience, robustness and adaptability have been brought into correspondence with each other to cover the domain of SC planning and execution under uncertainty and dynamics.

The proposed *adaptive* planning and control approach is based on a combination of the MPC and AC frameworks as well as of optimal control theory and OR. We took as a basis the adaptive planning in which the SC plan is modified periodically by a change of SC parameters or characteristics of control influences on the basis of information feedback about a current SC state, the past and the updated forecasts for the future. For the forecast updating, the MPC technique has been used; the AC application has been extended from the signal identification to the whole complex systems dynamics with the help of structure dynamics control theory.

By designing the controller, the delays between the deviation's identification and adjustment decision making are handled within the SDC approach and a combined people–machine adjustment system is used for SC adaptation in the case of different disruptions. A hierarchy of adjustment actions is brought into correspondence with different deviations in the SC execution. As such, the controller serves both for the deviation's identification and for the adjustment measures' generation, taking into account the distributed system nature and delays in *managerial decisions*. Besides, this allows the transit from continuous control models that are characterized for the process industry and the application of adaptive planning and control to many other branches with discrete operations.

We interpreted planning and scheduling not as discrete operations but as a continuous adaptive process. We considered *planning and scheduling* as an integrated function within an adaptive framework. In the adaptation framework, we interpreted SC functioning as SC operations dynamics. Plan adaptation is connected to the model adaptation. The parametric adaptation is enhanced by a structural adaptation. This basic feature will allow us to apply this model simultaneously to (1) a SC functioning model under uncertainty, (2) a SC planning and scheduling model, and (3) a simulation model to run a "what-if" scenario. A particular feature of the model is that not only control u but also a number of conjunctive variables can be adapted to a current execution environment.

In the presented approach to structural–operational SC dynamics, planning and execution problems in SCs are tightly interlinked. First, the planning and execution models are inter-reflected. This means that, in both of the models, the decision-making principles of the other model are reflected. Second, the planning and execution stages are interlinked through the monitoring level.

In this study, we extended the control frameworks for SCs by taking into account the particular features of SCs as multi-structural systems with *active* independent elements and managerial (unlike automatic control) adjustment actions. A generic adaptation framework, operation dynamics model and controller framework have been developed.

The explicit integration of the external and internal *adaptation control loops* and the operation dynamics model makes it possible to integrate the monitoring and control models and to connect the measured and controlled parameters explicitly. In integrating the monitoring and control models, the rational extraction of only the current necessary execution parameters for monitoring and control from the whole high-dimensional parameter vector becomes possible.

Another result of the integration of adaptation control loops and the operations dynamics model is that the parameters of the SC execution and operations dynamics model can be *tuned* simultaneously. This so-called dual control makes it possible both to fulfil the SC mission (the goals set by management) and to construct an adequate model of SC dynamics control.

The proposed approach provides the possibility to cover SC dynamics and permanent changes in SC processes and environment without a great necessity to accomplish a total "remodelling". In these settings, the integration of the planning and scheduling stages is possible. This means that not only a problem solution in a fixed environment (system under control) but also a simultaneous consideration of system formation and management problem solution are possible in this system. Hence, the goal-oriented formation of SC structures and the solution of problems in this system are considered as a whole.

To capture the issues of high dimensionality, non-linearity and non-stationarity, we proposed to modify the dynamic interpretation of operation control processes. We formulated the planning and scheduling models as *optimal control* problems, taking into account the *discreteness* of decision-making and *continuity* of flows. Here, the main idea is to implement non-linear technological constraints in sets of allowable control inputs rather than in the right parts of differential equations. In this case, Lagrange coefficients, keeping the information about economical and technological constraints, are defined via the local-sections method. Furthermore, we proposed to use interval constraints instead of relay ones. Nevertheless, the control inputs take on Boolean values as caused by the linearity of differential equations and the convexity of the set of alternatives.

The originality of the proposed dynamic scheduling model is composed of several features. The *first* feature is that the right parts of the differential equations undergo discontinuity at the beginning of interaction zones. The problems considered can be regarded as control problems with intermediate conditions. The *second* feature is the multi-criteria nature of the problems. The *third* feature is concerned with the influence of uncertainty factors. The *fourth* feature is the form of time–spatial, technical and technological non-linear conditions that are mainly considered in control constraints and boundary conditions.

The process control model is presented as a dynamic linear system while the non-linearity and non-stationarity are transferred to the model constraints. This allows us to ensure convexity and to use interval constraints. As such, the constructive possibility of discrete problem solving in a continuous manner occurs. The modelling procedure is based on an essential reduction of a *problem dimensionality* that is under solution at each instant of time due to connectivity decreases. The problem under solution can be presented with a polynomial complexity rather than with an exponential one. This results in the possibility of solving high-dimensional problems with computational complexity in a dynamic manner.

The narrowing of Pareto's set in discrete models is performed in the interactive mode by means of eliminating elements from this set. The elimination is based on the mathematical investigations of Pareto's set features and the consideration of decision-makers' opinions (evaluation of the set's power, of the quality measures' range, and of the quality measures' contradictoriness).

If the power of *Pareto's set* becomes acceptable, then the SC structures selected on the basis of static models can be checked by means of queuing theory models and then by means of simulation models. If the constraints characterizing the models are not fulfilled, then corresponding structure variants are no longer considered.

The Pareto-optimality-based *multiple objective* problem formulation allows us to take into account individual managers' preferences, SC strategies, etc. The model is scaleable to other management levels of SCs, i.e. orders and operations can be presented as SC configuration elements and orders correspondingly. The transformation of parameters and goal criteria is also possible, i.e. the lead time can be considered as the SC cycle time. Hence, the SC strategic configuration and tactical planning can be optimized.

To implement the simultaneous analysis of both the *performance and stability* of the SC, the method of *AS* can be used. By constellating different SC execution scenarios with regard to the scope and scale of perturbation influences and the corresponding control influences, the decision makers can analyse different constellations of SC performance and stability, and select the most preferable one from a number of alternatives in accordance with the individual risk perception. The final choice of a SC configuration or plan occurs on the basis of managerial individual preferences and the risk perception.

The *SC stability* analysis model addressed the problem of the direct connection of business processes' stability estimation and analysis with problems of estimation and the analysis of their economic performance. We formulated the SC global stability as a *dynamic* SC property that emerges through controlled adaptability on the basis of feedback loops. Hence, stability can be considered as a *dynamic property* of system behaviour that should be maintained despite perturbation influences by means of corresponding control actions in the feedback loops.

As such, stability becomes interconnected with *adaptability*. Such an approach to stability allows us to reveal other possible ways for decision-making in business complex systems rather than to consider stability of SCs merely at the beginning of planning. Besides, the stability indicator meets the SCM nature to a greater extent. Increases in sales and cost reductions may be related to operational logistics improvements at local knots of SC. But the stability of the whole SC is even the direct performance indicator of SCM.

Within the proposed global stability concept and formal model, we showed that the mutual influence of perturbation and control (adjustment) impacts affects SC stability immediately. Hence, it can be concluded that SC stability depends on a balance of control and perturbation areas. This opens up new perspectives for solving numerous problems of SCM under uncertainty, e.g. the problem of balancing SC reliability and flexibility.

The approach to stability analysis allows multi-criteria estimation and stability analysis; the consideration of the different variants of initial data on perturbation influences (the determined, fuzzy, stochastic, interval data and their combinations) is possible. The approach allows direct connection of the problem of estimation and the analysis of the stability of SC business processes with the problem of estimation and the analysis of their performance. Due to a dynamic interpretation of the SC functioning process, it is possible to use it at the constructive level received earlier in the classical theory of automatic control fundamental and applied scientific results for the estimation of SC stability. The simplicity of the results' presentation and interpretation is to be underlined due to their representation in the form of geometrical figures.

In general, the proposed model complex (see Chaps. 10–15) is based on the dynamic interpretation of SC processes and allows the use of fundamental and applied results gained in modern optimal control theory, systems analysis, OR and multiple-criteria optimization for solving different classes of SC synthesis and analysis with regard to SC economic performance and stability.

In concluding this paragraph, we would like to emphasize that the main motivation of the mathematical part of the research approach is to combine the possibilities of different decision-making techniques, such as OR, control theory, systems analysis, and agent-based modelling to achieve new quality of the decision-making support, e.g. in applying the proved fundamentals of the control theory to the SCM domain, the conventional OR-based modelling techniques for SCM can be enriched by new viewpoints on the dynamics, stability, adaptability, consistency, non-linearity and high-dimensionality of the complex system.

The mathematics of optimal control can help in revealing new conformities to natural laws that have yet to be revealed within the OR field. Hence, the conventional SCM problems may be considered from a different viewpoint and new problems may be revealed and formulated.

Summarizing, this is to say that the results gained suggest quantitative methods to transit from a simple open time slots incremental planning to a dynamic, feedback loop-based adaptive SC planning and scheduling to implement value chain adaptability, stability and crisis-resistance throughout. The findings contribute to the understanding of SC from the perspectives of adaptable and stable processes that provide the achievement of management goals with a sufficient degree of stability and crisis-resistance instead of "ideal" optimal plans and schedules that fail in a real perturbed execution environment. The modern optimal control theory, in combination with systems analysis, OR and agent-oriented modelling, is a powerful technique to handle dynamics and uncertainty in SCs.

Future Research Outlook

Let us discuss the limitations of the proposed approach. In the management concept A-SCM, some particular aspects of the approach show limitations regarding branch independence, i.e. with regard to flexible suppliers' structuring, the approach can be applied especially to the cases where there is the possibility to attach alternative suppliers to a number of operations in the value-adding process. For example, this flexible structuring could not be implemented in the automotive sector because of the strict quality policies of OEMs. Hence, the proposed concept can be applied in two cases: (1) for unique products (in a special machinery industry) or (2) for products without strict technical quality policies (i.e. the textile branch).

Another very important point is the *trust and collaboration* in the network. Before automation, a huge amount of organizational work should be carried out to convince the OEMs and suppliers to collaborate within a common informational space, share the data, actualize the data and ensure financial trust. While automating, it is important to elaborate and to maintain throughout product and process technological documentation and classification. Last, but not least, the firms themselves should perceive the necessity for such collaboration.

Certain elaborations in this study are still in the form of theoretical frameworks and hypotheses that must be proved and improved in concrete case studies. For concrete application cases, it is nevertheless almost impossible to take into account the whole variety of complex interrelated constraints and parameters. As the success of the modelling with the proposed dynamical complex depends greatly on the formulation of constraints, this becomes a bottleneck in the proposed approach. For each concrete problem, we should actually build a new complex of constraints. This is a very time-consuming process as we should achieve the strong monosemantic problem formulation to ensure only one solution is gained. In some cases, certain model simplifications will be needed.

A great challenge is the calculation precision both in the planning model itself and at the interface between the adaptation loops and the operations dynamics model. In the developed mathematical model, calculation precision is subject to individual decision-maker perceptions. The operations dynamics model itself is implemented in a software prototype, but the adaptation loops are still undergoing programming. Constructive ways to achieve calculation precision with regard to the interrelations of measured and calculated (control) parameters are to be developed further. Finally, the mathematical formalization of uncertainty factors is complicated by a high complexity of stochastic dynamic models.

With regard to overcoming these shortcomings and further developments, the following future research needs in adaptive SCs can be indicated. In our future research, we are going to investigate in depth some referenced SCM problems with the use of optimal control theory. Furthermore, a promising area for using control theory is SC monitoring and event management. Both problem areas will lead us to a deeper understanding of the interrelations between control theory, systems analysis, OR and agent-oriented modelling.

In answering the indicated challenge to interrelate decisions at different management levels and in different structures, we will prove the efficiency of integrating the elaborated structure–operation dynamics approach with the system dynamics approach. In general, the developed fundamentals must be further proved in different SC environments. Only this proof will provide the basics to develop a theory of operation dynamics in complex production-logistics networks (like SCs) as a further development of system dynamics and structure dynamics theories.

In future, the investigations into the triangle "manageability–optimality–complexity" may potentially provide new insights into SCM and engineering. Furthermore, with increasing transportation, the minimizing of negative impacts on ecology may become one of the primary objectives in SCM and logistics. Actually, this is already stated by global automotive companies. Hence, we are moving towards a triangle of goals: profitability, stability and ecological goals.

In future, the adaptive SC should evolve into self-organizing and self-learning SCs. In self-organizing SCs, both the system and its goals would evolve, unlike in adaptive SCs where the system's design and goals are predetermined. The system's borders would become fuzzy, the system can broaden by "acquiring" a space from the environment, or the system can narrow in the reverse way. However, to approach the self-organizing SCs, time is needed to rethink many crucial management paradigms.

Glossary

Term	Explanation
Adaptability	The ability of a supply chain to change its behaviour for the prevention, improvement, or acquisition of new characteristics for the achievement of supply chain goals in environmental conditions that vary in time and the aprioristic information about which dynamics is incomplete
Adaptation	The property of a system consisting of the continuous changes of its functioning and the abilities to function in unpredicted conditions by a goal-oriented adjustment of the process parameters and/or structures
Adaptive management	A management method of a supply chain with varying unknown environmental characteristics, in which for the final time are reached defined (satisfactory, wished for, or optimum) goals of supply chain management by means of a change of the supply chain parameters, processes , and structures or characteristics of control influences on the feedback loop driven basis
Adaptive planning	A method of planning and scheduling in which the plan of a supply chain is modified periodically by a change of parameters of the supply chain or characteristics of control influences on the basis of information feedback about a current condition of the supply chain, the past and the updated forecasts for the future
Adaptive supply chain	A networked organization wherein a number of various enterprises • collaborate (cooperate and coordinate) along the entire value-adding chain and product life cycle to: acquire raw materials, convert these raw materials into specified final products, deliver these final products to retailers, design new products, and ensure post-production services; • apply all modern concepts and technologies to make supply chains stable, effective, responsive, flexible, robust, sustainable, cost-efficient, and competitive in order to increase supply chain stability, customer satisfaction and decrease costs, resulting in increasing supply chain profitability
Adaptive supply chain management	A business concept, a technology, and a scientific discipline that studies the resources of enterprises and human decisions with regard to stability, adaptability and profit-

ability of cross-enterprise collaboration processes to transform and use these resources in the most rational way along the entire value-adding chain and product life cycle, from customers up to raw material suppliers, based on cooperation, coordination, agility and sustainability throughout

Attainable set	The set of all the ends of different supply chain execution scenarios at $t = t_1$, which begin at $t = T_0$ at the point $\vec{x}(T_0)$ and result from different variations of managerial actions $\vec{u}(t)$ within the time interval $(T_0, t_1]$. The economic sense of the attainable sets consists of the following: the attainable set characterizes a set of supply chain plans and the values of the supply chain potential goals corresponding to them
BIBO stability	A property of the supply chain functioning; the state of a supply chain that is in a planned mode of functioning is stable, if it is considered that the fixed set of admissible control influences and limited and small perturbation influences lead to limited and small changes of goal variables
Catastrophic (disaster) situation	A supernumerary mode of functioning during which the supply chain passes from an efficient to a disabled catastrophic state in which the transition to an efficient state is essentially excluded and/or is economically inefficient. The liquidation of a catastrophic situation is carried out on the basis of the changes in SCM goals and financial plans, i.e. changes in the strategic plan change. Actually, this situation leads the strategic management to the formation of new supply chains and their management systems
Control (in the narrow interpretation)	Input influence to regulate the supply chain plan
Control (in the wide interpretation)	Input influences on the object of the planning and execution, intended for the achievement of the management goals. The set of control influences can be divided into two categories: the construction of supply chain plans, taking into account uncertainty, and the regulation of (adaptive) control influences at a stage of a supply chain realization
Critical situation	A supernumerary mode of functioning in which the supply chain performance metric or the environment indicators are out of the intervals of a regular mode in such limits that there is a real threat of disruption of the plan or a catastrophe
Dangerous situation	A supernumerary mode of functioning in which the indicators of supply chain performance or the environment indicators are out of the intervals of a regular mode in such lim-

	its that the plan disruption or a catastrophe is almost inevitable.
Deviation	The short-term transition of a supply chain from an efficient state to a disabled state, which does not lead to a loss of manageability. Deviation withdraws without external influences. Deviation is characterized by a supernumerary situation
Disruption	The transition of a supply chain from a planned state to an unplanned state in which the achievement of the SCM goals without additional control influences is impossible
Disturbance	The impossibility of the realization of the planned event (or a critical amount of events) according to the supply chain plan
Dynamics	A system's change and evolution in the form of changes in object and process states in space and time as driven by perturbation influences and control influences of both planned control actions to transit a system from a current state to a desired one and adaptation control actions to adapt a system (structures, processes, and operations) to a changed execution environment.
Effectiveness	Approaching the process goal
Efficiency	Fulfiling the process with minimum costs
Execution	The process of control realization for the achievement of the supply chain goal at all three levels of planning
Flexibility	A property of a supply chain concerning its ability to change itself quickly structurally and functionally depending on the current execution state, reaching SCM goals by a change of supply chain structures and behaviour
Macro-state	General supply chain state in which one or a number of supply chain objects can occur
Manageability	A general system property to generate, implement, analyse and adjust managerial actions to lead the system to the achievement of its goals
Multi-structural macro-state	A supply chain macro-state that reflects the current states of objects and structures in supply chains as well as interrelations between them
Non-purposeful perturbation impact	An external input influence on a supply chain of the casual nature (for example, demand fluctuations, resource failure)
Order penetration point	A part in the supply chain where a strategic inventory is held in as generic a form as possible
Performance	A complex characteristic of the potential and real results of the supply chain functioning, taking into account the con-

	formity of these results with the goals set by management. Performance is measured with certain metrics (indicators)
Plan correction	The end result of a dangerous situation; the adaptation by means of changes in plans (for example, changes of the delivery time of production)
Plan disruption	A supernumerary mode of functioning during which the supply chain passes from an efficient to such a disabled state that it is necessary to execute replanning for the transition to an efficient state. The liquidation of the situation of the plan disruption is carried out on the basis of replanning, i.e. a change in the tactical plan
Planning	A purposeful, organized, and continuous process including the synthesis of a supply chain's structures and elements, the analysis of their current state and interaction, the forecasting of their development for some period, the forming of mission-oriented programmes and schedules, and the development of supply chain structure-dynamics control programs for the supply chain's transition to a required (optimal) structural macro-state
Process	A content and logic sequence of functions that are needed to create an object in a specified state
Process correction	The end result of a critical situation; the adaptation by means of operative changes in processes with the use of reserves (for example, safety stocks)
Purposeful perturbation impact	An external input influence on a supply chain goal to harm or damage a supply chain
Reliability	A complex characteristic of a non-failure operation, durability, recoverability, and the maintenance of the supply chain elements and the supply chain as a whole
Replanning	The end result of a situation of disruption of the plan; the adaptation by means of changes in tactical plans (for example, changes of manufacturing volumes)
Resilience	A property of supply chains consisting of their ability to maintain a regular mode of functioning in predicted conditions of the purposeful influence of destabilizing factors and to exclude the possibility of transition from a regular mode to a situation of plan disruption or a catastrophe in unpredicted conditions of influence of predicted destabilizing and/or unpredicted risk factors
Risk	A multi-meaning term that includes the following: 1. The risk is a likelihood estimation of a negative outcome of the event leading to losses/losses (the technological approach)

	2. The risk is an individual estimation by the person of the danger of a negative outcome of the event leading to losses/losses; the risk is ultimately a property of any entrepreneurship (the psychological approach) 3. The risk is an integral property of any process or system, the management of which is key to economic performance and stability maintenance (the organizational approach)
Robustness	A property of a supply chain consisting of its ability to continue its functioning at a certain level of perturbation influences
Security (in the narrow interpretation)	Resistance to the external, unauthorized actions developed to cause damage or to break a supply chain; a set of measures to protect the supply chain assets (product, facilities, equipment, information, and personnel) from theft, damage, or terrorism, and to prevent supply chains against unauthorized people or weapons of mass destruction
Security (in the wide interpretation)	A general system property characterizing the uninterrupted performance of a supply chain's functioning to achieve its goals under protection against external purposeful threats
Stability (in the wide interpretation, the global supply chain stability)	A complex property of a supply chain, characterizing the ability of a supply chain to maintain, realize, and restore goal-oriented functioning in an ever-changing execution environment under the influence of perturbation factors of unauthorized purposeful and non-purposeful nature
Structural state	A supply chain macro-state that reflects the current states of objects in a supply chain structure as well as interrelations between these objects
Structure dynamics	The process of supply chain structure transition from one to another planned macro-state
Supernumerary situation	A mode of supply chain functioning in which several indicators of a supply chain or environment indicators are out of the intervals of a regular mode in such limits that there is no disruption or catastrophe threat
Supply chain (SC)	A network of organizations, flows, and processes wherein a number of various enterprises (suppliers, manufacturers, distributors, and retailers) collaborate (cooperate and coordinate) along the entire value chain to acquire raw materials, to convert these raw materials into specified final products, and to deliver these final products to customers
Supply chain management (SCM)	A scientific discipline that studies human decisions in relation to cross-enterprise collaboration processes to transform and to use these resources in the most rational way along the entire value chain, from customers up to raw material

	suppliers, based on functional and structural integration, cooperation, and coordination throughout
Uncertainty	A property characterizing the incompleteness of our knowledge about the system's environment and conditions of its development
Virtual enterprise	A number of organizations that collaborate to develop a common working environment or virtual breeding environment with the goal of maximizing flexibility and adaptability to environmental changes and developing a pool of competencies and resources
Vulnerability	Resistance to external perturbation influences (planned and unplanned) of a casual character

Notation

Relations, Maps, Sets, Subsets and Elements of Sets

B	An area of allowable parameter values
B	A set of internal objects, e.g. enterprises that are embodied in an SC and are necessary for its functioning
$\overline{B}$	A set of external objects (customers, share holders, creditors, logistics service providers) interacting with the SC (the interaction may be informational, financial or material)
$\widetilde{B}$	A set of the objects in the SDC
$\overline{B}^{(i)}$	An element of the set $\overline{B}$
$B^{(j)}$	An element of the set B
C	A set of internal objects channels
$\overline{C}$	A set of external objects channels
$\widetilde{C}$	A set of channels that are used for informational interaction
$CatD$	A category of dynamic models
$Cat\Phi$	A category of digraphs
D	A set of interaction operations with the object $B^{(i)}$
$\mathbf{D}$	An attainable set (AS)
D^{+}	The external approximation of an AS
D^{-}	The internal approximation of an AS
$D_{x}^{(\xi)}$	An AS of the SC under interval disturbances
$\overline{E}$	The convex capsule
$E_{\widetilde{\mu}}$	The given point of an extended state space $\mathbf{x}_{<\widetilde{p},\widetilde{\mu}>}(T_f)$
F	A sigma-algebra over the space Ω
$F^{t}_{\chi'}$	A set of arcs of the DAMG G^{t}_{χ} representing relations between the DAMG elements at time t
G	A set of structures that are being formed within the SC
G_{χ}	An element of set G
GF	A functor (a set of maps)
$\hat{I}$	An index set of the SCM models
$\widetilde{K}$	The initial class of allowable control inputs
$\widetilde{\widetilde{K}}$	An extended class of allowable control inputs
$K_{i}^{(f)}$	An index set of flows (informational, financial of material) produced when the objects $B^{(i)}$ and $B^{(j)}$ interact
$\widetilde{K}_{i}^{(f)}$	An index set of flows produced by or necessary for the object $B^{(j)}$

$K_i^{(o)}$	An index set of interaction operation with the object $B^{(i)}$
$\widetilde{K}_i^{(o)}$	An index set of interaction operation with the object $B^{(i)}$
$K_j^{(p,1)}$	An index set of non-storable resources of the object $B^{(j)}$
$K_j^{(p,2)}$	An index set of storable resources of the object $B^{(j)}$
$\widetilde{\widetilde{K}}_t^{(o)}$	An extended class of allowable operations' control inputs
$\widetilde{\widetilde{K}}_t^{(k)}$	An extended class of allowable channel control inputs
$\overline{L}$	A set of models' types
L_χ	An index set of elements of the structure G_χ^t
$\overline{\overline{M}}$	A set of SCM models
$\overline{M}$	A set of SC possible elements
M	A dynamic model of SC control
M_θ	An element of the set $\overline{\overline{M}}$
$M_{\overline{g}}$	A simulation model describing the SC functioning under perturbation impacts
$M_\Theta^{(rr)}$	An element of the SC model set at iteration number "rr"
$MM^t_{<\chi,\chi'>}$	A map which characterizing interrelation between the structure G_χ^t and $G_{\chi'}^t$
N	An index set of internal objects
$\overline{N}$	An index set of external objects
NS	An index set of an SC structure type
P	A set of SC flows that are under consumption at different resources (financial flows, material flows, and information flows)
P_J	A set of allowable values of the performance metrics
$\overline{P}_{cs}$	A set of the model characteristics
$\overline{\overline{P}}_{cs}$	A set of possible values of the SC characteristics
$P^{(i,j)}$	A set of flows (informational, financial of material) produced when the objects $B^{(i)}$ and $B^{(j)}$ interact
$P^{(j)}$	A set of flows produced by or necessary for the object $B^{(j)}$
$\overline{P}_{\overline{g}}^{(cs)}$	An element of the set $\overline{P}_{cs}$
$P^{(i,j)}_{<\overline{\mu},\overline{\rho}>}$	An element of the set $P^{(i,j)}$
$P^{(j)}_{<\overline{\mu},\overline{\rho}>}$	An element of the set $P^{(j)}$
$\widetilde{\widetilde{\mathbf{Q}}}(\mathbf{x}(t))$	An extended domain of allowable control inputs
$\mathbf{Q}_\Theta$, $\mathbf{V}_\Theta$, $\mathbf{\Xi}_\Theta$	The corresponding sets of allowable areas for program control, real-time regulation control inputs, and perturbation input

$Q^{(\Theta)}$	A set of a mathematical structure
$R_{\tilde{r}}$	A set of business and information processes constraints
$\mathrm{R}^{(\bar{n}+\bar{n}n)}$	The dimension of the space of the SC aggregate model state
$\bar{r}_{i_1}^{\alpha(\Theta)}(\omega)$	A set of preference relations to be used for selection of the best alternatives via the structures
S_δ	An element of the set S
S_δ^*	The preferable element of the set S
$S_\delta^{*T_f}$	The preferable element of the set S at the moment T_f
$\tilde{s}_t^{(o)}$, $\tilde{s}_t^{(k)}$	The function-theoretic constraints imposed on the classes of allowable controls
$\mathbf{U}(\mathbf{x}(t))$	A set of allowable control inputs
U^t	An allowable scenario of transition from the initial SC multi-structural macro-state to the final one
$W^{(3)}$	A set of allowable values for the vectors of the structure-adaptation parameters
$\mathbf{W}_{<k-1>}^{(u)}$	A set of the parameter vector of the simulation model of SC execution under perturbation impacts
$\tilde{X}$	A space of the SC aggregate model state
$\mathbf{X}(\xi(t),t)$	An area of the allowable states of the SC structure dynamics
X_χ^t	A set of elements of the structure G_χ^t
Z_χ^t	A set of parameters characterizing relations numerically
$\hat{\Gamma}_2$	An index set of rules for constructing the resulting choice functions and the preferences relations
Γ_{1i}^-, Γ_{i2}^-	The sets of processes which immediate precede the process $\bar{B}^{(i)}$ (interaction operation with $\bar{B}^{(i)}$)
$\Gamma_{i\mu_1}^-$, $\Gamma_{i\mu_2}^-$	The sets of operation which immediate precede the operation $D_\mu^{(i)}$
Δ	A set of allowable areas for programme control and real-time regulation control inputs
$\Delta^{(d)}$	A set of allowable dynamic alternatives of SCs
$\Delta^{(nd)}$	A Pareto's set
$\Delta_{\hat{\eta}}^{0(\Theta)}$	A collection of the main basic sets of alternatives
$\Delta_{\hat{\rho}}^{(\Theta)}$	A set of auxiliary alternatives to be used mostly in coordination choice tasks
$\Pi_{<\delta,\delta'>}^t$	A map describing allowable transitions from one multi-structural macro-state to another one
$\bar{\bar{\Phi}}$	An operator of iterative construction (selection) of the model $M_\Theta^{(rr)}$
$\tilde{\varphi}$	A transition function of SDC model
$\bar{\Phi}_1$	An index set of the construction formed of basic sets via Gartesian products and the generation of subsets

$\overline{\Phi}_2$	An index set of the construction which corresponds to the output scale
$\Phi N_{\pi'}^{(j)}$	A set of storable resources of the object $B^{(j)}$
ΦR	A set of SC resources
$\Phi S_{\pi}^{(j)}$	A set of non-storable resources of the object $B^{(j)}$
$\widetilde{\psi}$	An output function of SDC model
Ω	A space of events (the set of uncertainly)

Continuous Variables and Functions

AD	A functional characterizing the adequacy of the model $M_{\Theta}^{(rr)}$ for the SC
$f(K-k)$	A monotone decreasing function of "forgetting"
F_1	A given functional characterizing a "distance" between $\hat{\mathbf{y}}(t_{<k-1>})$ and $\mathbf{y}(t_{<k-1>})$
F_2	A given functional characterizing the adequacy of the planning model
$g^{(K-k)}$	An adaptation coefficient that "depreciates" the information received at the previous step
H	A Hamiltonian function
$\mathbf{h}_0$, $\mathbf{h}_1$	The known vector-functions that are used for the state $\mathbf{x}$ end conditions at the time points $t=T_0$ and $t=T_f$
$\mathbf{h}_0^{(f)}$, $\mathbf{h}_1^{(f)}$	The known differentiated vector-functions which determine the end conditions for the variable vector $\mathbf{x}^{(t)}$ at the time points $t=T_0$ and $t=T_f$
$\mathbf{J}$	A vector of a quality functional
$J_G(t)$	The generalized SC performance functional, which is constructed by the multi-criteria procedures
J_{ζ}	The SC performance metrics (costs, service level, etc.)
$\mathbf{J}_{\Theta}$	The vector quality measure for different models
J_{ϑ}	The SC performance metrics within an SC model
$J_{\overline{\vartheta}}$	The components of generalized performance functional within an SC model
J_G	An SCalar form of the vector quality measure $\mathbf{J}_{\Theta}$
$\mathbf{J}_{\tilde{h}}^{(pl)}$	A vector of SC performance metrics for the case of zero perturbation influences
$\mathbf{J}_{\tilde{h}}^{(\xi)}$	A vector obtained as a result of simulation replicating the conditions of plan realization
$\mathbf{q}^{(1)}$, $\mathbf{q}^{(2)}$	The vector-functions, defining the main spatio-temporal, economic, technical and technological conditions for the SC functioning process

$q^2_{<k>}$	A possible variant of the functional F_1 for the parametric adaptation case
$Q^{<\Theta>}_{<k>}$	A possible variant of the functional F_1 for the structural adaptation case
$\overline{S}$	An objective function which used for performance estimation during SDC problem decision
$S_{\tilde{h}}(\mathbf{u}_{\tilde{h}}(t))$	The area of the sets' $\overline{\overline{D}}{}_J^{(\xi)}(T_f, T_0, X_0, \boldsymbol{\Xi}, \mathbf{u}_{\tilde{h}})$ and P_J intersection
$S^*_{\tilde{h}}$	The minimal value of the area of sets' $\overline{\overline{D}}{}_J^{(\xi)}(T_f, T_0, X_0, \boldsymbol{\Xi}, \mathbf{u}_{\tilde{h}})$ and P_J intersection
t_{st}	The total time of SC SDC models' structure adaptation
$\mathbf{u}$	An allowable control input
$\mathbf{u}(t)$	A control vector representing the SC control programms
$\overline{\mathbf{u}}(t)$	An arbitrary allowable control vector
$\tilde{\tilde{\mathbf{u}}}(t)$	The optimal control in an extended class of allowable control inputs
$\mathbf{u}(t_{<k>})$	A control vector of the SC at the control cycle $<k>$
u_{ij}	A control input action ($u_{ij}(t)=1$, if the resource $B^{(j)}$ is used for process $B^{(i)}$, $u_{ij}(t)=0$ otherwise)
$\mathbf{u}_{pl}(t)$	A vector of SC program control
$\mathbf{u}^*(t)$	An optimal control vector of an SC
$\mathbf{u}^{(c)\mathrm{T}}(t)$	A control vector of the SC structure dynamics
$\mathbf{v}$	A vector of SC real-time regulation control inputs
$\mathbf{w}$	A subvector of the parameters being adjusted through the SC external/internal adapter or defined within the structural adaptation
$\mathbf{x}(t)$	The general state vector of the SC
$\Delta\mathbf{x}(t)$	An indicator of a difference between the planned state trajectory and the real one
$\mathbf{x}(t_{<k>})$	The state vector of an SC at the control cycle $<k>$
$\mathbf{x}(t, \boldsymbol{\lambda}')$	The state vector of an SC
$\mathbf{x}_{\tilde{a}}(t)$	Additional elements of the SC state vector
x_i	A variable characterizing the state of the dynamic process $\overline{B}^{(i)}$
$\mathbf{x}_p(t)$	An extended state vector characterizing SC multi-structural macro-state
$\mathbf{x}_{pl}(t)$	The planned SC state trajectory
$\mathbf{x}^*(t)$	The optimal state vector of an SC
$\mathbf{x}^*(T_f)$	The optimal value of the general state vector of an SC at the time point $t=T_f$
$\mathbf{x}^{(c)\mathrm{T}}(t)$	A vector characterizing the state of the SDC

$\tilde{x}_{ij}^{(k)}$	A variable characterizing aggregate state of readjustment process of the SC
$x_{i\mu}^{(o)}$	A variable characterizing the state of the operation $D_{\mu}^{(i)}$
$\mathbf{x}_{\beta}^{*}(T_f)$	A point of the internal approximation of the AS
$\mathbf{x}_{\tilde{h}}^{(\xi)}(T_f)$	Perturbed SC vector obtained as a result of simulation replicating the conditions of plan realization number "$\tilde{h}$"
$\mathbf{x}_{\tilde{h}}^{(pl)}(T_f)$	Planned SC state vector at the time point $t = T_f$ for plan realization number "$\tilde{h}$"
$\hat{\mathbf{x}}_{\tilde{h}}^{(\xi)}(T_f)$	Stochastically perturbed SC vector obtained as a result of simulation replicating the conditions of plan realization number "$\tilde{h}$"
$\mathbf{y}(t)$	A vector of the output characteristics
$\int_{T_0}^{T_f} f_{\vartheta}$	A functional part of the SC performance metrics
$\boldsymbol{\beta}$	A vector of parameters of SC SDC models
$\varepsilon_{ij}(t)$	An element of a preset matrix time function of time-spatial constraints
$\boldsymbol{\lambda}'$	A vector of coefficients which are used for multi-objective decision-making
λ'_{ϑ}	Coefficients which are used for the multi-objective decision-making
$\boldsymbol{\xi}(t)$	A vector of perturbation influences
$\tilde{\Pi}_{<k>}^{(\xi)}$	An index of the total losses caused by the necessity for correction inputs at control cycle $<k>$
$\tilde{\boldsymbol{\varphi}}_{\Theta}$	A transition function of the SC SDC models
$\varphi_{\vartheta}(\mathbf{x}(t_f))$	A terminal part of the SC performance metrics
$\boldsymbol{\psi}^{*}(t)$	An optimal conjugate vector
$\tilde{\boldsymbol{\Psi}}_{(r)}$	A running value of the conjugate vector on the iteration r
$\tilde{\tilde{\boldsymbol{\Psi}}}_{(0)}$	A running value of the conjugate vector at the moment $t = T_0$
$\tilde{\boldsymbol{\psi}}_{\Theta}$	An output function of the SC SDC models

Parameters

a_i	The given quantities (end conditions) values of which should have the corresponding variables $x_i(t)$ -
$c_{i\eta\rho j}^{(f)}$	A potential productivity of the resource $B^{(j)}$ with regard to the flow ρ and subject to the operation $D_{\mu}^{(i)}$

$c_{i\mu j}^{(f)}$	The total potential productivity of the resource $B^{(j)}$ subject to the operation $D_{\mu}^{(i)}$
$\mathbf{c}^{\mathrm{T}}$	A given vector which is used for the external approximation of AS
$\overline{\overline{D}}$	The total amount of the approximation points
$d_{ij\eta}^{(1)}$	The maximal productivity of the resource $B^{(j)}$ subject to the collaboration operation $D_{\mu}^{(j)}$ with the customer $\overline{B}^{(\mu)}$
$d_{ij\rho}^{(2)}$	The maximal productivity of the resource $B^{(j)}$ subject to the collaboration operations to deliver products ρ to customers $\overline{B}^{(\eta)}$
$\widetilde{e}$	The total amount of the preferable multi-structural macro-state
$h_{\eta i}^{(j)}$	A given time of channel revamping
$\overline{G}$	The total amount of model characteristics
$\widetilde{I}_1$	The total amount of the constraints $\mathbf{q}^{(1)}(\mathbf{x},\mathbf{u})$
$\widetilde{I}_2$	The total amount of the constraints $\mathbf{q}^{(2)}(\mathbf{x},\mathbf{u})$
$\mathbf{J}_a^*$, $\mathbf{J}_b^*$	The given vectors defined respectively the lower and upper bounds of $\mathbf{J}^{(pl)}(T_f)$ and $\mathbf{J}^{(\xi)}(T_f)$
J_{a1}, J_{b1}	The given constants defining respectively the lower and upper bounds of the values of the performance metric J_1
K	The total amount of control cycles
K_{σ}	The total amount of SC multi-structural macro-states
$k_j^{(p,1)}$	The total amount of non-storable resources
$k_j^{(p,2)}$	The total amount of storable resources
$\widetilde{m}$	The total amount of the disturbance scenarios at the stage of SC execution
n	The total amount of the internal objects
$\overline{n}$	The total amount of the external object
p_i, $\overline{p}_i$	The total amount of flows
$\widetilde{R}$	The total amount of business and/or information processes constraints
$\widetilde{\widetilde{R}}_{1j}^{(f)}$	The maximal total productivity of the resource $B^{(j)}$ with regard to the product flows
$\widetilde{\widetilde{R}}_{1j\eta}^{(f)}$	The maximal channel intensity to deliver products to the customer $\overline{B}^{(\eta)}$
$\widetilde{\widetilde{R}}_{\widetilde{r}}$	A known constant
s_i $\overline{s}_i$	The total amount of interaction operation
T_0	The start instant of time of the planning horizon
T_f	The end instant of time of the planning horizon
t_{st}	The total time of SC models' structure adaptation

$\bar{t}_{st}$	The maximal allowable time of the structural adaptation
$\tilde{\tilde{\Delta}}$	A step of integration
ε_1, ε_2	A known constants characterizing the accuracy of the iterative solution of a boundary-value problem
$\bar{\bar{\varepsilon}}_2$	A given constant establishing an allowable level of the SC SDC model $M_{\Theta}^{(rr)}$ adequacy
$\varepsilon_1^{(f)}$, $\varepsilon_2^{(f)}$	The given constants characterizing the stability conditions of SC plans in the space of state vectors and the space of performance metrics
$\tilde{H}$	The total amount of plans (schedules)
θ	The total amount of SC structural-dynamics control models
λ	Coefficients which are used for multi-objective decision-making
$\xi_j^{(1)}$, $\xi_j^{(2)}$	The given vectors functions for minimal and maximal disturbances
$\Im$	The total amount of performance metrics

Indices

$\hat{e}$	A running number of the basic set corresponding to the input choice scale
$\bar{g}$	A running number of model characteristics
$\tilde{h}$	A running number of a plan (schedule)
i	A running number of an external object (e.g., a customer)
i_1	A running number of the preference relation, which is used for the selection of the best alternative
i_2	A running number of the relation, which is satisfied when an alternative is selected
j	A running number of an internal object resource
$\tilde{j}$	A running number of a disturbance scenario at the stage of SC execution
k	A running number of a time interval
$\hat{k}$	A running number of the basic set corresponding to the output scale
$\tilde{k}(\Theta)$	A type of rule for constructing the resulting choice functions and preferences relations
l l'	A running numbers of a SC structure element in the set $B^{(j)}$
$\tilde{\tilde{l}}$	A running number of conjugate vector element
$\tilde{p}$	A running number of Pareto-optimal points
r	A running number of an iteration during the plan construction
$\tilde{r}$	A running number of a business and/or information process constraint
rr	A running number of an iteration during the model adaptation
$\tilde{\alpha}$	A running number of the constraints $q^{(1)}$

$\alpha(\Theta)$ A type of a preference relation

$\beta(\Theta)$ A type of a satisfied relation

$\overline{\beta}$ A running number of an approximation point

$\widetilde{\beta}$ A running number of the constraints $q^{(2)}$

$\widetilde{\gamma}$ A running number of the coordination level

δ A running number of SC multi-structural macro-state

$\widetilde{\delta}$ A running component number of the performance metric vector

ε A running number of a process (operation) with regard to the constraint "and"

ζ A running number of a performance metric

η A running number of an external object (customer) with regard to the end product delivery

$\hat{\eta}$ A running number of a coordinate alternative

μ , $\overline{\mu}$ A running number of an interaction operation

$\overline{\overline{\mu}}$ A running number of a measure class (e.g. fuzzy measures, probabilistic measures, etc.)

$\widetilde{\mu}$ A running number of the preferable multi-structural macro-state

π A running number of a non-storable resource

π' A running number of a storable resource

ρ A running number of a flow produced when the objects $\overline{B}^{(i)}$ and $B^{(j)}$ interact

$\hat{\rho}$ A running number of an auxiliary alternative

$\overline{\rho}$ A running number of a flow with regard to the object $B^{(j)}$

ϑ A running number of a SDC model type

$\hat{\vartheta}$ A running number of a SDC model type without consideration of the operation control model

φ A running number of a process (operation) with regard to the constraint "or"

χ A running number of a SC structure

Θ A running number of a model

Index

FSC
www.fsc.org
MIX
Papier aus verantwortungsvollen Quellen
Paper from responsible sources
FSC® C105338

If you have any concerns about our products,
you can contact us on
ProductSafety@springernature.com

In case Publisher is established outside the EU,
the EU authorized representative is:
Springer Nature Customer Service Center GmbH
Europaplatz 3, 69115 Heidelberg, Germany

Printed by Libri Plureos GmbH
in Hamburg, Germany